The Hunter's Guide to Ballistics

Practical Advice on How to Choose Guns and Loads, and Use Them Effectively

Wayne van Zwoll

THE LYONS PRESS
Guilford, Connecticut
An Imprint of The Globe Pequot Press

to Alice, for all her help

The Lyons Press is an imprint of the Globe Pequot Press.

Printed in the United States of America

10 9 8 7 6 5 4 3 2 1

Design by Compset, Inc.

Library of Congress Cataloging-in-Publication Data

Van Zwoll, Wayne.

The hunter's guide to ballistics : practical advice on how to choose guns and loads, and use them effectively / Wayne van Zwoll.

p. cm.

ISBN 1-58574-575-5

1. Hunting guns. 2. Ballistics. I. Title.

SK274. V35 2001

799.2'028'3--dc21

2001029899

Acknowledgments

A book is never a solo project. This one required many hands over many years. Some of the shooters and hunters who tutored me are no longer here to read this book. Many of those who are will wonder why I'm not spending my time more profitably. I can't thank all the people who, by giving of their experiences, have directed my work. They come from companies such as Speer, Sierra, and Nosler, Remington and Winchester. Larry Werner, with his career at DuPont, and Dave Emary, a Hornady ballistician with a distinguished competitive record, were of special help. The people at Hodgdon, and Lee Reed at Swift Bullets, lent a hand, as did D'Arcy Echols, one of the most talented gunmakers I've ever met. Garth Kendig at Leupold knows about long shooting with special optics, and Art Alphin delivered information on interior ballistics. Target shooters Gary Anderson and Lones Wigger, Vic Fogle, Rich McClure and Johnny Moschkau all have beaten my scores on the line—and told me how. Hunters Dan O'Connell, Vern Woosley, Ken Nagel, and Leonard Giles showed me what can be done with modest rifles, good shooting and a willingness to hike where the game is. Fellow journalists Bill McRae and Dave Henderson have contributed their seasoned perspectives and technical knowledge. Dr. Ken Oehler, "Mr. Chronograph," has shown me a great deal about ballistics through his equipment and computer software. I've learned so much from so many that I'm forced to close here with the standard apology: If you're not mentioned, it isn't because you haven't helped. You know who you are, and you have my sincere thanks.

—Wayne van Zwoll
March 2001

Contents

Acknowledgments . . . iii
Preface . . . vii
Introduction . . . 1

Part I: Ballistics Background
The First Projectiles . . . 3
From Arrows to Ammunition . . . 9
Rifles for Our Manifest Destiny . . . 13
Loading from the Breech . . . 17
Bolt Actions and New Cartridges . . . 23
Naming Ammunition . . . 29
A Close Look at Primers . . . 33
Powder Makes Things Happen . . . 37
DuPont: Cornerstone of an Industry . . . 45
Bruce Hodgdon's Big Gamble . . . 51
A Look Back at Bullets . . . 57
Bullet-Making . . . 63
Bullets for Tough Game . . . 69

Part II: Ballistics to Measure
The Violence Inside the Gun . . . 77
Headspace . . . 81
Figuring Bullet Flight . . . 87
Chronographs and Bullet Speed . . . 91
Bullets on Leashes . . . 97
Zero . . . 103
How Not to Miss . . . 111

Getting the Drift . . . 117
Wind, Mirage, and Luck . . . 123
Getting the Angles Right . . . 127
Aiming Where They Aren't . . . 133
Recoil . . . 137

Part III: Ballistics Afield
Extending the Rifle's Reach . . . 145
Choosing a Big Game Cartridge . . . 149
The Deadliest Big Game Cartridges . . . 153
The Knockdown Myth . . . 157
What About the .30–06? . . . 161
Cartridges Outside the Mainstream . . . 167
Small Bores, Long Reach . . . 173
The Biggest .30's . . . 177
Short but Fast . . . 183
.33s: Coming Back Strong . . . 187
The Biggest Bores You Should Need . . . 193
Shotguns and Shotshells . . . 199
Buckshot: Tests Are Mandatory . . . 203
Shotguns for Big Game . . . 207
The New Slugs . . . 211
Handloading, Simply . . . 215

Part IV: Terms and Charts
Glossary . . . 221
Comprehensive Ballistics Tables . . . 229

Preface

This book is for hunters who want to know more about ballistics—not just the numbers for their particular loads, but enough to understand the bullet's flight. For example. . . .

A bullet is moving so fast that air can't easily get out of its way. Like water shoved to the side by a diver's body, air cleaved by a bullet exerts pressure on that bullet. It's pneumatic pressure, as opposed to hydraulic pressure, but pressure nonetheless. There's pressure against the nose and shank, and a vacuum forming behind the heel. Drag can exert many times the force of gravity on a bullet in flight. As that drag overcomes inertia, the bullet slows down, and gravity pushes it to earth. The slower the bullet goes, the more time gravity has to work on it, and the steeper its arc. That's why a bullet travels on a parabolic course, not one shaped like a rainbow. Add wind, and that arc bends.

This book tells you why bullets behave like they do and how to aim so they go where you want them under field conditions. You'll get the information you need to shoot better at long range, in the wind, up and down hills, and at moving game. You'll find out what makes some rifles kick harder than others, and why some bullets shoot tighter groups farther from the rifle.

Here, too, you'll find ballistic data for all factory cartridges currently loaded in the United States, and for many popular wildcats. You can use that data to improve the performance of your hunting rifle—or to choose your next one. You'll also discover here a condensed but detailed history of ballistic study, and explanation, in plain language, of the most important ballistic principles.

There's more. You'll find out what causes mirage and how to read it, and why bullets that move left also rise, while right-hand drift lowers point of impact. You'll get straight talk on bullet deflection in brush. If you want to know which bullets are best for big game, and why, or how gunpowder is made, or which cartridges give you a ballistic edge, turn the page. There's a chapter on shotguns, and one on recoil too. Whatever kind of shooting you do, you should find this book useful. It isn't an academic work, or a technical treatise for ballisticians. Rather, it was written for hunters to help them hit. I hope you find it readable, entertaining, and above all, practical.

Having studied ballistics for 40 years, I still have a lot to learn about what is, in the details, a complex, highly technical discipline. Fortunately, you don't need a graduate degree in physics to predict how a bullet will behave in flight, or to put that bullet where you want it. But getting in touch with the forces that control your bullet will help you aim accurately and give you the confidence that leads to better shooting.

Introduction

Ballistics is the study of hurtling objects. Throw a rock, and you give it ballistic properties. The label "ballistic missile" seems to me a redundancy, because every missile becomes "ballistic" as soon as it is launched.

Most commonly, though, ballistics has to do with bullet behavior. If you're a hunter, you're smart to know something about ballistics, because each shot you take is really a launch. As with the rock, you're hurling that projectile into the atmosphere where the laws of physics quickly render it impotent. Once it leaves the muzzle, it has only momentum to carry it—unlike a guided missile, which gets both power and direction during flight. Understanding what happens to a bullet after you pull the trigger is your first step toward more effective shooting.

Ballistics is a three-part study. *Interior* ballistics concerns the turbulence inside the rifle: ignition, gas buildup, bullet release and acceleration, the pressure curve during barrel time. *Exterior* ballistics takes over when the bullet exits the muzzle; it is bullet flight described by numbers. *Terminal* ballistics tells you about the expansion and penetration of the bullet after it contacts the target.

Interior ballistics is a laboratory science. Sophisticated instruments measure very high pressures over very small slices of time. Charting pressure curves enables ballisticians to formulate propellants and design bullets for commercial loads. Though it represents a tiny fraction of the period between striker fall and the impact of a bullet on a distant target, the instant of launch has a profound effect on bullet trajectory and accuracy. However, few shooters are equipped to measure what goes on before bullet exit, so beyond an understanding of the forces at work, there's not much to be gained by dwelling on interior ballistics. Better to focus on exterior ballistics—the bullet's flight. You can clock a bullet with a chronograph to determine its speed. Plugging bullet speed and weight into a formula will give you energy. It's easy to measure bullet drop and drift. These are the variables that matter to hunters. By changing loads, you affect both interior and exterior ballistics, but except for pressure signs on cartridge cases, the measures of bullet performance are all taken forward of the muzzle.

Speed, energy, and trajectory are the variables commonly used to compare cartridges, loads, and big game bullets. Knowing these numbers, you're able to pick the right cartridge and components for the hunt. Without them, you can't tell how close you are to reaching your rifle's potential, or where to aim for shots at long range. Without them, you can't discern differences between cartridges as closely matched as, say, the .270 and .280, the .35 Whelen, and .338–06. Without them, you'll still make lethal hits on game because most big game gives you a big target, is shot quite close to the rifle, and dies right away when properly hit with a bullet that has far less energy than most bullets that are shot at game these days. On the other hand, if you're an avid hunter, you'll eventually have the chance to shoot at long range or in adverse conditions. In that case, you'll benefit from knowing a lot more about your bullet than its weight and diameter.

Long ago, on a windy Wyoming plain, I crawled over a rise and saw a pronghorn buck in the basin below me. He looked far away. I got excited, held high and fired too quickly. He ducked and ran hard. I'd find later that the 90-grain Remington bullet from my .244 had grazed his back. I jacked another cartridge into the spout of the 722 rifle and dug my elbows deeper into the sand. He stopped. I'd have only one more chance. Taking a measured look at the yardage, I became conscious of the strong wind from nine o'clock. And this time the buck, now facing left, would be close to 400 yards out. I held the crosswire of the 4X Lyman Challenger on his nose. The bullet fell 18 inches, drifted about the same and sailed through the antelope's heart. Great planning.

It was also great luck. Even when you know a little about ballistics, you can make bad shots. Sometimes they're bad because you err in calculating bullet drop and drift; sometimes you just don't hold the rifle still when the bullet leaves the barrel. And sometimes things happen that you can't explain at all.

Last year, for example, I crawled over a Wyoming rim to get a closer look at a small band of pronghorns. A half dozen does lay only 30 yards away, alert. Nearly a quarter-mile beyond them stood two bucks. One was a big buck. I had no way to get closer and little time to wait. The late afternoon sky hung heavy with dark purple clouds. The stillness that kept my scent from the does meant a storm would come soon. Even now, the bucks were moving nervously, as if wondering whether to abandon this draw and head for another with taller sage.

I slid the rifle, a Magnum Research .280, forward over my hand and snugged the Latigo sling. Almost as steady as on a bench, I settled the crosswire of the Burris scope on the big buck. It was very far. The range-finding reticle, with a modest ladder of cross-tics below center, told me 400 yards. I'd shot at 400 yards the day before, to find out where the 140-grain Barnes X-Bullet would strike. Holding the second tic where I wanted to hit, I squeezed the trigger. The shot looked good.

But even the best planning and execution sometimes fails. The antelope dashed away as if hit, then stopped. I knew the bullet had missed, but didn't know where. The question now: Shoot again?

Think about that for a minute. I did.

There was no wind. The reticle had told me the range (I later confirmed it with a Leica laser rangefinder at 394 yards). I'd shot the rifle repeatedly at 400 yards and knew the bullet's point of impact there. My position was solid. I'd called the shot a good one. Firing again without knowing what I'd done wrong seemed irresponsible. Instead of shooting with confidence, I would be launching a second bullet with only the hope that it would land where I wanted it to.

I almost didn't shoot.

Finally, I decided that my first shot had struck low. The does watched me warily as my lungs relaxed and I blocked out everything but the image of the buck in the scope. Elevating the reticle slightly, I pressed the trigger again. The scope rose in recoil and came back down. The pronghorn buckled.

The X-Bullet had struck him between the shoulders, in the spine, right where I'd held the second cross-tic. It was a higher hit than I'd planned. Why hadn't the first bullet gone to the middle of the second rib, where I had placed that cross-tic for the first shot? I don't know. I may have horsed the trigger or given the forend a twitch at the final instant. Possibly the bullet had not flown true. A rifle that shoots a 5-inch group at 400 yards, as this one had the day before, could be expected to plant any bullet within 2½ inches of point of aim. But not all minute-of-angle rifles shoot minute-of-angle every time. Occasionally a shot from a cold, clean barrel will go wild.

My decision to shoot again was based more on experience than on empirical knowledge. And in the field of ballistics, experience can give you an edge. Even career ballisticians tell me that not all goes as expected in the firing lab. While exterior ballistics should be predictable for any given load out of a rifle with an established record, some bullets behave inexplicably. The more you know about ballistics, the fewer inexplicable shots you have. And the better you'll be able to make smart decisions when something happens that you don't expect.

It isn't always smart to shoot at an animal again when you miss. If I were on that Wyoming ridge today, I might decline that second chance. It's part of our responsibility as hunters to take shots we're sure to make. If you miss, you can just as easily cripple and lose the animal. Confidence in a hit is a big part of any successful shot. It's a mark of character when you decline a shot. A miss is, at best, an embarrassment.

PART I

Ballistics Background

The First Projectiles

A rock is a rock until you throw it. Then it becomes a projectile with kinetic energy. In the air, it has what we call ballistic properties: speed and energy, and a track known as a trajectory. Long before gunpowder, hunters were using projectiles to kill game—first rocks, then spears, boomerangs, arrows, darts and bolts. The rocks, spears, and boomerangs had limited range because they depended directly on the power of the human arm. Blowguns put a new source of power to work. Air would later be employed in gunbarrels. The bow and the crossbow gave soldiers great reach.

Dating as far back as 15,000 years to early Oranian and Caspian cultures, the bow helped the Persians in their conquest of the civilized world. Around 5000 B.C., Egypt managed to free itself from Persian domination, at least in part because the Egyptians became skilled in the use of bows and arrows. By 1000 B.C. Persian archers had adapted the bow for use by horsemen. Short recurve bows date to at least 480 B.C., and the Turks are credited with launching arrows half a mile in flight contests with their sinew-backed recurves.

Apparently, it was the Greeks who first started studying ballistics, about 300 B.C. As has been the

Primitive rifles still kill game. The study of ballistics began well before the advent of muzzleloaders.

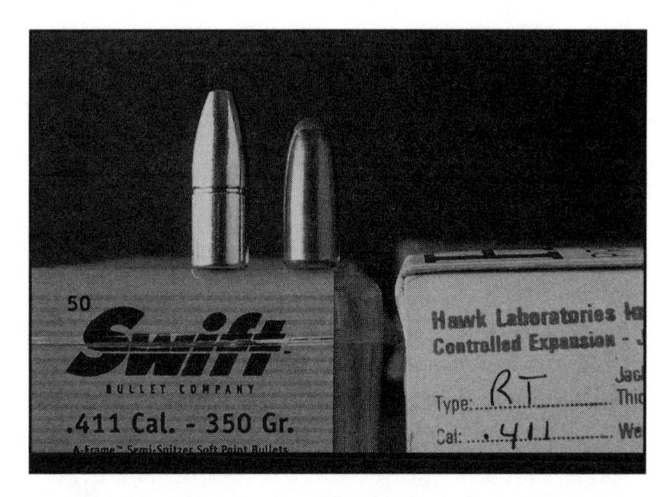

Modern bullets don't look like the rocks hurled by early hunters, but they're subject to the same forces in flight.

case since, their inspiration was no doubt the development of better armament for battle. At that time, little was known about gravity and air resistance, the two main forces acting against a projectile in flight. Later, such brilliant people as Isaac Newton, Leonardo da Vinci, Galileo, Francis Bacon, and Leonard Euler would examine how these forces could be measured and countered.

Bows and arrows surely helped people understand the principles of ballistics. The arrow's arc was visible, and for years after gunpowder was invented, archers ruled the battlefields. In 1066, at the Battle of Hastings, the Normans drew their English foes onto the field with a false retreat, then drove arrows en masse toward the oncoming troops, inflicting heavy casualties and winning the day. Unlike the Turks, English archers preferred a one-piece bow. It was called a longbow not for its tip-to-tip measure, but for the manner of the archer's draw, with an anchor point at the ear or cheek. Short bows of the day were commonly drawn to the chest. The English adopted the Viking and Norman tactic of arcing a hail of arrows into distant troops. In this way, archers resembled a wall of Swiss pikemen but with fearsome reach. Unfailingly, a charge toward well-positioned archers resulted in huge casualties, even when the attackers had armor. The armor might turn arrows, but it also wearied the advancing foot soldiers who wore it. English bowmen aimed for the joints in the armor, and for the exposed head and neck of any man so foolish as to shed his helmet on a hot day. They also shot deliberately at horses during a cavalry assault, the effect being not only to cripple and kill but to make the steeds unmanageable.

At Crecy (1346) and Agincourt (1415), England's archers vanquished the French army with volleys of arrows. In those days, English conscripts were required to practice with their bows. Royal statutes dictated that anyone earning less than 100 pence a year had to own a bow and arrows, which could at any time be inspected! Poachers in Sherwood Forest were even offered a pardon if they agreed to serve the king in battle as archers. Many did, and a contingent of these outlaws won a spectacular victory at Halidon Hill in 1333. Their arrows killed 4000 Scots in a conflict that left only 14 English dead.

The bow was not only a tool of war. Poachers used it with telling results, and if caught with iron arrowheads on the King's hunting ground were summarily hung with their own bowstrings.

The author with a favorite deer rifle: a modified Springfield in .30–06 improved.

The English longbow became an everyday tool. Specimens were not embellished or displayed as were early firearms or even swords. Bow wood and its utility deteriorated with age and weather. Only a handful of examples remain: unfinished staves in the Tower of London and a bow recovered in 1841 from the wreck of the *Mary Rose,* sunk in 1545. The "war" bow averaged 6 feet in length, with a flat back and curved belly. Yew was the preferred wood, but the English yew couldn't match that from Mediterranean countries for purity and straightness. Some of the best English bows derived from Spanish wood. After the longbow gained its fearsome reputation, Spain forbade the growing of yew on the premise that it might find its way to England and thence to battlefields in which Spain might feel its sting. The English, desperate for staves, cleverly required that a certain number be included with every shipment of Mediterranean wine. Stag hunters as well as the King's archers benefited.

In North America, bows varied regionally in shape and construction. Generally, flat, wide limbs were fashioned from soft woods like the yew preferred by Indians in the Pacific Northwest. Hardwoods, like ash and hickory in the East and Osage orange in the Midwest, gave better service in slender bows, rectangular in cross-section. Sinew backing was popular among some tribes, as it protected the back of the bow at its extreme arch. Strips of horn occasionally found their way onto the bellies of bows. Unlike sinew, horn has no stretch, but it performs well under compression.

Before the advent of the horse, bows of plains Indians averaged nearly 5 feet from tip to tip, almost as long as bows used by forest Indians east of the Mississippi. Mounted warriors soon switched to bows between 3 and 4 feet in length, like those favored by Northwest tribes. Arrows from the Eastern Indians were long and beautifully made, with short feathers to clear the bow handle while the hunter was stalking. Accuracy was of paramount importance, because one shot was all that could be expected. On the plains, where hunters rode alongside bison and elk and shot several arrows quickly at short range, accuracy didn't matter. Fletching was long because the crude shafts needed strong steering and because the feathers didn't rest against the bow handle long before the shot. Raised nocks on the arrows of the plains Indian aided the "pinch" grip preferred by horsemen under pressure to shoot quickly. Steel arrowheads replaced the equally deadly but more fragile obsidian after white traders came. Most big game heads were small, to aid penetration.

The plains Indian pushed the bow as much as he pulled the string. This technique allowed him to

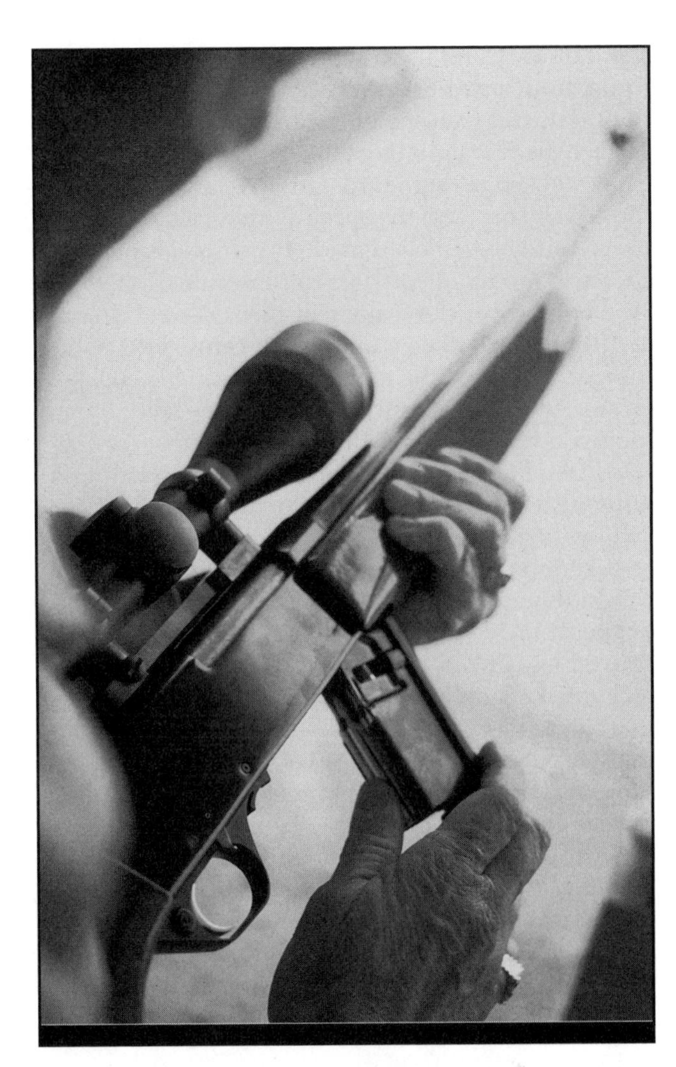

Modern magazine rifles evolved as hunters and soldiers sought more reach and faster loading.

Some modern rifle competition requires the use of cast lead bullets similar to those of early pioneers.

shoot quickly and use a short draw to get lots of thrust from a strong bow. He normally drew only to the chest, and well shy of the arrowhead. A typical 24-inch arrow might be pulled 20 inches. That short draw stacked enough thrust to drive arrows through bison. And the speed with which repeat arrows could be launched made firearms second choice to mounted warriors until Samuel Colt's Walker revolver came along in 1839.

"Bigfoot" Walker, a nineteenth-century Texas Ranger, respected the plains Indian and his bow. "I have seen a great many men in my time spitted with 'dogwood switches' . . . [Indians] can shoot their arrows faster than you can fire a revolver, and almost with the accuracy of a rifle at the distance of fifty or sixty yards."

Even in the East, firearms did not help American Indians kill game for some time after the musket became available, mainly because it often proved less effective than a bow and arrows! Flintlock muskets delivered poor accuracy—besides making a frightful noise and spewing thick smoke that obscured the animal. Several arrows could be loosed in the time needed to recharge a musket once. Powder and ball had to be bought or stolen, but arrows could be fashioned in the field. Bows were lightweight and easily maneuverable on horseback, while a 12-pound musket was unwieldy. The effective range of eighteenth-century long guns edged by only a few yards that of a bow and arrow in practiced hands.

Bowhunters borrow from the 15th-century technology of English archers.

From Arrows to Ammunition

Chemical propulsion probably occurred to early scientists long before the first firearms. It seems odd that the explosive "Chinese snow" appeared in fireworks a couple of centuries before Roger Bacon, an English friar and philosopher, described gunpowder in 1249. But those first compounds were hardly what we'd call reliable. Also, the idea of bottling the pressure and directing a projectile fom a barrel had yet to be explored. It would be another 60 years before crude guns appeared in England, following experimental work on propulsion by Berthold Schwarz. In 1327, Edward II used guns as weapons during his invasion of Scotland.

The first gunpowder contained about 41 percent saltpeter, with equal proportions of charcoal and sulphur. In 1338, French chemists changed the composition to 50–25–25. In 1871, the English settled on a mix of 75 percent saltpeter, 15 percent charcoal and 10 percent sulphur. That composition of black powder was commonly accepted as standard until the development of guncotton in 1846.

Powder manufacture in the United States antedated gun building. A powder mill at Milton, Massachusetts, near Boston, was probably the first such facility. By the beginning of the Revolution, enterprising colonists had amassed, by manufacture or capture, 40 tons of black powder. Half went to Cambridge, where it was wasted before George Washington took charge of the Revolutionary Army. In short order, the Continental Army had no powder!

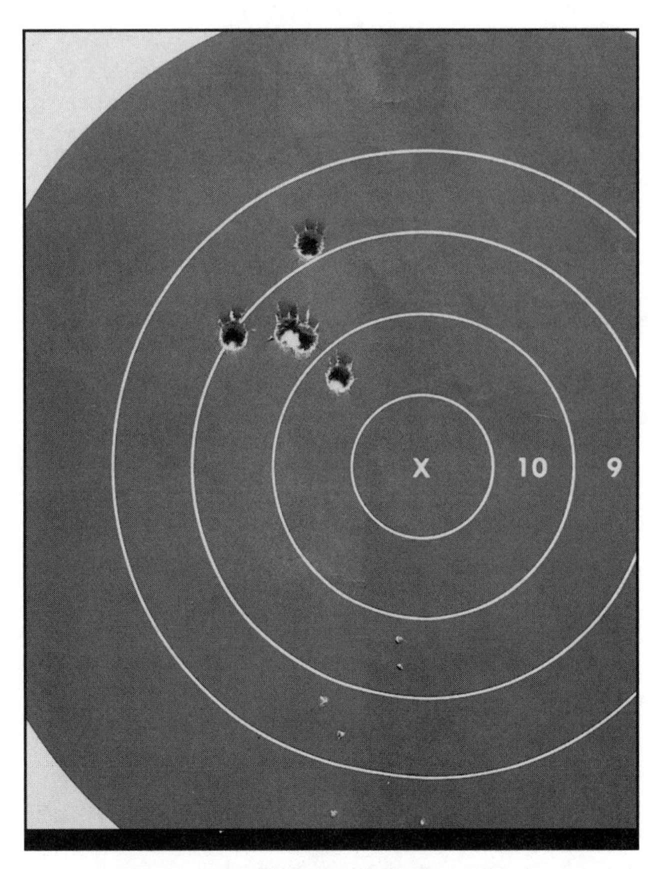

This 100-yard group measures 1½ inches—mediocre by current standards for hunting rifles but well beyond the capabilities of firearms used for hundreds of years by both hunters and soldiers.

New mills became a top priority, and by war's end the American forces had stocks of powder totaling 1000 tons. By 1800, the new nation's powder mills were producing 750 tons annually.

Torching black powder was easy in the open air. Igniting it in a chamber that bottled the violence to launch a ball was not so easy. The first guns, developed in Europe a century and a half before Columbus sailed for the New World, were heavy tubes that required two attendants. The Swiss called these firearms culverins. The culveriner held the tube, while his partner, the "gougat," applied a priming charge then lit it with a smoldering stick or rope. Culverins were clumsy and inaccurate. Though they often misfired, they were nonetheless popular. The noise and smoke they generated could unnerve an enemy armed with spears or pikes or even bows. Culverins were eventually lightened so that one soldier could load and fire unassisted. Many barrels were also fitted with ax heads, to make them useful when ignition failed. The heavy barrels were hard to steady by hand, so shooters commonly used mechanical rests. A forked brace adapted from fourteenth-century artillery supported the petronel, a hand cannon held against the breast for firing. Forks could be made to support infantry guns and were even used on the saddle of a mounted warrior.

A stationary cannon, whose muzzle was aimed at a wall or a camp or a mass of men, could be fired without regard to timing because gun and target had a fixed relationship. But soldiers on the move could ill afford to wait while an assistant caught up with a burning wick and then wait some more while the cannon fuse burned, all the while pointing the muzzle at the enemy. A soldier needed a lock, a mechanism to cause instant ignition. The first lock was simply a crude lever by which a smoldering wick was lowered to the touch-hole in the barrel. The long wick was later replaced by a short wick or match that got help from a cord or long wick that was kept smoldering on top of the barrel. The shooter eased the serpentine into that wick until the match caught fire. Then he moved the mechanism to the side and lowered the match to the touch-hole. A spring was eventually added to keep the match from the touch-hole. A trigger adapted from crossbows gave the shooter more control.

Guns with this crude mechanism were known as matchlocks. Among them, the Spanish arquebus gained particular notoriety. Arquebusiers carried extra wicks smoldering in perforated metal boxes on their belts. Even the best of these weapons was unreliable. In 1636, during eight hours of battle at Kuisyingen, one soldier managed only seven shots! At Wittenmergen two years later the rate of fire doubled to seven shots in four hours. Eliminating the wick became the priority of sixteenth-century German gun designers, who developed the "monk's gun" with a spring-loaded jaw that held a piece of pyrite (flint) against a serrated bar. To fire, the shooter pulled a ring at the rear of the bar, scooting it across the pyrite to produce sparks. The sparks fell in a pan containing a trail of fine gunpowder that led into the touch-hole in the barrel.

A more sophisticated version of this design appeared around 1515 in Nuremberg. The wheellock featured a spring-loaded sprocket wound with a spanner wrench and latched under tension. Pulling the trigger released the wheel to spin against a fixed shard of pyrite held by spring tension against the wheel's teeth. Sparks showered into the pan. Wheellocks were less affected by wet weather than were matchlocks. They also gave quicker ignition, were faster to set, and proved more effective as cavalry guns. As with the matchlock, the name of the mechanism became the name of guns that featured it.

In the *Lock a la Miquelet,* the roles of pyrite and steel were reversed. Named after the Spanish *miquelitos* (marauders) operating in the Pyrenees, this design seems, oddly, to have Dutch origins. It would later be modified to become what Americans know as the flintlock. Guns of this type have a spring-loaded cock that holds a piece of flint and swings in an arc when released. At the end of its travel, the flint in the jaws of the cock hits a pan cover or hammer, knocking it back to expose the primed pan. Sparks shower into the pan, igniting a priming charge of black powder, which conducts flame through the barrel's touch-hole. The cock eventually became known as a hammer, the hammer a frizzen. Flintlocks were less costly to build than were wheellocks, and in time became more reliable.

The common weakness of matchlock, wheellock, and flintlock mechanisms was exposed priming. Moisture could render any of them useless. Weak spark could fail to ignite even dry priming, and if the priming did ignite, the flame might not finish its trip to the main charge, yielding only a "flash in the pan." Generating a spark inside the gun, an inventor's dream, became possible early in the eighteenth

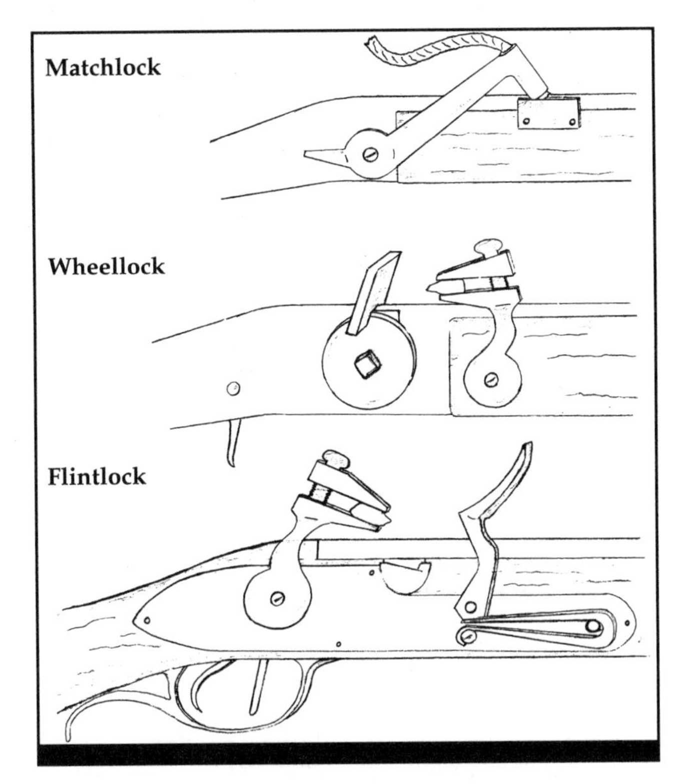

These illustrations represent the three basic firing mechanisms used by early armsmakers. The matchlock was essentially a tube with a hole to accept the lighted end of a wick. Triggers made shots controllable.

century, with the discovery of fulminates. Chemists found that fulminic acid (an isomer of cyanic acid) produced shock-sensitive salts. Percussion caused them to release their energy immediately, and more reliably than flint generated sparks. In 1774, the chief physician to Louis XV wrote about the explosiveness of fulminate of mercury, but more than a decade passed before Antoine Fourcroy and Nicolas Louis Vauquelin followed up with experiments. French chemist Claude Louis Berthollet found that substituting chlorate of potash for saltpeter in gunpowder made it unstable. Two huge explosions almost ended Berthollet's efforts.

Adding saltpeter to fulminates of mercury produced a shock-sensitive but manageable explosive. Called "Howard's powder" after Englishman E. C. Howard who discovered it in 1799, this compound may have influenced the work of Scotch clergyman Alexander John Forsythe. In 1806, Forsythe became the first experimenter on record to ignite a spark inside the chamber of a gun. Two years later the Swiss gunmaker Johannes Pauly designed a breech-loading percussion gun that used a cartridge with a paper percussion cap on its base. A spring-loaded needle pierced the cap, detonating the fulminate. The Lefauchex needle gun came later.

Internal combustion was clearly a landmark development, and even in its infancy drew enormous interest from military and civilian shooting circles. New ammunition and the guns to fire it were developed simultaneously by legions of inventors. In 1818 Joseph Manton, an Englishman, built a gun with a spring-loaded catch that held a tiny tube of fulminate against the side of the barrel and over the touch-hole. The hammer crushed the fulminate, and breech pressure blew the tube off to the side. The Merrill gun, 14,500 of which were bought by the British government, employed this mechanism. In 1821 the London firm of Westley Richards designed a percussion gun that used fulminate primers in a flintlock-style pan. The pan cover, forced open by the falling hammer, exposed a cup of fulminate. The hammer's sharp nose pierced it, sending sparks through a touch-hole at the cup's base. Two years later American physician Dr. Samuel Guthrie found a way to make fulminate pellets, which proved much more convenient than loose fulminate or paper caps.

Many people have claimed credit for inventing the copper percussion cap, but the honor is most commonly attributed to sea captain Joshua Shaw of Philadelphia. In 1814 Shaw was denied a patent for a steel cap because he was British-born and yet to become a citizen. He persevered with a disposable pewter cap, then one made of copper. The hollow nipple appeared soon after and eventually became standard on all percussion guns. It provided a tunnel that caught sparks at their origin and transmitted them directly to the chamber. In 1822 Shaw patented his own lock. In 1846 Congress awarded the 70-year-old inventor an honorarium for his work.

Between 1812 and 1825, the United States patent office issued 72 patents for percussion caps. Only a few of these designs proved out. Some caps fragmented, spattering the shooter. Others had so little priming mixture that they failed to ignite the main charge—or so much that they started the ball down the bore before the powder could build pressure. To throttle primer blast, an Englishman named Nock designed an antechamber perpendicular to the bore and behind the chamber. Powder burning there ignited the main charge through a short tunnel.

Percussion rifles and shotguns were slow to catch on. Part of the reason: fulminates were chemicals, and in the early nineteenth century, chemistry was still viewed with suspicion by many people. Also, early caps had performed inconsistently. Governments, traditionally wary of new things, resisted replacing pyrite with primers. By this time, flintlocks had been refined, both mechanically and esthetically, so there was no apparent need to change ignition types. Rumors circulated among shooters unfamiliar with the new caps. Percussion guns were said to kick harder but deliver less punch downrange. Even Britain's Colonel Hawker, a firearms authority of the day, qualified his praise of percussion ignition: "For killing single shots at wildfowl rapidly flying, and particularly by night, there is not a question in favour of the detonating system, as its trifling inferiority to the flint gun is tenfold repaid by the wonderful accuracy it gives in so readily obeying the eye. But in firing a heavy charge among a large flock of birds the flint has the decided advantage."

Eventually percussion caps would win over the doubters. Meanwhile, firearms were changing in other ways. Long abandoned were the cumbersome firearms that had come from Europe with the Pilgrims in the early seventeenth century. These smoothbores were typically 75-caliber flintlocks, 6 feet long. Though the superior accuracy of rifled bores was well known by this time (rifle matches had been held as early as 1498 in Leipzig, Germany and 1504 in Zurich, Switzerland), rifled barrels were expensive and slow to load. Early firearms were mostly for military use, and the fighting tactics of the day made fine accuracy unnecessary. Quick reloading by green recruits mattered more.

But in the New World, battles between settlers and Indians did not follow the traditional European pattern. There was no wall of uniforms, squarely presented as a collective target. The enemy was usually a single antagonist, partly hidden behind vegetation. Accuracy mattered. For hunters, long shots became more common and second shots less important. The huge lead balls used in British muskets constituted a waste of valuable lead. For these reasons, Americans favored the French-style flintlock common in Europe at the beginning of the eighteenth century; from it evolved the *jaeger* (hunter) rifle. The typical jaeger had a 24- to 30-inch barrel of 65 to 70 caliber, with seven to nine deep, slow-twist grooves. Most wore a rectangular patchbox on a stock with a wide, flat butt. Double set triggers were common. In the interest of conserving lead, frontier gunsmiths started making jaegers with 50-, 45-, and even 40-caliber bores. (A pound of lead will yield 70 balls of 40 caliber, but only 15 of .70-inch diameter.) They lengthened the barrel, replaced the jaeger's sliding patchbox cover with a hinged lid and trimmed the stock, giving it a "crescent" butt to fit comfortably against the shooter's upper arm. The result was what we know as the Kentucky rifle (though most of the changes were wrought in Pennsylvania by riflesmiths of German extraction).

The jaeger's rifled bore made it much more accurate than the Brown Bess musket British troops brought to the Revolutionary War. Precision, however, came at a cost in reloading speed. Balls—even patched balls—fit the bore tightly. They required careful attention and some effort to seat. Pounding them home was not only slow, hard work; it made noise that gave away the shooter's position. To a hunter in trouble with Indians, or hurrying for a second shot under difficult conditions, loading tight balls was more than a bother. To a mounted man, it proved prohibitively difficult. (Years after rifles became available to Indians through trade and warfare, many preferred bows and even muskets for hunting, because rifles took too long to load.)

To speed loading, Americans learned early on to swath undersize balls in greased patches that took the rifling. Crack *Jaeger* troops, against which they fought in the Revolutionary War, loaded their rifles with tight-fitting balls. The colonists beat the *Jaegers* almost as handily as they defeated British regulars. The patched ball soon gained favor among hunters, who appreciated the cleaning action of the patch and its protection of the bore against leading.

Rifles for Our Manifest Destiny

In 1816, before Jethro Wood of Cayuga County, New York, had invented the first all-metal plow, Eliphalet (Lite) Remington II was going on 23. Lite lived with his wife Abigail in his father's stone house on Staley Creek in Litchfield, four miles from New York's Mohawk River. It was there that he borrowed from the jaeger rifle design to build his own rifle in the family shop. He made all its parts, because in those days, you had to. Pumping the bellows, Lite heated the rod he'd chosen for his barrel. When it glowed red, he hammered it until it was half an inch square in cross-section. Then he wound it around an iron mandrel that was not quite as big in diameter as the finished bore (in this case, 45 caliber). Heating the barrel again until it was white-hot, he sprinkled it with Borax and sand. He held one end with his tongs and pounded the other on the stone floor to seat the near-molten coils. After it cooled, Lite checked it for straightness and hammered out the curves. Then he ground and filed the eight flats that would make the tube octagonal, as were most barrels of the day. Lite traveled to Utica to pay a gunsmith four double reales (about a dollar in country currency) to rifle the bore. At the time, $200 a year was a living wage.

Returning home, Remington bored a touch-hole and forged a breech plug and the lock parts. He shaped them with a file, then brazed the priming pan to the lockplate. Lite used uric acid and iron oxide, a metal preservative known as hazel-brown, to finish the steel. Starting with a stick of walnut, Lite fashioned a stock with draw-knife and chisel.

Flintlocks were simpler, sturdier and less expensive than their wheellock forebears and remained in use long after the advent of percussion ignition.

He smoothed it with sandstone and sealed it with beeswax. Screws and pins, also hand-made, brought the various parts together. Lite had a rifle. Shortly, the young man placed second at a local shooting match. So impressed was the match winner with Remington's rifle that he asked if Lite would build one for him.

"Sure."

"How much?"

"Ten dollars. And you'll have it in ten days."

As the frontier edged south and west, the needs of hunters changed. Grizzly bears, bison, and elk were not easily felled by the light charges fired from the svelte Kentucky rifle. Neither was the Kentucky's long barrel well suited for carrying horseback, nor its slender stock for rough treatment in the saddle. By the late 1700s, when Daniel Boone was probing the Cumberlands, gunmakers were already redesigning the Kentucky rifle. They gave it iron hardware, a bigger bore. The result: the "mountain" or "Tennessee" rifle.

Shortly after the turn of the century, General W. H. Ashley, head of the Rocky Mountain Fur Company, successfully promoted the rendezvous as a way to collect furs from trappers in the West. Tons of pelts funneled from frontier outposts to St. Louis, Missouri, which soon came to be a gateway to the West. Among the many Easterners drawn to this boom-town was gunsmith Jacob Hawkins. In 1822 his brother Samuel closed a gunshop in Xenia, Ohio to join Jake. The two changed their name to the original Dutch "Hawken" and started building rifles for frontiersmen.

As Youmans in North Carolina had become a pre-eminent maker of Tennessee rifles, the Hawken brothers would define the plains rifle. It borrowed from Youmans' design but had a shorter, heavier barrel for horseback carry and to accommodate bigger powder charges. The full-length stock was replaced by a half-stock, typically maple with two keys (market price for sugar maple in 1845 was only $2 per hundred board feet!). The traditional patchbox was often omitted. Until about 1840, the standard firing mechanism was a flintlock, sometimes purchased but often fabricated in-house. The Hawken brothers used Ashmore locks as well as their own. A typical Hawken weighed just under 10 pounds, with a 38-inch, 50-caliber octagonal barrel fashioned of soft iron with a slow rifling twist. Hawken barrels were less susceptible to fouling problems than were the quick-twist, hard-steel barrels common to "more advanced" English rifles of that era. Double set triggers became standard equipment on Hawken rifles. Other makers—notably Henry Lehman, James Henry, and George Tryon—turned out plains rifles that looked like (and in some cases were patterned after) Hawkens.

Black powder produced low pressures and velocities, so a big ball was used to add energy. Not until the development of progressive-burning smokeless powder would we get fast bullets from small bores.

A flintlock firing. The billows of black-powder smoke come not only from the muzzle but from the breech, where a priming charge in the pan has caught sparks from the flint striking the frizzen.

Though they got their start by repairing rifles, the Hawken brothers were soon kept busy building them. Their reputation reached all corners of the frontier, and their rifles were credited with extraordinary performance. Francis Parkman told of killing a pronghorn at 204 paces and of watching another hunter drop a bison at nearly 300.

Charge weights in Hawken barrels typically ran 150 to 215 grains. Bore size increased as ammunition became easier to get and buffalo became more valuable at market. The secret to the Hawken's legendary accuracy seems to have been its soft barrel iron, which didn't "slick up" like ordnance steel. Instead, it absorbed a trace of bullet lube and delayed fouling. The slow rifling pitch controlled patched round balls better than did the fast-twist barrels of contemporary English rifles. Hawken rifles proved less finicky as to loads as well, demonstrating fine accuracy over a wide range of charges. Also, they reportedly dished out less recoil.

In an article for the *Saturday Evening Post* (February 21, 1920, as cited by Hanson in "The Plains Rifle"), Horace Kephart wrote of finding an unused Hawken rifle in St. Louis: "I found that it would shoot straight with any powder charge up to a one-to-one load, equal weights of powder and ball. With a round ball of pure lead weighing 217 grains, patched with fine linen so that it fitted tight, and 205 grains of powder it gave very low trajectory and great smashing power, and yet the recoil was no more severe than that of a .45 caliber breech loader charged with seventy grains of powder and a 500-grain service bullet. . . ."

This Thompson/Center replica of a Hawken plains rifle is a caplock. The author shot a caribou with it.

The Hawken brothers continued to repair rifles. On December 26, 1825, the Hawkens billed the U.S. Indian Department, through its agent Richard Graham, $1.25 for "Cutting Barrel & new brich" and 50 cents for "Repairing Rifle." During spring of the following year, Graham gave the shop more business, mostly repairing locks at 25 to 87 ½ cents apiece. For "Repairing Lock, bullet molds, ram rod, & hind sight" the charge totaled $2. A side pin cost 18½ cents, as did repairing spurs. The Hawkens billed 50 cents for shoeing a horse. Labor—even skilled labor—was cheaper than parts.

In 1849, when the California Gold Rush began, you could buy a Hawken rifle for $22.50. That year Jake Hawken died of cholera. Sam continued in the business alone. In 1859 he made his first trek to the Rocky Mountains, where so many of his rifles had gone. That journey, from Kansas City to Denver, took 57 days. Sam apparently worked in nearby mines for a week, then started back to Missouri, where upon his return he was quoted as saying ". . . and here I am once more at my old trade, putting guns and pistols in order. . . ."

William Stewart Hawken, Sam's son, became a mountain man and rode with Kit Carson's mounted rifles. O.P. Wiggins, a fellow trapper, told of seeing William for the last time September 23, 1847, during the battle of Monterey. Badly outnumbered by Mexican troops, General Henderson and his small group of frontiersmen fought to hold a bridge over San Juan Creek. The clash left only nine ambulatory men among

Clockwise from top center: bullet starter, box of Maxi-Ball bullets, capper, canister of percussion caps, powder measure. All are necessary (plus powder and ramrod) to charge a muzzleloader.

Many of today's big-game hunters prefer to hunt with black powder rifles. Special seasons give them extra opportunities.

43 Texas Rangers. William Hawken, age 30, was one of the wounded. He made his way back to Missouri.

When Sam went west he left William in charge of the Hawken facility. William got this letter:

> Evans Landing Nov. 27th 1858
> Mr. Wm. Hawkins
> Sir, I have waited with patience for my gun, I am in almost in a hurry 2 weeks was out last Monday I will wait a short time for it and if it don't come I will either go or send. If your are still waiting to make me a good one it is all right. Please send as soon as possible. Game is plenty and I have no gun. Yours a friend.
>
> Daniel W. Boon.

William must have traveled west shortly after Sam returned, because the Rocky Mountain News reported that the ". . . son of the worthy and well known Uncle Samuel, of Hawken rifle notoriety . . . has removed his shop to F Street, the sign of the Big Gun, where everybody can have their guns and pistols put in order. . . ." Nothing more was heard of William Hawken until some years later, when a 56-caliber muzzle-loader bearing his name was found under a pile of rocks in Querino Canyon, Arizona. It told nothing of the man's later years or of his fate, as it could easily have been stolen or fallen into Indian hands.

Sam hired a shop hand, J. P. Gemmer, after his trip west. Gemmer proved capable and industrious and in 1862 bought the Hawken enterprise. He is believed to have used the "S. Hawken" stamp on some rifles, but marked most of them "J. P. Gemmer, St. Louis." As the bison dwindled and cartridge rifles became popular, Gemmer turned to smaller bores and target options. Sam Hawken continued to visit his shop in retirement and once built a complete rifle. He outlived Jim Bridger and Kit Carson and many other frontiersmen who had depended on Hawken rifles. When he died at age 92 the shop was still open for business. It remained in St. Louis but changed locations in 1870, 1876, 1880, and 1912. Three years after the last move it closed. Mr. Gemmer died in 1919.

The author sets the trigger on a Thompson/Center caplock black powder rifle. The front trigger will then drop the hammer, crushing the cap and shooting sparks through the nipple to the main charge.

Loading from the Breech

Percussion ignition spurred inventors to develop a breechloading rifle. Guns with a hinged breech date to at least 1537, but went nowhere without the percussion cap. A seventeenth-century French musket featured a cylindrical breech plug that dropped when the trigger guard was turned. After inserting the charge and raising the block, a shooter was ready to fire. Sadly, raising the block sometimes pinched the powder and caused premature detonation. The most famous example of this type, designed in 1776 by British Major Patrick Ferguson, had a threaded breech plug that retracted by rotating the trigger guard as one might loosen a screw. A charge was inserted in the barrel through an open breech plug hole and the plug spun back into place. The Theiss breechloader of 1804 had a sliding block activated by a button in front of the trigger guard. The block was raised for loading. This mechanism, a flintlock, leaked lots of gas. Captain John Harris Hall of Maine designed one of the first successful breechloaders in this country by employing the same principle. Hammer, pan, and frizzen all rode on the movable block.

European inventors Lepage, Pauly, Potet, and Robert pioneered in fashioning breechloaders for cartridges. The first cartridges—assembled in 1586—were paper. They had no priming, of course, and the guns were still loaded from the muzzle. Biting or ripping off the cartridge base was necessary before loading to expose the powder charge. The

Among the most effective and popular of early cartridge rifles was the Winchester 1885 High Wall, a rifle designed by John Browning. This is a modern version.

case burned to ashes upon firing. Replacing pyrite with the percussion cap did away with the biting and tearing because the cap's more powerful spark could penetrate thin paper.

A major problem with early breechloaders was making one stout enough to withstand the heavy powder charges used for hunting big game and reliable enough to keep functioning while hot and dirty. The first American breechloader that won popular acceptance was developed by John Harris Hall in 1811. Six years later the U.S. government issued a limited number of these rifles to soldiers. But the Hall was weak, heavy and crude, a flintlock firing paper cartridges. As war with Mexico threatened to drain U.S. arsenals in 1845, factories instead began producing the proven Harpers Ferry muzzleloading rifle, a design dating all the way back to 1803.

Johann Nikolaus von Dreyse was among the first inventors to put a primer in a cartridge. The bullet in his paper cartridge had a pellet of fulminate on its base. A long striker penetrated the paper cartridge from the rear and lanced through the charge to pinch the pellet against the bullet. The von Dreyse "needle gun," developed in 1835, became very popular. About 300,000 were built for the Prussian army over the next 30 years. (Incidentally, the "needle gun" mentioned by post-Civil War writers was not the European Dreyse. It was the .50–70 Springfield that became in 1873 the .45–70 "trap-door" rifle used in the last of the Indian Wars. The long breech block required a long, needle-like firing pin.)

These .22 rimfire cartridges haven't been on the market long, but their lineage dates back 150 years. Horace Smith and Dan Wesson were instrumental in the development of .22 rimfires.

Meanwhile, Eliphalet Remington bought (for $2581) the gun company and services of William Jenks, a bright Welsh engineer who had designed a breechloading service rifle but had wearied of awaiting a reply from the U.S. government. Remington adapted the Jenks rifle to use Edward Maynard's percussion lock, which advanced caps on a strip of paper. Later, a new breech mechanism designed by J.H. Merrill improved the Jenks rifle. In government tests, its tallow-coated cardboard cartridges fired reliably, even after a one-minute submersion in water.

Rail transport, meat markets, and floods of immigrants across the Great Plains would themselves have exacted a heavy toll from native wildlife. But development in the mid-nineteenth century of self-contained rifle cartridges accelerated the demise of big game as no single event had before.

In 1847 Stephen Taylor patented a hollow-base bullet with an internal powder charge held in place by a perforated end cap that admitted sparks from an external primer. The following year, Walter Hunt, an inventor from New York, devised a similar bullet. This one had a cork cap covered with paper. Primer sparks shot through the paper to ignite the charge. What made Hunt's development of "rocket balls" exciting, however, was his concurrent design of a repeating rifle to fire them.

Hunt's rifle had a brilliantly conceived tubular magazine, and a pillbox mechanism to advance the metallic primers. But the action, operated by a finger lever under the breech, was prone to malfunction. Lacking the money to promote or even improve his "Volitional" repeater, Hunt sold patent rights to fellow New Yorker and machinist George A. Arrowsmith. At that time, Arrowsmith had in the shop a talented young engineer named Lewis Jennings, who, with his boss, corrected several design deficiencies in the Hunt repeater.

In 1849, after receiving patents for Jennings' work, Arrowsmith sold the Hunt rifle for $100,000 to railroad magnate and New York hardware merchant Courtland Palmer. With Palmer's financial backing, gun designers Horace Smith and Daniel Wesson began in 1852 to develop a metallic cartridge like that patented in 1846 and 1849 by the Frenchman Flobert. Rather than using a ball atop a primer as Flobert had, Smith and Wesson modified a rocket

Breech loading is something most hunters take for granted. Many inventors worked for decades to perfect it.

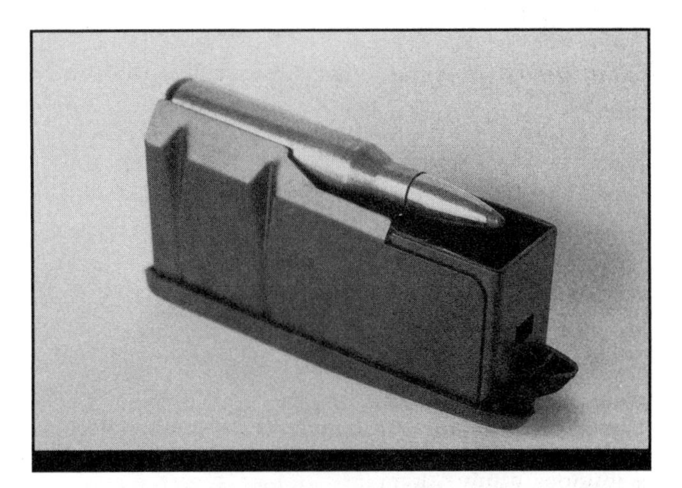

A stack of centerfire cartridges in a detachable magazine looks simpler than it is!

ball to include a copper base that held the fulminate priming mix. The pistols they built for this ammunition failed to sell. In 1854 Courtland Palmer joined his designers in a limited partnership, putting up $10,000 for new tooling in a company that would become known as Smith and Wesson.

In 1855 a group of 40 New York and New Haven investors bought out Smith and Wesson and Palmer to form the Volcanic Repeating Arms Company. The investors chose one of their own as company director: a shirt salesman named Oliver F. Winchester. Winchester soon moved the firm from Norwich to New Haven. When sluggish sales of Volcanic guns sent the firm into receivership in 1857, Winchester bought all assets for $40,000. He reorganized the company into the New Haven Arms Company, hiring Benjamin Tyler Henry as chief mechanic.

Oliver Winchester knew success hinged on the development of better ammunition; so that became Henry's first assignment. In 1860 Henry received a patent for a 15-shot repeating rifle chambered for .44 rimfire cartridges. The brass-frame Henry would father Winchester's first lever rifles: the 1866, 1873, and 1876. Confederates called the Henry "that damned Yankee rifle you loaded on Sunday and fired all week." Underpowered, prone to leak gas, and marginally reliable, it was nonetheless coveted by hunters as well as soldiers because it could be recharged from the shoulder with the flick of a hand.

The middle of the nineteenth century was the most active period in the history of firearms design. While many entrepreneurs struggled to perfect a repeating rifle, a young machinist named Christian Sharps decided to build a better breechloading single-shot. Sharps, a New Jersey native, had worked under John Hall at the Harpers Ferry Arsenal. In 1848 he received his first patent, for a rifle mechanism with a sliding breechblock. Fitted to an altered 1841 Springfield, the prototype held great promise for big game hunters because it could handle powerful loads. But Sharps failed to market the action. He was bailed out by businessman J.M. McCalla and gunsmith A.S. Nippes.

The first Sharps rifle, Model 1849, had a priming wheel that held 18 primers and advanced mechani-

Early lever-action breechloaders carried "rocket balls" in a tube under the barrel. Modern lever guns like this Marlin owe their design to Hunt, Henry, Jennings and Winchester, who persevered early on.

cally. In 1869, five years before Christian Sharps succumbed to tuberculosis, the Sharps Rifle Manufacturing Company courted hunters with the New Model 1869, its first rifle in metallic chamberings. The New Model 1874 appeared the next year to serve market hunters. George Reighard was one of those hunters. In a 1930 edition of the *Kansas City Star,* he explained how he shot bison:

> In 1872 I organized my own outfit and went south from Fort Dodge to shoot buffaloes for their hides. I furnished the team and wagon and did the killing. (My partners) furnished the supplies and did the skinning, stretching and cooking. They got half the hides and I got the other half. I had two big .50 Sharps rifle with telecopic sights. . . .
>
> The time I made my biggest kill I lay on a slight ridge, behind a tuft of weeds 100 yards from a bunch of a thousand buffaloes that had come a long distance to a creek, had drunk their fill and then strolled out upon the prairie to rest, some to lie down. I followed the tactics I have described. After I had killed about twenty-five my gun barrel became hot and began to expand. A bullet from an overheated gun does not go straight, it wobbles, so I put that gun aside and took the other. By the time that became hot the other had cooled, but then the powder smoke in front of me was so thick I could not see through it; there was not a breath of wind to carry it away, and I had to crawl backward, dragging my two guns, and work around to another position on the ridge, from which I killed fifty-four more. In one and one-half hours I had fired ninety-one shots, as a count of the empty shells showed afterwards, and had killed seventy-nine buffaloes. . . . On that trip I killed a few more than 3000 buffaloes in one month. . . .

The Sharps Rifle Company folded in 1880, having failed to market its hammerless Model 1878 rifle or design a practical repeater. A peacetime military budget and the decimation of game herds by commercial hunters contributed to the company's financial crisis. By the early 1880s, so many bison had been killed (many with Sharps rifles) that human scavengers would glean more than three million *tons* of bones from the plains.

The only thing deadlier than a Sharps rifle at that time would have been a repeater stout enough to handle the powerful cartridges that made the

In Africa, breech-loading rifles gave hunters quick repeat shots, a comfort when hunting buffalo.

This under-hammer percussion rifle (c. 1855), a crude but functional repeater, was made by Jonathan Browning; it featured a sliding bar with several chambers.

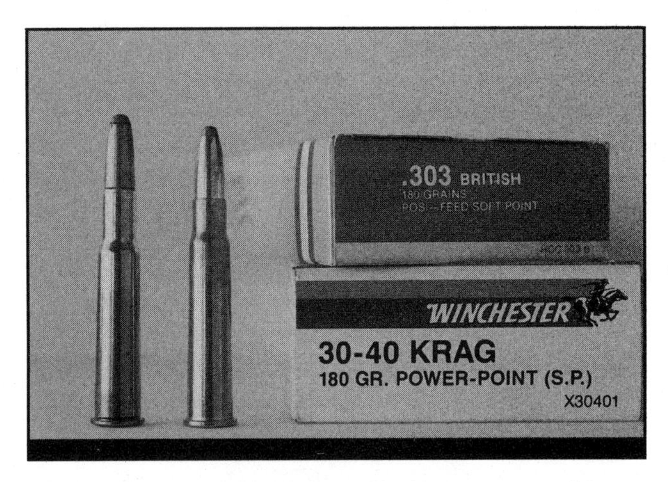

Breech loading didn't bring with it modern cartridges. Development of smokeless powder did. These two rounds were among the first smokeless cartridges in the 1890s.

Sharps popular. In Ogden, Utah, such a rifle evolved, albeit too late for commercial hunters to give it much play. The remarkable fellow who developed this repeater was John Moses Browning, son of gunsmith Jonathan Browning, who arrived on the Wasatch Front in 1852, five years after the first of his Mormon brethren completed the trip from Nauvoo, Illinois. John was born in 1855, one of Jonathan's 22 children. He quickly became interested in firearms, and by the time he was 11 had built a working gun. Later he took over the shop. In 1878, not long after his twenty-third birthday, John designed a breechloading single-shot rifle. He fashioned the prototype by hand, with files, chisels, and a foot-lathe his father had brought from Illinois. In May 1879, he received a patent for the rifle, which would one day find its way to market as the Winchester Model 1885.

At this time, the Winchester Repeating Arms Company was a behemoth, with a net worth of $1.2 million. By 1875 it had reached an output of a million cartridges per day. Nonetheless, declining demand for military rifles and ammunition prompted the company to shift its focus in the late 1870s. For the next 36 years, it would cater primarily to hunters. To that end it developed the Model 1876 lever rifle, a big, iron-frame version of the popular Model 1873, whose roots reached to Walter Hunt's Volitional repeater. But the Model 1876, while massive, lacked the strength to handle the pressure generated by the .45–70 and similar rounds—favorites not only of ex-soldiers but of elk and bison hunters. For long shooting at big animals, you still could not beat a Sharps.

The double-rifle design predates cartridges. It's still favored by some hunters in Africa. This rest lets the shooter fire while standing, so his body can move with the recoil. Less pain!

Then in 1883 a Winchester salesman showed Thomas Bennett, the company president, a single-shot rifle he had bought and used during a trip west. The action was of clever design and evidenced great strength. A competitor selling rifles that could accept .45–70 cartridges had to be taken seriously. Bennett immediately booked passage by train to Ogden.

He arrived to find a group of brothers barely out of their teens, laboring in a small, crude shop under a sign proclaiming it the "Largest Arms Factory Between Omaha and The Pacific." The boss, John, was 28. Bennett came straight to the point: He wanted to buy all rights to the single-shot. John said very well; he could have them for $10,000. Bennett countered at $8000, and John Browning accepted. It was the start of a relationship that would last 17 years and produce for Winchester 40 gun designs, among them several flagship models. Thomas Bennett was on the train back to New Haven six hours later.

Barely had Winchester dubbed Browning's rifle the Model 1885 when John and his brother Matt arrived at company headquarters with a new idea: a repeating lever-action rifle that would handle the .45–70. Bennett bought the design, reportedly for $50,000, or, as John would say later, "more money than there was in Ogden." The rifle later became Winchester's Model 1886. John Browning returned from his two-year Mormon mission in March 1889 and worked furiously designing firearms. During the next four years, he garnered 20 patents. His primary contribution to Winchester was the Model 1892, a petite lever-action with the Model 86's vertical locking lugs. The '92 would spawn the Model 1894, initially designed for black powder but chambered in 1895 for the .30–30, America's first centerfire smokeless round.

Development of the metallic cartridge accelerated the pace of rifle design in Europe too. But lever-action rifles did not appear. Double rifles became a British specialty, while a talented German contributed the bolt action.

Bolt Actions and New Cartridges

Peter Paul Mauser was born in 1838 in the Swabian village of Oberndorf. The youngest of 13 children, he would later work with his older brother Wilhelm. Paul's first gun project went nowhere, but in 1872 the single-shot 11mm Mauser Model 1871 became the official arm for Prussian troops and drew the attention of other governments. Its simple, rugged turn-bolt action derived, legend has it, from Paul's inspection of a door latch.

Mauser Bros. and Co. was established in February 1874. After Wilhelm's premature death, it became a stock company, whose controlling shares were purchased by Ludwig Lowe & Co. of Berlin. In 1889 Fabrique Nationale d'Armes de Guerre (FN) was founded near Liege to build Mauser rifles for Belgium's military. The 1889 rifle, designed for smokeless-powder cartridges, established Paul Mauser as the dominant rifle engineer in Europe. Over the next six years he improved the action. A staggered box magazine appeared in 1893, and two years later the Model 1895 incorporated most of the features that would make Mauser's Model 1898 the top choice of ordnance officers from eastern Europe to the tip of South America. By the late 1930s, Mauser was selling rifles to hunters in the U.S. through A.F. Stoeger of New York, which at one time listed 20 types of Mauser actions, differing not only in length (four were available), but in magazine configuration. These rifles did not come cheap. In 1939 retail prices ranged from $110 to $250. Since the

Lever-action rifles like the Savage 99 remained popular long after Paul Mauser came up with the door-latch approach to securing cartridges in the chamber.

This hunter used an iron-sighted Savage 99 in .300 Savage to kill a fine Quebec-Labrador caribou.

second world war, modifications of Paul Mauser's Model 1898 action have appeared in all the world's game fields. A recent survey I took of elk hunters in the U.S. brought responses from 598 riflemen. Of these, 522 used bolt actions, the familiar mechanism that a young German engineer brought to firearms years before the first successful automobile.

The shift among American hunters to the centerfire bolt rifle began with the introduction of the Mauser-inspired 1903 Springfield. The Springfield's fine accuracy, and the great reach of its .30–06 cartridge, quickly won converts. The metallic case, combined with smokeless powder and jacketed bullets, offered new possibilities to experimenters seeking better ballistic performance.

Among the most active cartridge designers during the Springfield's heyday was Charles Newton. Educated as a lawyer, Newton designed the .250–3000 Savage, so named because its 87-grain bullet sped away at a reported 3000 feet per second. He also fashioned, around 1912, the .25–06. With Fred Adolph, he designed several powerful hunting cartridges. The .30 Adolph Express had a big case with a rebated rim to fit Mauser bolt heads. The Adolph line later included a .280, .35, and .40 on the

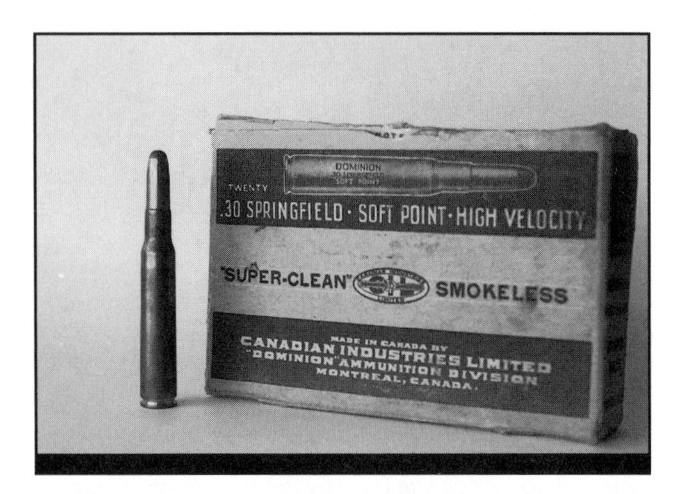

The .30–06 followed the .30–40 Krag as a U.S. service round. Both were designed for bolt rifles.

Autoloading rifles like this Remington appeared early in the 20th century. The first ones were heavy and less reliable than bolt rifles and could not handle as much breech pressure.

Winchester lever rifles remained popular with hunters until scopes earned their allegiance. This .348 Model 71 was made from 1935 through the mid-50s.

German 11.2x72 case. Newton developed a 7mm-06, a short 6.5–06 he called the .256 Newton, and a .276 on a bigger case. Many Newton rounds performed like modern magnums.

Charles Newton started his own Newton Arms Company in August 1914, intending to fashion his own rifles from DWM Mausers. His timing couldn't have been worse: Germany went to war a day before Newton's first shipment of Mausers was due. It was an omen. Not to be deterred, Newton designed a new rifle from scratch and loaded his own rounds (he even designed and built bullets that would stay together when driven into tough game at high speed). But he depended on Remington for the brass. The first production run of Newton rifles was finished in January 1917—just as America entered the war and the government seized control of all cartridge production.

The courageous Newton tooled up to make cartridge cases. But in August 1918, the banks supporting his venture sent it into receivership. He persevered with the Chas. Newton Rifle Corporation in April 1919. A deal with Eddystone Arsenal to supply equipment fell through, however, and Newton sold only a few rifles, built on imported Mauser actions. Later he formed the Buffalo Newton Rifle Corporation, which manufactured rifles of his design. It failed in 1929, shortly before Newton's death.

Established American firearms companies began offering commercial bolt rifles soon after World War I. Remington's Model 30, based on the 1917 Enfield action, was followed in 1922 by Winchester's Model 54. The versatile, familiar .30–06 proved the most popular chambering in both. Then in 1925, Winchester chambered its 54 for a new cartridge, the .270. Blessed with a receptive press, the .270 prospered. The .300 Holland & Holland Magnum, introduced at the same time, wooed some big game hunters; but its long case and broad, belted head were not suited to standard-length military rifle actions. In 1937 Winchester replaced the

Charles Newton used his genius to design both rifles and cartridges. The .250 Savage is his idea. It was dubbed the .250–3000 by Savage, to show its velocity with 87-grain bullets.

Mauser bolt rifles are still the preferred raw material in many custom gunshops. This commercial 98 Mauser was stocked by Fred Zeglin in quilted maple, and chambered for the .411 Hawk.

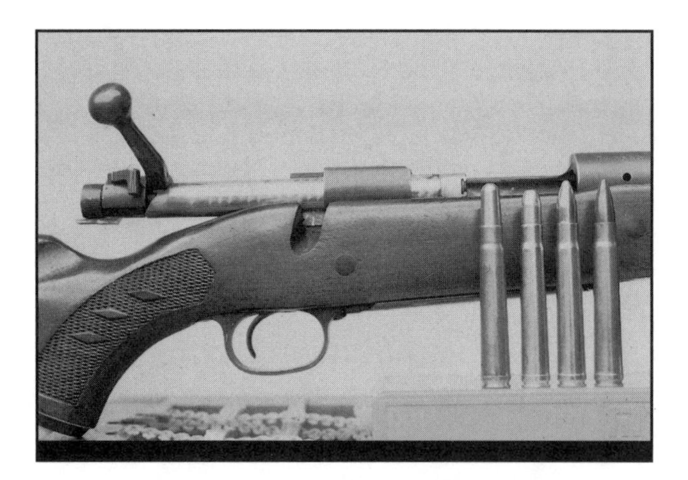

A professional hunter in Zimbabwe owns this .375 Model 70 Winchester, a Mauser derivative.

Model 54 with the new Model 70, destined to become the most popular centerfire rifle of the century. Chambered for every useful big game round (and for some that were not very useful at all), it had an action long enough to accommodate the .300 and .375 H&H Magnums, both of which were offered. The strong Mauser-type extractor and a rugged but finely-adjustable trigger helped boost Model 70 sales.

About this time Charles Newton's quest for more powerful hunting cartridges and innovative rifle designs was taken up by another experimenter. In 1937, insurance salesman Roy Weatherby moved from his native Kansas to California. To pursue his interest in firearms design, he bought a drill press and a lathe from Sears. He rebarreled or rechambered surplus 98 Mausers, 1903 Springfields, and 1917 Enfields to cartridges he fashioned at home. His first publicized round, the .220 Weatherby Rocket, was a blown-out Swift. His next, the .270 Weatherby Magnum, became the first in a stable of high-velocity hunting rounds based on .300 Holland brass.

During the years immediately following World War II, when deer and elk hunting in the western U.S. reached a zenith, the .30–06 and .270 became increasingly popular. The .30–06 had already supplanted the .30–30 as the most popular U.S. deer cartridge. In a 1947 survey reported by *The American Rifleman,* author J.S. Rose stated that "Hunters, 1948 model, want bolt actions and fast loads!" Roy Weatherby was there to accommodate riflemen who craved even higher performance.

Charles Newton developed rimless cartridges that, with modern powders, would have matched the ballistic performance of today's best, like this .300 Remington Ultra Mag. He was ahead of his time.

Weatherby quickly made a business of his avocation, promoting high velocity as a way to make lightning-like kills at long range. His .270 Magnum drove a bullet 300 fps (feet per second) faster than a .270 Winchester. The .257 and 7mm Weatherby Magnums that followed were, like the .270 Magnum, based on a shortened Holland & Holland case to fit .30–06-length actions. The .300 Weatherby Magnum, a 1946 offering, had a full-length case. All Weatherby rounds were distinguished by minimum body taper and a double shoulder radius. Roy Weatherby marketed these cartridges (and the rifles he built for them) expertly, making sure that prominent politicians and

The .270 Winchester appeared in 1925. It has sold well since, a tribute to its versatility and reach.

The .300 H&H Magnum (left) was introduced in the U.S. in 1925, the .348 Winchester a decade later. The .300 could use pointed bullets prohibited in the .348's tube magazine. The .300 set a trend.

movie stars hunted and were photographed with Weatherby products. In 1948 he standardized the Weatherby rifle, using a commercial Fabrique Nationale (FN) action. Ten years later he and company engineer Fred Jennie completed work on a new rifle action they called the Weatherby Mark V. Its bolt had nine locking lugs in three rows of three, and a low lift. The stock gave the rifle a futuristic look. But Weatherby rifles were expensive, and so was Weatherby ammunition, loaded only by Norma of Sweden. Weatherby's cartridges did not appear as chamberings in any other rifles. Winchester took note and in 1956 gambled on a short magnum with its .458. Two years later the New Haven firm followed with .264 and .338 Magnums on the .458 case.

Remington countered with a cartridge especially for mule deer and elk. The 7mm Remington Magnum, based on the same case as Winchester's new magnums, appeared in 1962. It got a boost from the concurrent introduction of Remington's Model 700 rifle. Intelligent advertising and performance tests on game accessible to the working-class hunter also helped. Les Bowman, a Wyoming big game guide who had helped develop the cartridge, praised it. In short order Remington's belted 7mm was far more popular than either of Winchester's new rounds—or the 7mm Weatherby Magnum, which had a market lead of nearly 20 years. In 1963 Winchester re-entered the magnum market with a .300. Since then, many high-performance cartridges from various makers have given big game hunters ever more reach.

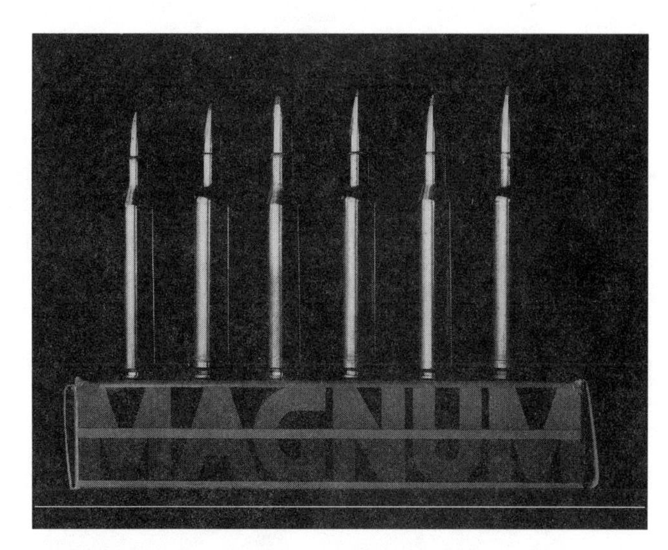

Roy Weatherby developed his magnum cartridges from the .300 H&H case in the early 1940s.

Naming Ammunition

The numbers used to describe ammunition have to do mostly with dimensions. But they don't all refer to the *same* dimensions. Also, the unit of measure can vary. And sometimes the numbers are rounded up or down so as not to duplicate numbers already in use. Names tacked onto the numbers don't always clarify. To a beginning shooter, it may seem odd that the .308 Winchester and the .30–06 use the same bullets. And that not every factory-loaded .300 Magnum will fit in his .300 Magnum rifle. How is it that there's a .220 that's bigger than a .221 and a .222? Why is there also a .218 and .219, a .223, .224, and .225 but only one common bullet size for all these ".22s"? The .30–30 is the same as the .30 WCF but not the same as the .30 Remington. And despite the popularity of .270 rifles, you can't find a bullet that measures .270 in diameter! Cartridge nomenclature remains a fascinating study that dates back to early muzzle-loading times, before there were any cartridges at all.

Labels on boxes of rifle and handgun cartridges usually specify bore diameter and parent (originator), as well as the name of the manufacturer. So you'll see .243 Winchester ammunition produced by Remington and .257 Roberts cartridges by Federal. This is important to know because a box of .300 Winchester Magnum rounds was not always the same as a box of "Winchester .300 Magnums." For some time after its debut in 1925, the .300 Holland & Holland Magnum was the only commercial .300 magnum around, so "Winchester" on the box meant

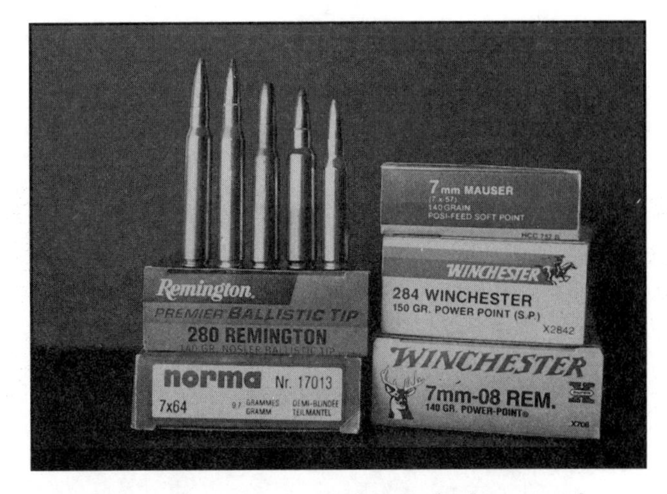

Cartridge nomenclature can be confusing, as evidenced here by cartridges that all use the same bullets. Decimal calibers are American, metric numbers European. "-08" refers to the parent .308 case.

only that the .300 H&H cartridges inside were manufactured by Winchester. Roy Weatherby came up with his own .300 magnum in 1944, Winchester its version in 1963.

Numbers designating caliber can be confusing because in rifle and handgun bores there are two diameters: one for the bore and one for the grooves cut as rifling to spin the bullet. The bore diameter (measured across the lands between grooves) is less than the groove diameter, and either number could be used as "caliber." The .250 Savage and .257 Roberts, for example, have bullets of the same diameter: .257 inch. That's groove diameter. But both .250 and .257 rifle bores have a .250 land diameter. The .270 Winchester is .277 across the grooves; .300s are .308 across the grooves, as are the .308 Winchester and .308 Norma Magnum. Centerfire .22s, from the .218 Bee to the .225 Winchester, use the same .224 bullets. Bullets for the .303 British make .311, for the .338 Winchester and .340 Weatherby Magnums .338, and for the .350 Remington and .358 Norma Magnums .358. The .243 Winchester, .375 H&H Magnum, and .458 Winchester Magnum, are named for bullet, or groove, diameter.

What about two-digit designations like the .22 Hornet, .30 Carbine, and .45 Colt? These numbers simply indicate bore diameter in hundredths, not thousandths of an inch. If you add a zero, you'll often come up with the name of a similar cartridge and a good idea of the bullet diameter. Sometimes the rule doesn't hold, though. The .38 Special uses the same .357 bullet as the .357 Magnum! *Two* two-digit numbers usually mean the cartridge dates to black-powder days, the second pair of digits indicating grains of black powder in an original load. The .30–30 (.30 Winchester Centerfire), was charged with 30 grains of black powder before it was loaded with smokeless powder. Of course, "grains" here is a weight measure, not a physical description as in "grains of sand." There are 437.5 grains in one ounce. The best-known exception to the paired double-digit tradition is the .30–06. This is a *30*-caliber

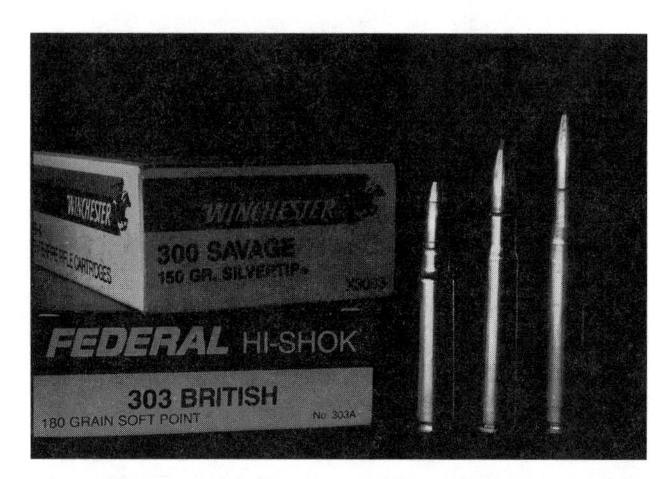

Numbers don't tell you much about cases or bullets. The .300 Savage (left) is smaller than .300s that dwarf the .30–06 (right). These two rounds use .308 bullets; the .303 British uses .311 bullets.

cartridge adopted by the U.S. Army for its new Springfield rifle in 19*06*. It replaced the .30–03. Another anomaly: the .250–3000 Savage. Developed by Charles Newton for Savage around 1913, it featured an 87-grain bullet at an advertised speed of 3000 feet per second. The company wanted the velocity (high for that day) in the cartridge name. Some old cartridges wear a third set of numbers, designating bullet weight in grains, as in .45–70–405, a designation for the official U.S. Army cartridge from 1873 to 1892. Three-number sequences can also show caliber, bullet weight, and case length, as in .45–120–3¼ Sharps. The 3¼-inch case is exceptionally long.

"Wildcat" cartridges are those not commercially manufactured but fashioned by handloaders from existing cases. The .338–06, for instance, is the .30–06 case "necked up" to .33 caliber. Some wildcats, notably the .25–06 and .22–250 (on the .250 Savage case), have gone commercial. For years, hunters handloaded the .35 Whelen, a .30–06 derivative named after Colonel Townsend Whelen, a firearms authority in the decades between the world wars. Remington began producing .35 Whelen rifles and ammunition in 1988.

Old cases have given us many modern cartridges. The 7x57, circa 1892, was revamped to produce the .257 Roberts and, later, the 6mm Remington. The .308 Winchester case fathered the .243 and .358. Remington's 7mm-08 later filled out that group. The 7–30 Waters is a .30–30 case blown forward and necked to .284.

We have many .22 rimfire cartridges (the priming compound in the case rim instead of in a separate, central primer). Perhaps the earliest was the .22 BB Cap, designed in 1845 for what were then known as parlor or saloon (salon) rifles. The "BB" stands for "bulleted breech." This round initially held no powder; the priming alone had enough thrust for indoor target shooting. In the 1880s the CB (conical bullet) Cap appeared, loaded with 1½ grains of black powder. The .22 Short and Long differ in case length. Originally, the Long Rifle had the Long's case but a much heavier bullet: 40 grains, compared to 29. A 37-grain hollowpoint followed. Now some hyper-velocity .22 Long Rifle rounds feature bullets of 33 to 36 grains.

In Europe, rifle and pistol bores are commonly measured in millimeters, not hundredths of an inch. Some American rounds have been labeled this way. The 6mm Remington uses a .243 bullet, the 6.5 Remington Magnum a .264 bullet (like Winchester's .264 Magnum). The various 7mms fire .284 bullets, the

The .30–06 is taller than the .300 Winchester Short Magnum, but not as powerful. Both use .308 bullets. The "-06" refers to that round's year of introduction to military service.

8mm Remington Magnum a .323 bullet. European cartridge designations also include case length in millimeters: The 7x57 Mauser has a 7mm groove diameter and a case 57mm long. Our .308 Winchester cartridge is known in military circles as the 7.62x51 NATO (51mm is the case length).

The term "magnum" means that the cartridge generates extra-high energy or velocity. But some magnums are not as potent as others with the same bore diameter. Some rounds not called magnums (though they may come from magnum cases) outperform cartridges that have a magnum label. The 7mm STW (Shooting Times Westerner) derives from the 8mm Remington Magnum and has a bigger case than the 7mm Remington Magnum, for example. "Belted magnum" refers to the belt in front of the extractor groove, present on cartridges based on the .300 H&H case. Almost all belted magnum rifle cartridges were fashioned from this case, which was itself an offshoot of the .375 H&H, circa 1912. The belt is a headspacing device and, like the "high brass" on heavy-load shotshells, has essentially nothing to do with structural reinforcement. The belt is a "stop" like the shoulder, case mouth and rim (flange, in British parlance) on other rounds. It prevents the case from entering the chamber too far. End play in the chambered round can result in case separations on firing.

Magnum handgun rounds like the .357 and .44 magnums are much smaller than the .300 H&H and are not belted. Magnum shotshells may be longer (a 3-inch Magnum 12-gauge) than standard shells, or may just have a heavier charge of shot. Standard-length magnum shotshells commonly produce *lower* velocity than non-magnums, but additional shot means denser patterns and greater effectiveness on game.

The British equivalent to "magnum" is "Nitro Express." It brings to mind nitroglycerine, the explosive used in some gunpowders, and evokes the image of a fast-moving train. The term came into common use early in the twentieth century. Mainly, it applies to rifle cartridges designed for thick-skinnned African game. "Black Powder Express" predated it, in the late 1800s. By the way, dual numbers on British cartridges signify parent cartridge and bore, in *reverse* order from American custom. A .450/.400 fires a .400 (actually .405)-diameter bullet.

Shotshell nomenclature amounts to another study altogether. Gauge is an early measure of bore diameter, telling the number of lead balls the size of the bore that equal a pound. So the bigger the bore, the smaller the gauge. Balls that fit a 16-gauge shotgun bore thus weigh an ounce apiece; those that fit a 20-gauge weigh less because the bore is smaller in diameter and more small balls are required to equal one pound. An exception is the .410, which is actually a caliber designation: .410-inch bore diameter. The .410 is equivalent to 67½-gauge. You can substitute "bore" for "gauge" when referring to a shotgun, as in "a 12-bore double." Common shotgun gauges are 10, 12, 16, 20, and 28. But during the late nineteenth century, "punt guns" of 4-gauge and even larger were mounted in small boats to kill waterfowl

The 7x64 uses .284-diameter (7mm) bullets in a case 64mm long.

Shotshell gauges indicate the number of lead balls the size of the bore that weigh a pound. Thus, a 12-gauge shell is bigger than a 16.

for market. Early African explorers used 1250-grain bullets in 8-gauge muskets, delivering 3 tons of energy to big game, even at the sedate speed of 1500 feet per second.

Labels on shotshell boxes show powder charge in "drams equivalent," a designation held over from when the propellant was black powder. A dram is a unit of weight; 16 drams equal one ounce. When smokeless powder supplanted black powder at the turn of the century, it was of a type known as "bulk powder" and could be loaded in place of black powder "bulk for bulk" (not by weight). "Dense" smokeless powders came later. They took up less space in the shell, so neither bulk nor weight measures of earlier loads applied. A "3¼ dram equivalent" charge is a smokeless charge that approximates the performance of a 3¼-dram black-powder charge. It has nothing to do with the amount of smokeless powder actually loaded.

The other numbers on a shotshell box refer to shot charge weight and pellet size. For example, a series that reads "3¼ 1¼ 6" gives you the drams equivalent, then tells you there's an ounce and a quarter of No. 6 shot. Lead birdshot ranges in size from No. 12 ("dust shot") to BB. A No. 12 pellet measures .05 inch in diameter, a BB .18. (This is not the same as a steel .177 air rifle BB.) Buckshot, in sizes 4, 3, 2, 1, 0, 00, and 000 range from .24 to .36 inch. One ounce of No. 9 birdshot contains 585 pellets, while an ounce of 00 buck has only 8. Since steel shot became mandatory for waterfowl, several new shot sizes have appeared (some resurrected). Steel pellets in sizes 7 to F were developed to duplicate the performance of lead pellets commonly used for ducks and geese.

These turkey loads, and other shotshells, bear numbers showing the powder charge in drams equivalent (dram measure is a term from black-powder days). There's also shot size and charge weight (ounces).

A Close Look at Primers

Modern centerfire primers derive from the percussion cap of the mid-nineteenth century. Drawn-brass, solid-head cartridge cases have no anvil (a solid surface against which the firing pin strikes to ignite the priming powder), so beginning in 1880, primer cups have been manufactured with anvils, to be installed in cases with a central flash-hole. Called Boxer-style primers, they now come in two sizes for rifle cartridges, two for handgun. The large and small rifle primers are of the same diameter as the large and small pistol primers, respectively, but the primer pellets differ. A large rifle primer weighs 5.4 grains and carries .6 grain of priming compound.

European cartridge designers followed their own path, choosing to incorporate the anvil in the case and to punch two flash-holes in the primer pocket on either side of the anvil. The Berdan primer, without an anvil, has a couple of advantages over the Boxer: There's more room in it for priming compound and the flame can go straight through the holes rather than having to scoot around the anvil and its braces. The Boxer primer owes its popularity largely to handloaders, who can pop the old primer out while sizing the case in a die with a decapping pin. Berdan primers must be pried out with a special hook or forced free with hydraulic pressure. In a strange twist, Edward Boxer, for whom American-style primers have been named, was British. Europeans named their primer design after

Modern rifle cartridges, like this .416 Remington Magnum, have a central primer with its own anvil. It is called a Boxer primer. Berdan-primed cases feature an integral anvil, double flash-holes.

RCBS now supplies plastic strips that hold primers (and a special tool) that makes priming easier.

Hyram Berdan, an American. Both men were military officers.

The first Boxer primers ignited black powder easily but sometimes failed to fire smokeless. When more fulminate was added to the priming mix, cases began to crack. Blame fell on the propellant, but the culprit was really mercury residue from the primer. This residue, which had been largely absorbed by the fouling from black powder, accumulated in harmful concentrations in the wake of clean-burning smokeless. It attacked the zinc in the case walls, causing them to split. A typical primer mix at this time was 41% potassium chlorate, 33% antimony sulfide, 14% mercury fulminate, 11% powdered glass, 1% gelatin glue.

The first successful non-mercuric primer for smokeless loads was the military H-48, developed in 1898 for the .30–40 Krag. The main detonating component was potassium chlorate, whose corrosive salts did not damage the case but could ruin the bore by attracting water. Vigorous cleaning with hot water and ammonia, followed by oiling, kept rifling shiny, but even short-term neglect could result in pitted bores.

In 1901 the German Company Rheinische-Westphalische Sprengstoff (RWS) announced a new primer, with barium nitrate and picric acid instead of potassium chlorate. These compounds did not cause rust. Ten years later the Swiss had a non-corrosive primer and German rimfire ammunition featured *Rostfrei* (rust-free) priming. *Rostfrei* contained neither potassium chlorate nor ground glass, the common element used to generate friction when the striker hit. The solution proved imperfect, because barium oxide turned up as a residue, scouring the barrel as aggressively as had the glass. Glass was purged from American military primers before World War I, when the FH-42 supplanted the H-48.

Wartime primer production overloaded drying houses in the U.S., causing sulfuric acid to build up in the priming mix. Misfires resulted. The FH-42 gave ground to the Winchester 35-NF primer, later known as the FA-70. This corrosive primer remained in military service through World War II. Its composition: 53% potassium chlorate, 25% lead thiocyanate, 17% antimony sulfide, 5% TNT.

Non-corrosive priming arrived in the states as Remington's "Kleanbore" in 1927. Winchester followed with its "Staynless," Peters with "Rustless." All contained mercury fulminate. It would fall to German chemists Rathburg and Von Hersz to remove both potassium chlorate and mercury fulminate from primers. Again, Remington took the lead with the first U.S. version of a non-corrosive, non-mercuric primer. The main ingredient was lead trinitro-resorcinate, or lead styphnate. Then comprising up to 45 percent of the priming mix, lead styphnate remains an important component in small arms primers today. The U.S. Army adopted non-

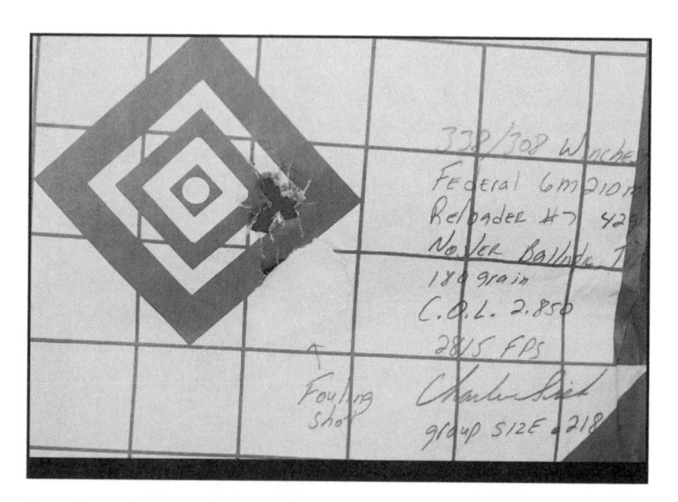

Loads this accurate reflect attention to detail. The shooter noted not only powder charge and bullet, but the primer, a Federal 210.

The author shot this Quebec caribou with an iron-sighted rifle after a long stalk. Reliable primers are a relatively recent blessing.

corrosive, non-mercuric primers in 1948 and currently uses one designated FA-956. It comprises 37% lead styphnate, 32% barium nitrate, 15% antimony sulfide, 7% aluminum, 5% PETN, and 4% tetracene.

During the 1940s, development of big, broad-shouldered rounds and extra-slow-burning powders called for a stronger ignition flame. More priming compound was not the answer, because it might shatter powder directly in front of the flash-hole, causing erratic pressures and performance. Dick Speer and Victor Jasaitis, a chemist from Speer Cartridge Works, came up with a better idea: the addition of boron and aluminum to the lead styphnate base to enable a primer to burn longer. The result would be more heat, more complete ignition before primer fade. This first successful magnum primer, was just what Roy Weatherby needed for his magnum cartridges. Other munitions firms now offer similar sparkplugs. Speer primers, still made in Lewiston, Idaho, wear a new label these days: CCI (Cascade Cartridge Industries).

Shotshell manufacture was plagued with the same primer problems visited on the production of rifle and pistol cartridges, and it came by the same salvation. But the shotshell cup assembly is different. A deep battery cup holds the anvil and a smaller cup containing the detonating material. As with rifle primers, a foil cover protects the pellet. The primer pocket of a shotshell has no bottom and, hence, no flash-hole. The flash-hole is in the battery cup. Made of thin, folded brass like old balloon-head rifle cases, shotshell heads are reinforced with a dense paper base wad or a thick section of hull plastic. Battery cups seal the deep hole in the base wad.

Primer manufacture is much the same now as it was at the outbreak of the second world war. Equipment is more sophisticated, but huge batches of primer cups are still punched and drawn from sheet metal and indexed on large perforated metal tables. A second perforated plate is smeared with wet, dough-like priming compound and laid precisely on top of the first so the little dabs in the holes can be punched down into the open-faced cups, or the cups are filled by a worker brushing the compound across the face of the table. Next comes the thin foil disc (or a shellacked paper cover). Anvils, punched from another metal sheet, follow. They're inserted as the plates line up. Priming mix is stable when it is wet, extremely hazardous to work with when dry. No mix is ever allowed to dry in the primer room.

Primer choice can influence pressures, velocity, and accuracy.

Powder Makes Things Happen

Propellants have certainly influenced rifle and cartridge design. Black powder fueled all firearms until the mid-nineteenth century. Mild steels effectively bottled the pressures it generated. Nitroglycerine, discovered in 1846 by Ascanio Subrero in Italy, promised higher performance but also metal-splitting pressures. "Nitro" is a colorless liquid comprising nitric and sulphuric acids plus glycerin. Unlike gunpowder, it is not a blend of fuels and oxidants; rather it is an unstable, oxygen-rich chemical compound. It can almost instantly rearrange itself into more stable gases. All it needs is a little jolt. As nitroglycerine ages it can become more unstable, more dangerous.

In 1863 Swedish chemist Emmanuel Nobel and his son Alfred figured out how to put this touchy substance in cans. That made it easier to handle but no less hazardous. Several shipments blew up, as did the Nobel factory in Germany. Alfred eventually discovered that soaking the porous earth *Kieselguhr* with nitro rendered the chemical less sensitive to shock. This process led to the manufacture of dynamite, patented in 1875. Dynamite in its original form is now rarely used, but the term has become a generic name for more effective, more manageable explosives.

During the 1840s, Swiss chemist Christian Schoenbein had discovered that nitric and sulfuric acids applied to cotton formed a compound that burned so fast as to consume a cotton patch without setting fire to a heap of black powder on top of it! Schoenbein obtained an English patent to cover his work and sold the procedure to Austria. Shortly, John Hall and Sons built a guncotton plant in Faversham, England. It promptly blew up. So did most of the other guncotton plants that followed. The substance had little use as a propellant because it burned too fast, almost detonating. But as an explosive designed to tear other things apart, guncotton excelled.

Much of powder development has proceeded through trial and error. Some of the errors have resulted in tragedy. Chlorate powders, pioneered by

Gunpowder, black or smokeless, does not explode when ignited in open air. It just burns fast.

Development of smokeless powder was a boon to soldiers. Not only did it give bullets greater speed; it was nearly invisible to the enemy. Black powder signaled their positions with clouds of white smoke.

Canister powders are blends ready for retail sales to shooters. They meet rigid specifications.

Berthollet in the 1780s, had great energy. But when a French powder plant at Essons blew up in 1788, the chemist concluded potassium chlorate was too sensitive for use as a propellant. Combined with other compounds, it was later put to harness. During the 1850s J.J. Pohl developed a substance he called "white powder." It comprised 49% potassium chlorate, 28% yellow prussiate of potash, and 23% sulfur. A second-rate propellant, it proved valuable to the Confederacy during the Civil War when black powder became unavailable. War-time backyard powder mills turned out propellants of varying colors, shapes, and potencies. Some formulas were sold through the mail by con artists who baited innocents with claims that coffee, sugar, and alum, combined with potassium chlorate and a lead ball, gave them ammunition!

Despite frequent setbacks, experimenters persevered in the late nineteenth century, keen to improve upon black powder and heartened by Nobel's success with nitro. In 1869 German immigrant Carl Dittmar built a small plant to make "Dualin," sawdust treated with nitroglycerin. A year later he came out with his "New Sporting Powder." In his native Prussia, he had tried to make smokeless powder from nitrated wood but ran out of money. By 1878 he was able to start up a mill in Binghampton, New York. An explosion destroyed the mill a few years later, however, taking part of Binghampton with it. Then Dittmar's health failed, and he sold what was left of his firm. A foreman, Milton Lindsey, wound up at the King Powder Company, where he worked with company president G.M. Peters to develop "King's Semi-Smokeless Powder." They patented their formula in 1899. Dupont's "Lesmoke" appeared soon after, with roughly the same components and proportions: 60% saltpeter, 20% wood cellulose, 12% charcoal, and 8% sulfur. One of several semi-smokeless powders that turned up around the same time, "Lesmoke" proved a fine propellant for .22 rimfires. Fouling remained a problem, but the residue didn't harden as with black powder. Its soft residue gave semi-smokeless an advantage over the first smokeless powders, which left nothing to carry away residue from the corrosive primers of the day. "Lesmoke" was more hazardous to produce than smokeless, however. It was discontinued in 1947.

French engineer Paul Vielle is generally credited with developing the first successful smokeless powder. His "Poudre B" comprised ethyl alcohol and celluloid. But in the mid-1870s, a decade before Vielle's achievement, Austrian chemist Frederick Volkmann had patented cellulose-based (single-base) powders. Unfortunately, Austrian patents were not acknowledged world-wide, and in 1875 Austria's government began enforcing its monopoly on the domestic powder supply. Volkmann closed his plant, then disappeared. His claims of transparent smoke, less barrel residue, more power and safer handling showed that this experimenter was on track to improve his propellant.

In 1887 Alfred Nobel increased the proportion of nitrocellulose in blasting gelatin and found he could use the new compound as a propellant. A year later he introduced "Ballistite." This double-base powder (it contained both nitrocellulose and nitroglycerin) was similar to a new powder developed concurrently by Hiram Maxim, of machine-gun fame. About this time, the British War Office started a search for a more effective rifle powder. The committee came up with "Cordite." This propellant, a modification of Nobel's formula and Maxim's, was at one stage in manufacture a paste that could be squeezed through a die to form spaghetti-like cords. The early formula included 58 percent nitroglycerine and 37 percent guncotton. The proportions were later changed to 30 percent nitroglycerine and 65 percent guncotton. The remainder of this most recent blend was mineral jelly and acetone. Nobel and Maxim sued the British government for patent infringement but got nowhere.

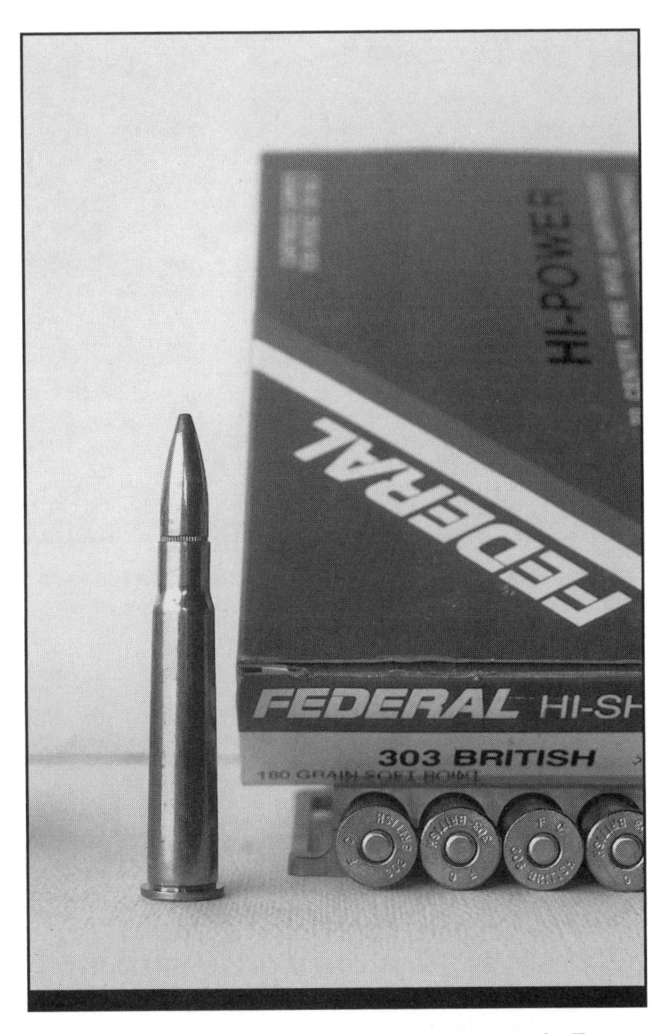

The first smokeless powder of consequence in England was cordite. Its long strands were loaded, spaghetti-like, into cases like the .303 British. Steep case tapers were popular for that reason.

Their nitrated lignin gave early smokeless powders a lumpy, fuzzy appearance. Variations in density put handloaders on notice. "Bulk" powders could be substituted, bulk for bulk, for black powder. "Dense" or gelatin powders could not be safely measured by bulk, because their energy-per-volume ratios were higher. The shooting industry responded by marking shotgun loads in "drams equivalent." That is, the smokeless powder used gave the same performance as black-powder shotshells loaded with the marked number of drams. Boxes of rifle and pistol cartridges did not traditionally carry propellant specifications, so handloaders had to mind new recommendations for the dense powders.

The 8mm Lebel, adopted in 1886, became the first military rifle designed for smokeless powder. Other countries quickly followed suit: England with the .303 British in 1888 and Switzerland with the 7.5x55 Schmidt-Rubin in 1889. By the mid-1890s

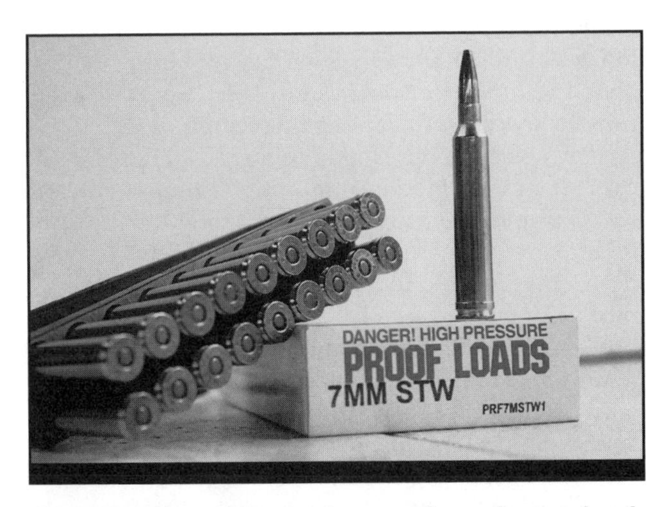

Special types and heavy charges of powder are developed for proof loads, used in testing guns and ammunition at the factory.

nearly all nations that could muster an army were equipping their soldiers with small-bore bolt rifles firing smokeless cartridges. The new propellant boosted bullet speed by a third and did not give away a rifleman's position with white clouds of spent saltpeter.

Most powder companies established in the 1890s failed. Fierce competition and flawed product, combined with the hazards of powder manufacture, made this a tough business. But most firm that buckled managed to sell their specialized equipment to larger, profitable companies. Mergers were common. In 1890, an English physician practicing in San Francisco formed the United States Powder Company to produce ammonium nitrate propellant. Samuel Rodgers joined his business that year with the Giant Powder Company. His "Gold Dust Powder" comprised 55 percent ammonium picrate, 25 percent sodium or potassium nitrate and 20 percent ammonium bichromate. This foul-smelling shotgun powder did not lend itself to bulk measure. The Giant Powder Company plant was destroyed by an explosion in 1898.

Tennessee's Leonard Powder Company, maker of "Ruby N" and "Ruby J" powders, folded in 1894. It was succeeded by the American Smokeless Powder Company, a New Jersey firm producing propellant under contract for the government. A creditor, Laflin & Rand, acquired ASPC in 1898. Laflin & Rand had once sought American rights to Ballistite but considered Nobel's price of $300,000 plus royalties too steep. Ballistite manufacture later came under the control of DuPont, which contracted that job to Laflin & Rand. The famous "Lightning," "Unique," "Sharpshooter," and "L&R Smokeless" powders were Laflin & Rand developments.

Vermont's Robin Hood Powder Company became the Robin Hood Ammunition Company before it was purchased in 1915 by the Union Metallic Cartridge Company. The American E.C. & Schultz Powder Company sold to DuPont in 1903, then became part of Hercules when DuPont was split by court order in 1912. Such corporate restructurings resulted mainly from under-capitalization. Many ambitious entrepreneurs who saw the huge profit potential in smokeless propellants failed to procure government contracts, believing mistakenly that private sales could sustain a young company. As with rifles, one military contract could herald salvation. Rejection by ordnance boded ill—even if, as one powder firm claimed, American sportsmen burned ". . . not less than three million pounds per annum."

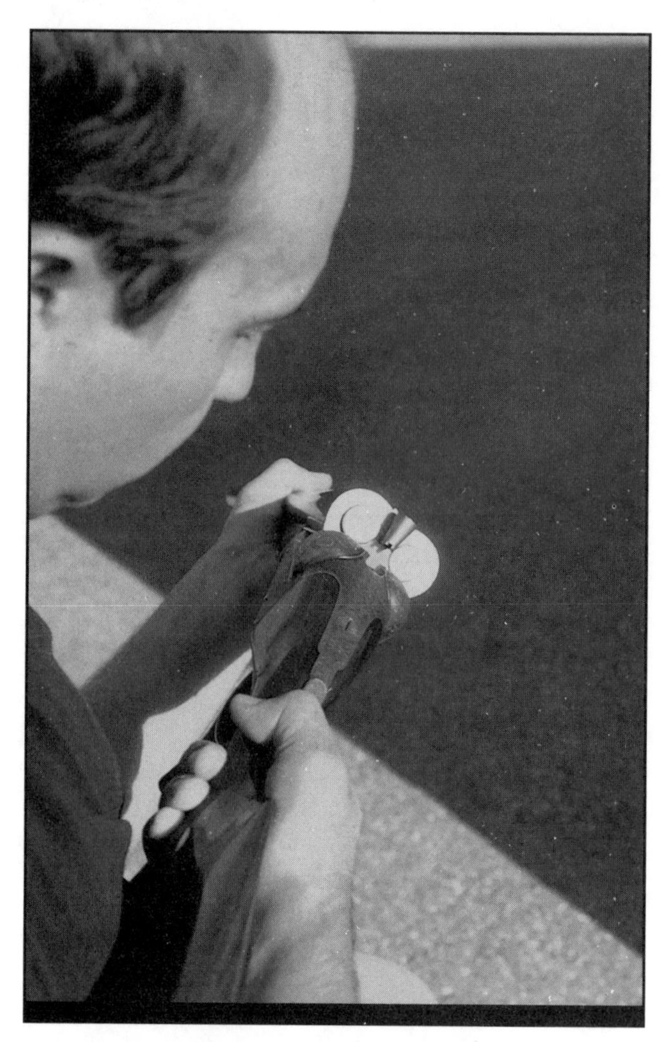

The muscular look of this double rifle belies the fact that many cartridges designed for bolt-action rifles operate at higher pressures.

During the late 1890s, "Peyton Powder" by the California Powder Works fueled .30–40 Krag ammunition for the U.S. military. This double-base powder had a small amount of ammonium picrate. Laflin & Rand manufactured a double-base powder for the Krag: the "W-A," named for developers Whistler and Aspinwall. "W-A" contained 30 percent nitroglycerin, which contributed to high burn temperatures and erosive tendencies.

DuPont and California Powder Works began filling military contracts in 1997, with nitrocellulose

These slow-burning powders excel in small-bore magnum cases like the .264 Winchester, and with heavy-bullet loads in the big .300s.

Winchester was the first to market ball powders, in the 1930s. These days all its ball powders are double-base, containing nitroglycerine as well as nitrocellulose.

powders that looked somewhat like Cordite. Guncotton dissolved in ether and alcohol formed a colloid. Pressed into thin, hollow strands, the colloid was then chopped into short tubes. "Government Pyro" became one of the first of these powders to see military use, in the 1903 Springfield's .30–06. DuPont later fueled that cartridge with 1147 and 1185 powders. IMR (Improved Military Rifle) 4895 became the powder of choice for the .30–06M2. It is still among the best choices for handloaders wanting maximum velocity and accuracy from middleweight bullets in the .30–06.

DuPont continued to dominate the powder industry after Hercules started up its own operation in Kenville, New Jersey. It manufactured dynamite and up to a ton of small arms powder daily. By the onset of World War I, Hercules had established a plant in Parlin, New Jersey, there producing nitrocellulose and popular rifle and pistol powders, including Bullseye, Infallible and HiVel. But it was DuPont that got the big war contracts. It built two new plants in Old Hickory, Tennessee and Nitro (really), West Virginia. Combined capacity: 1.5 million pounds per day! After the war, DuPont bought the town of Old Hickory to establish a rayon factory. But the powder magazines remained. One day in August 1924 they caught fire. More than 100 buildings and 50 million pounds of powder vanished. Hercules manufactured Cordite powder for the British government—up to 12,000 pounds a day. Wartime Cordite production totaled 46 million pounds. In addition, Hercules sold 3 million pounds of small arms propellants and 54 million pounds of cannon powder. It was a quick answer to an immediate need. Before 1914, Hercules had made no artillery powder and only token amounts of small arms powder for sportsmen.

The Great War brought new ideas to domestic powder makers. Plagued by copper residue fouling cannon bores, U.S. munitions experts took a tip from the French and added tin to their propellants. Soon rifle powders got the same treatment. DuPont's No. 17 became No. 17½ with a dose of 4 percent tin. No. 15 became No. 15½. Tin levels were halved when dark rings appeared in the bores of National Match rifles, a result of the tin cooling near the muzzle. Black powder, still in use for detonating charges of high-energy propellants in artillery shells, and to propel ejection seats clear of aircraft, is made pretty much as it was in the nineteenth century. It still comprises sulfur, charcoal, and saltpeter, ground fine and mixed at 3 percent moisture. The powder "meal" is pressed into cakes, which are fed through a granulating machine where toothed cylinders chop them up. Screening segregates particles by size; then they're polished in revolving wooden barrels. Black powder is labeled by granule size: A-1, Fg, Ffg, FFFg, FFFFg, and FFFFFg, in decreasing order of size. Bigger grains generally burn slowest and work best pushing heavy balls or bullets. Very

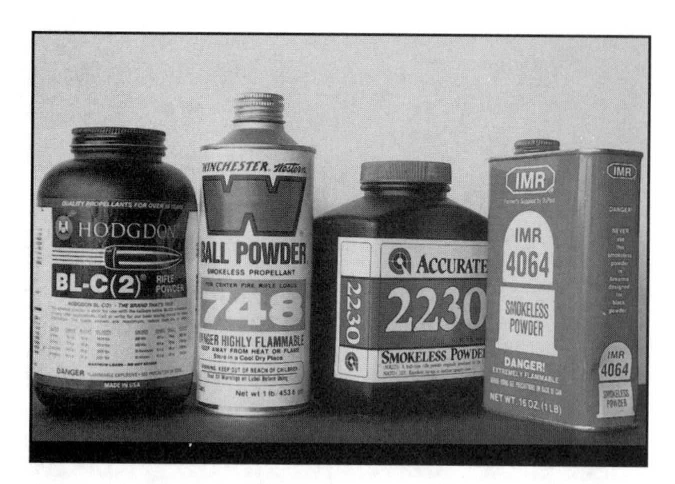

Medium-fast propellants like these work best in cases of modest capacity, like the .308 Winchester.

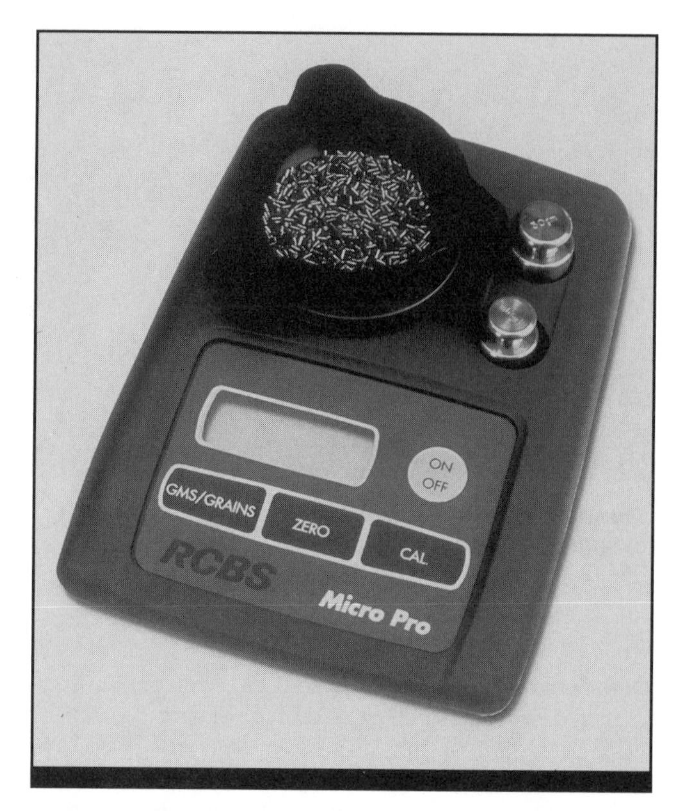

Powder charges are expressed in grains weight, 437.5 grains to the ounce.

fine black powder is suitable only for priming charges and pyrotechnics.

Smokeless powder results from a different process. It starts as nitrocellulose (a vegetable fiber soaked in nitric and sulfuric acids). "Pyrocellulose" is an industry name that refers to artillery powders used during the first world war. Guncotton, a special kind of nitrocellulose, has slightly higher nitrogen content (13.2 percent compared to a standard 12.6) and lower solubility in ether-alcohol solution. Like other forms of nitrocellulose used in powder production, it comes from short crude-cotton fibers or "linters," which are first boiled in caustic soda to remove oils. Water formed by the nitric acid during nitrating is absorbed by the sulfuric acid, thus preventing decomposition by hydrolysis. A centrifuge next strips excess acid. Following a rinse, the linters are boiled for 48 hours to remove all traces of acid, which can cause spontaneous combustion. The cotton is beaten and boiled again, five more times. Agitators fluff it. The nitrocellulose is washed in solvent, then heated to evaporate the solvent. Only hard grains of powder (and water) remain. "Cooking" the solvent off with heat applied to a wet solution is less dangerous than the old air-dry method. Ether is used to dissolve the fibers in nitrocellulose marked for single-base powders, acetone for double-base. Nitroglycerin is then added to form the double-base powder.

At this stage, the powder is a soup, and unstable. More mixing turns the soup to a jelly, which is squeezed through dies (extruded) to form slender tubes of precise dimensions. Rollers run these "noodles" through a plate, where a whirling knife shears them off in measured increments. The sections or grains of single-base powders still contain ether, so are sent to a warm solvent recovery room and soaked in water for about two weeks. Wet single-base and freshly-sheared double-base powders are then air-dried and sieved. Next, the granules are polished in drums that coat them with graphite. The tumbling smooths edges that might produce heat if allowed to scrape as the powder is jostled. The graphite further reduces friction. Gunpowder's slate color, incidentally, derives from that graphite coating. Uncoated powder is yellow.

Extruded powder granules are tube-shaped to mitigate the "burn down" effect that reduces gas production as the grains shrink in size. Perforated sections burn from both the inside (increasing burning surface) and outside. The burning rate of a powder is partly determined by the ratio of inside diam-

The Hercules series of Reloder powders includes these medium-speed (RL-15) to very slow (RL-22) fuels.

eter to outside diameter, partly by grain size and wall thickness.

Grain shape and composition determine not only burn rate but the shape of the burning curve. If powder grains are spherical or rod-like, the powder's surface quickly diminishes as the charge is consumed. This type of powder is called regressive. Powder formed in thin strips or flakes, or in tubes, loses little of its surface relative to time during the burn. It can be termed "neutral-burning" fuel. Progressive powder has been fashioned to increase in surface area during gas production. Tubular cannon powders with multiple perforations are an example of a progressive-burning powder; however, other grain shapes can be made to burn progressively with the addition of chemical deterrents in the powder. Deterrents like dibutylphthalate are commonly used to change the burning rates and pressure curves of sporting powders.

In 1933 Western Cartridge company produced the first successful ball powder. Ball powder manufacture differs from the production of extruded powders, though the raw materials are essentially the same. Nitrocellulose intended for ball powders goes through a hammer mill that grinds it to a fine pumice. Blended with water and pumped in slurry form into a still, the nitrocellulose combines with chalk added to counteract the nitric acids still present. Ethyl acetate dissolves the nitrocellulose, producing a lacquer. Heat and agitation then break the lacquer into small particles; or it is pressed through plates much like extruded powders and chopped to pieces by whirling knives. Tumbling and heating leave the grains round. When they're of proper size and roundness, the ethyl acetate is distilled off and salt is added to draw out moisture. In a slurry of fresh water, the powder rushes through sizing screens. A heated still adds nitroglycerin. Coatings of burning deterrents come next, to smooth the pressure curve by controlling burn rate. A centrifuge removes excess water. The grains are tumbled in graphite, then sized again. Some ball powders are measured blends of sizes. Ball powder meters better than stick powder, and its manufacture is quicker and simpler.

DuPont: Cornerstone of an Industry

E. I. DuPont de Nemours was a well-known company when I started shooting. Since its inception in 1802 on Delaware's Brandywine River, it had grown with the country. By the end of World War II, any list of major U.S. corporations included the chemical giant DuPont. But the varied product line it fielded in my youth could hardly have been imagined by its founder, French immigrant Eleuthere Irenee DuPont.

"I can make better black powder than what your country has in its magazines," DuPont told Alexander Hamilton. Apparently the prospect appealed to our government; it helped the enterprising engineer build a plant in Wilmington. The new propellant satisfied U.S. ordnance officers, and DuPont put down roots. Gunpowder was the firm's primary product for most of the nineteenth century. In the 1880s DuPont built a plant at Carney's Point to boost capacity. During World War I, 25,000 people went to work at this facility on the Brandywine, providing more than 80 percent of the military powders used by the Allies (the British, French, Danes, and Russians, as well as U.S. troops).

Soon after the transition from black to smokeless powders at the close of the nineteenth century, "MR" began appearing on canisters of DuPont powders. It meant "military rifle." The IMR line of "improved military rifle" powders came along in the 1920s, when four-digit numbers replaced two-digit numbers in DuPont designations. MR 10 and the like died out. IMR fuels, beginning with 4198, supplanted them. The first had relatively fast burn rates, because in those days, rifle cartridges were small. In 1934, DuPont introduced IMR 4227. In the early 1940s the firm developed IMR 4895 specifically for the .30–06 in the M1 Garand service rifle. About that time the first slow IMR propellant made its debut. Developed for 20mm cannons, IMR 4831 would become one of the most popular powders for high-capacity rifle cartridges developed by wildcat-

The DuPont chemical company began as a powder company when our nation had just been declared independent. The "IMR" designation came along in the 1920s.

ters like Roy Weatherby and P.O. Ackley. Incidentally, the numbers have nothing to do with burning rate. According to long-time DuPont engineer Larry Werner, powder from DuPont is labeled chronologically. The highest numbers indicate the most recent propellants.

"IMR powders are introduced mainly to fill gaps in the product line," says Larry. "Gaps appear with the advent of new cartridges." IMR 7828, for example, is a super-slow powder that would have had little application in rifle cartridges before the 1950s. It suited some Weatherby magnums, but there weren't enough handloaders using the fast Weatherbys to justify the substantial expense of developing a powder. In 1958 Winchester's .264 Magnum added incentive. But the .264 flopped at market, and the more versatile 7mm Remington Magnum needed nothing slower than IMR 4831 for bullet weights to 160 grains. But more recent long-range cartridges have bigger case capacities relative to bore volume. The 7 STW, 7mm and .300 Remington Ultra Mag, and the .30–378 and .338–378 Weatherby can all use 7828 to advantage. By the way, IMR 7828 makes a top performer out of the undersung .264 Winchester, pushing 140-grain bullets up to 300 fps faster than the anemic 3030 fps claimed in catalogs.

IMR 7828 is a relatively recent powder, very slow. It's ideal for the .264 Winchester and .257 Weatherby.

Not all powders finish the trip from inception to store shelf. At DuPont, propellants under development have traditionally been tagged "EX" for "experimental." Larry, who started working with the company in 1954, recalls that "almost always, firearm design preceded cartridge design. Powder for a new round came last. We'd start with a cookbook formula, then tweak the mix until we had it right." Once the customer was satisfied and committed to a batch (generally 5000 pounds), the "EX" was dropped from the powder designation, and commercial labeling replaced it. Propellants made expressly for the military had no number prefix. "For instance, EX 7383 was developed for 50-caliber spotter rounds during the 1950s. DuPont made lots of it, but not for the civilian market. So you don't see the number in loading manuals."

The best-known IMR rifle powder may still be 4895. It remains one of the most versatile as well, a first pick for most medium-bore cartridges based on the .30–06 and .308 (7.62 NATO) cases. IMR 4320 and 4064 lie close in burning rate. You'll find differences in charts ranking the burn rates of IMR and other smokeless powders. The reason: powders can behave differently as you change case shape and bore diameter, fuel charge and bullet weight. IMR gives all its powders a Relative Quickness value, assigning IMR 4350 an arbitrary value of 100. According to Larry Werner, quick-burning IMR 4227 has an RQ of 180; IMR 4198 comes in at 165 and IMR 3031 at 135. IMR 4064, 4320, and 4895 are listed at 120, 115, and 110 respectively, though you'll find some loading manuals suggest a different order. "Closed bomb" tests are used to measure burn rate. A unit charge of powder ignited in a chamber of known volume produces a pressure curve that's then compared to the curves from other propellants.

Tiny groups like these are possible only when powder burns consistently, one caseful to the next.

There's no better powder for the .30–06 than IMR 4895. It shines in other mid-capacity cases too.

DuPont's old MR powders included single-base (nitrocellulose) and double-base (nitrocellulose with nitroglycerine) varieties. "The nitro gives you more energy per grain," explains Larry, "and it reduces the tendency for the grains to pick up moisture. Its drawback is more residue. Double-base powders generally don't burn as clean." Larry adds that to get the full effect of nitroglycerine, you really need 8 to 12 percent in the mix, but that some powders claimed to be double-base contain less. "All commercial ball powders are double-base," he says. The current IMR line includes only single-base propellants.

According to Larry, IMR's load data is developed using 20-round strings in the lab. The resulting pressure and velocity averages are a sound basis for load recommendation. However, loading and shooting conditions affect powder performance. Once, Larry recalled, Remington technicians on the commercial shotshell line left powder canisters uncovered when production was shut down for the weekend. The SR 7625 powder used in those 1¼-ounce 12-gauge loads did not perform up to par because the propellant had picked up moisture from the air. It acted like a slower powder. Cold temperatures on a hunt can also drain the energy from gunpowder. Conversely, ammo left in a hot car may produce higher breech pressures.

IMR powders are no longer made by DuPont. World events and a changing society have shaped

DuPont and delivered the IMR trademark to EXPRO, another chemical firm. Here's the historical sketch: The 1930s humbled many mighty corporations. DuPont weathered not only the Depression, but scathing political attacks from certain U.S. senators who accused the company of war-mongering. As Hitler revved up his war machine and the U.S. prepared to rearm, DuPont boosted its production capacity. "But the company was fed up with the treatment it had received from Congress," remembers Larry. "Rather than build new plants, it contracted to operate government facilities for a dollar a year. That way, it could not be said to have had a stake in the hostilities. Of course, the government had no powder works that could match DuPont's, so the firm supervised construction of seven factories modeled on the Carney's Point plant. Another went up in Canada. At the height of the second world war, these operations shipped *a million pounds of powder a day.* We had 16 million guys in uniform then, and they used a lot of ammo!"

After armistice, the government took control of the propellant plants. Carney's Point stayed in production for DuPont. So did the Belin Works at Scranton, Pennsylvania. "Belin made black powder for the government," says Larry. It was used as an igniter in artillery shells. A plant explosion in 1971 sent debris over the nearby interstate and littered runways at the local airport. The plant closed right away but was reopened for military production to supply troops in Vietnam. "In 1973 Marv Gearhardt and Harold Owens bought the Belin Works for $400,000 and contracted to make powder," continued Larry. "Their company, GOEX, later moved to Louisiana, and the Scranton facility shut down."

Explosions at powder plants occur infrequently, but they're predictably serious. In June 1969, an accident in the finishing area at Carney's Point destroyed the facility. A new finishing area was built at DuPont's Potomac River Plant near Falling Waters, Virginia. Powder granulated at Carney's Point was freighted there for finishing. In April 1978, a fire swept through the main mix house at Carney's Point, cancelling operations there. That summer, DuPont contracted with Valleyfield Chemical Products in Quebec to produce its commercial smokeless propellants. (The Valleyfield plant was the Canadian factory built during World War II on the Carney's Point model and had been operated by CIL, or Canadian Industries, Limited, a branch of the government.) DuPont helped the Valleyfield people start their new enterprise and even finished the first few batches of powder at the Potomac plant. In 1982, Valleyfield Chemical sold to Welland Chemical, which became EXPRO.

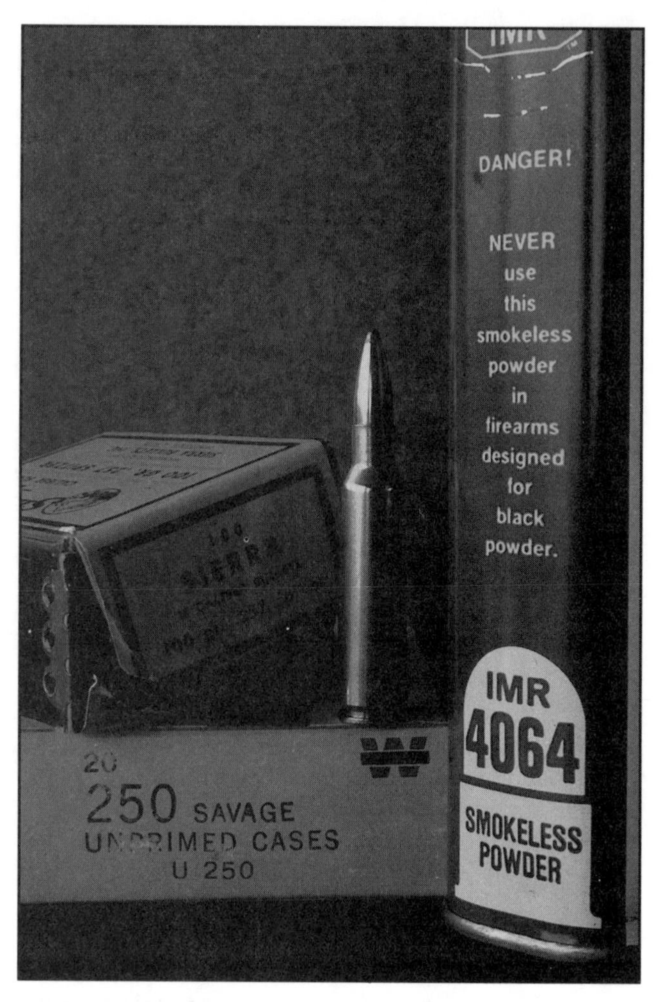

Like IMR 4320 and 4895, IMR 4064 is a well-established medium-burning "stick" powder.

In December 1986, DuPont sold its smokeless powder business to EXPRO. The IMR Powder Company established itself as a testing and marketing firm for EXPRO propellants. IMR's laboratory and offices in Plattsburg, New York, develop ballistics data for IMR powders and package and distribute them to dealers. EXPRO, with an annual manufacturing capacity of more than 10 million pounds, also produces other powders, including Alliant. Primex, another large U.S. powder firm, is owned by Olin and makes propellant for the Winchester, Ramshot, and Alliant labels. Though DuPont owned 70 percent of Remington for decades, it also provided powder for competing ammunition firms. Ironically, Hodgdon, a long-time IMR competitor, got its start with surplus powders, much of which had been produced by DuPont.

After 1945, huge stocks of surplus military powders remained in government arsenals. Bruce Hodgdon thought he could sell it to handloaders. He bought as much as he could store in underground rail cars and began parcelling it out at a profit. His business grew, and eventually his stocks of powder ran out. "A lot of that powder was 4831," says Larry Werner. "To get more of that powder for his customers, Bruce went to an Australian factory with a sample of the remnant. New production matched that sample. But the sample—and consequently the new 4831—did not quite equal the burning rate of original 4831 as it had been produced decades before for the war. Even proper storage can influence burning characteristics over time. That's why Hodgdon's 4831 is a tad slower than IMR's, and you're not wise to interchange the data." I've found in magnum cases like the 7mm Remington and .300 and .338 Winchester that a charge weight increase of about 4 percent is called for when substituting H4831 for IMR 4831, to get equal velocity.

Though the powders marketed with the IMR label have changed since the 1940s, and America's powder industry is nothing like E.I. DuPont de Nemours found it in 1802, shooters in these United States remain indebted to the enterprise of this young French immigrant. Twenty decades later, they can also thank the generations of men and women who produced gunpowder for the free world's armies at DuPont and EXPRO plants. Gunpowder isn't just for hunting.

A handloaded Barnes bullet from a Magnum Research rifle in .280 Remington reached nearly 400 yards to drop this fine buck for the author. IMR 4350 is a useful powder for the .280.

Bruce Hodgdon's Big Gamble

Years ago in somebody's basement I was shown a keg of Hodgdon 4831 powder. "It's old, but still gray," said the man, as he bent over to unwrap the tape holding the cardboard lid tight. He set the lid aside, plunged his hand into the silvery, slippery kernels of nitrocellulose and let them dribble through his fingers as he lifted his palm to me. "See. No red dust."

I didn't have enough .270 cases to use 50 pounds of powder, but the thought of owning that much appealed to me. How many other shooters had 50 pounds? And the price of powder was climbing. Who knew how far it would go? Besides, H4831 would work in almost any big rifle hull with a neck. "I'll take it," I said.

There's still some of that powder left. And because I've kept the keg sealed in a dry, cool room, the powder is still good, 60 years after it came off the line.

During the second world war, Bruce Hodgdon knew that gunpowder would keep far beyond the government's need for it. He also knew that huge stocks of powder were burned after the Great War, just to be rid of it. An avid handloader even during his Navy stint in the 1940s, Bruce winced at the thought of tons of precious powder going to waste. So after the war he bought 50,000 pounds.

Brewster E. Hodgdon was born in Joplin, Missouri in 1910. His father and grandfather were civil engineers, so he grew up as an apprentice. But one frigid day, riding in the back of an open 1917 Buick to survey a property, Bruce decided it was time to break the tradition. He studied business at Pittsburg State College, then at Washburn. He married high-school sweetheart Amy Skipworth in 1934, soon after he started work as a gas appliance salesman. A few moves later, he and Amy bought two acres with a house and garage, a chicken coop, and a small orchard in Johnson County. That site would later serve as headquarters for Hodgdon Powder Company.

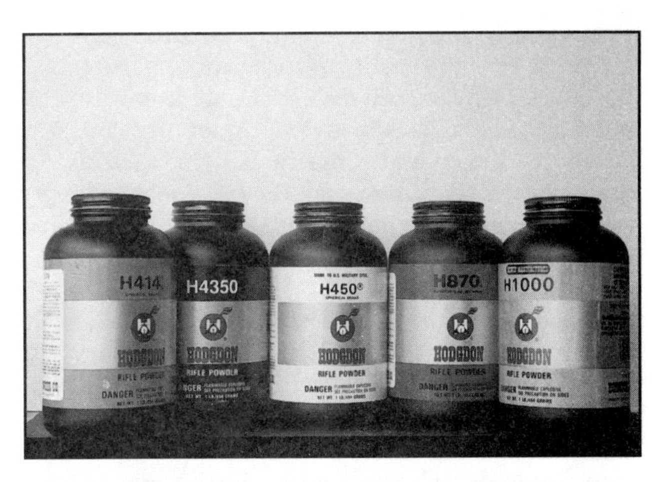

Bruce Hodgdon started his powder business buying war-surplus machine-gun powder from the army and storing it in rented rail cars. The Hodgdon line now includes a wide variety of propellants.

By all accounts, Bruce was a star salesman, but he never was happy hewing to increasingly narrow mandates from the gas company. An avid shooter and reloader, he figured there were enough of his kind in circulation that he could peddle 25 tons of surplus powder and make a little money on the side doing what he wanted to do.

It was a gamble. Sure, this was 4895, a popular fuel for the popular .30–06, and it was cheap. But Bruce had nowhere to store such a quantity and no market. He bought a derelict boxcar and moved it to a rented cow pasture. He placed a one-inch ad in *The American Rifleman.* The price for that first Hodgdon powder was $30—for 150 pounds!

Bruce and his two grade-school boys, Bob and J.B., fashioned shipping boxes from orange crates. They glued labels on the metal cans that held the powder. On their way to school, the boys drove thousands of pounds of 4895 to the REA and Merriam Frisco terminal in the trunk of a 1940 Ford. Business grew, and Amy became a bookkeeper. Mail orders came to include other reloading components as the Hodgdons devoted more and more time to selling powder. Eventually they were stocking rifles and ammunition.

By 1952, Bruce was up to his eyebrows in a venture that he'd started with a few dollars borrowed on the cash value of his life insurance policy. He quit his full-time job as appliance salesman in Kansas City, and devoted all his energy to the powder business. Sons J.B. and Bob joined him after finishing school in 1959 and 1961. The company name remained B. E. Hodgdon, Incorporated. In 1966 the family decided to separate the powder business from the wholesale firearms business. Hodgdon Powder Company was born. A year later, the company built what was then a huge indoor shooting range. The 75-foot facility accommodated 44 shooters at a time. It was open to the public from noon to 9 P.M. weekdays, 10 A.M. to 9 P.M. Saturdays.

Besides promoting the shooting sports, Bruce Hodgdon also lent a hand to help rid the industry of unnecessary legal restraints. In 1963 and '64, he and Ted Curtis, with Homer Clark and Dave Wolfe (founder of Wolfe Publishing), persuaded the Interstate Commerce Commission to downgrade certain types of smokeless powder to the classification of "Flammable Solid." This change made the powder a lot easier to ship. Now containers under 8 pounds each, boxed in shipments totaling under 100 pounds, can be sent by common carrier.

H4831 is one of the author's favorite powders. It's a first choice behind medium-weight bullets in big wildcat cases like this .300 Jarrett.

As Hodgdon Powder Company sales grew, so did the firm's Kansas City headquarters. Eventually it came to include powder magazines and packaging facilities on 160 acres six miles west of the main office—plus rented magazine space on closed military bases that are controlled by the municipalities of Topeka and Olathe. While the mail comes to a post office box in Shawnee Mission, Kansas, the offices are at 6231 Robinson, in Overland Park.

Bruce Hodgdon was always open to fresh ideas, and the company has stayed that way. In 1976 it agreed to distribute a new product called Pyrodex. Pyrodex had its start far from Missouri, in the Seattle-area laboratory of a gifted young shooter and entrepreneur named Dan Pawlak. His goal during the early 1970s was to come up with a substitute for black powder—a propellant that would be safer to manufacture, ship, and use. But it would have similar density and appearance, generate the same smoke and smell, and yield comparable velocities at pressures that any black powder gun would handle. Not everyone thought it could be done. Those who conceded the possibility argued that such a substitute would be far too costly to interest shooters.

Though Pawlak had little formal background in explosives, he was a quick learner and optimistic. During his preliminary work with propellants, he was granted several patents. His purpose in developing a black powder substitute was partly to release shippers and shooters from the onerous regulations governing black powder. Pawlak aimed to provide a fuel that would stay outside the Class A, High Explosive category, into which the Bureau of Explosives in the U.S. Department of Transportation had placed black powder. Ideally, his propellant would pass stringent Flammable Solids tests, so it could be handled like smokeless powder.

Pawlak teamed up with an experienced propellant engineer, Mike Levenson, and within two years the pair developed a steel-gray fuel that met some of Dan's criteria. The new substance was dubbed "Pyrodex," an abbreviation for "pyrotechnic deflagrating explosive."

Surplus H4831 enabled wildcatters like Roy Weatherby to develop cartridges requiring slow-burning propellants. Weatherby magnums like this 7mm still like H4831!

Meanwhile, Pawlak's work had caught the eye of Warren Center, at the gun firm of Thompson/Center. He phoned R. E. Hodgdon and urged him to consider a joint financial venture. In January 1975, R. E. Hodgdon traveled to Seattle to meet with Dan Pawlak. The strapping 6'4" entrepreneur hosted Hodgdon in his Issaquah mobile home, next to a private airstrip and a remarkably well-equipped ballistics laboratory. After examining the transducer pressure guns, oscilloscopes and other sophisticated devices, R.E. knew the young man was both serious and talented. A quick study of pressure curves and some test firings convinced him that Pawlak was onto something.

By this time Mike Levenson had left. With the financial help of Linn Emrich, Pawlak continued to accumulate equipment, most of it used. He refined Pyrodex, increasing its potency, reducing its range of pressure variation, making it burn cleaner, adjusting the components to make it smell like black powder. He even changed the color to black. Also, he began to remodel a former explosives facility, Excoa, on Taylor Mountain north of Issaquah for Pyrodex production.

Before submitting Pyrodex to government tests, Pawlak arranged his own. The detonator test was considered most severe. In it, an igniter is placed in one pound of 10 one-pound canisters in a case. The case is then covered and surrounded with similar cases of the same size and weight but containing inert material, usually sand. When ignited, there must be no detonation. Apparently, smokeless powders occasionally fail this test. But Pawlak was hard on himself: He put a detonator *in each can* in the center case. Noise and flame ensued when the Pyrodex burned, but there was no detonation. In fact, the subsequent Department of Transportation laboratory report stated that "Pyrodex is probably a little safer than conventional smokeless powder because the ignition temperature is considerably higher."

Pawlak targeted the first shipment from Pyrodex Corporation for May 1976, and he went to Hodgdon Powder Company for help with distribution. Hodgdon took on sole distributorship for Pyrodex, agreeing to market it as "the replica black powder." That name came about at the suggestion of Maxine Moss, editor of *Muzzle Blasts* magazine. The National Muzzleloading Rifle Association had first dictated that only original black powder rifles could be used at sanctioned matches. The limited supplies and steep prices of original muzzleloaders soon proved a hurdle in recruiting shooters to the sport. An exception was then granted for replica firearms. A replica fuel would likely receive the same welcome—and so it came to pass. In 1976 Pyrodex became legal for use in sanctioned NMLRA matches.

Then, on January 27, 1977, tragedy struck the man who had done what many thought impossible. At the Issaquah plant, three tons of Pyrodex flashed off, killing Dan Pawlak and three technicians. In the aftermath, Pyrodex was again tested. Again, it was found to be less hazardous than smokeless propellants.

The directors of Pyrodex Corporation, with Mrs. Pawlak and the board of Hodgdon Powder Company, met to consider the manufacture of Pyrodex. The venture was not without risk. Pyrodex had just begun to prove itself at market. Though initial demand had been strong, the Issaquah disaster would certainly affect shooters' perceptions. Besides, Hodgdon had little experience in the manufacture of small arms propellants. The company's background lay in ballistics and powder sales.

But the outcome of that somber January meeting in Issaquah may never have been in doubt. Dan Pawlak had believed firmly that his propellant would not only make money but encourage more people to take up black powder shooting. He also believed it would make the sport safer. Cathy Pawlak, Linn Emrich, Dave Wolfe, Neil Knox, and R. E. and J. B. Hodgdon agreed to back construction of a new Pyrodex mill on an abandoned B-29 base eight miles from Herington, Kansas. The facility was only about 130 miles from Hodgdon's distribution center. In May 1979, after three years and numerous delays, the 1.5-million-dollar plant began turning out canisters of Pyrodex. It became available in retail stores in early 1980.

H4350 is slightly faster than H4831, and better suited for cartridges like the .338 Winchester Magnum. So are these other powders.

Pyrodex is now available in three grades: RS, the equivalent to FFg black powder, works best as the main charge in most muzzleloading rifles; P, which mimics FFFg in burning characteristics, is for small-bore rifles and pistols; Pyrodex "Select" appeared in 1991 as a specialty powder for sabot and heavy conical bullets. In 1997 Hodgdon announced the Pyrodex pellet, a 50-grain (equivalent) pill you just pop into the muzzle. Two pellets equal a 100-grain charge. There's no measuring required; you just seat the bullet or ball over the pellets. A caveat: Pyrodex pellets are recommended only for in-line rifles.

Pyrodex, however, is just one product in an expanding Hodgdon stable. The smokeless propellants include 10 shotgun and pistol fuels and 15 rifle powders. Five of the latter are ball powders: BL-C2, H335, H380, H414, and H870. They represent a wide range of burning rates, for rounds as small as the .218 Bee and as large as those based on the .378 Weatherby. Hodgdon's extruded propellants also span a broad gamut. H4227 is very fast, H50BMB glacial. Between them you have H4198, H322, H4895, Varget, H4350, H4831, H4831 Short Cut, and H1000.

Over many years, I've emptied dozens of Hodgdon powder cans. I've certainly burned more H4831 than any other powder of any kind. It is a fine performer in the .270 Winchester and any short magnum through 35 caliber. The Short Cut version meters nicely and seems to yield the same velocities. H1000 burns a little slower in magnum cases and begins to edge out H4831 for heavy-bullet loads in full-length medium-bore rounds like the .300 Weatherby.

For medium-bore hunting rounds of modest capacity, H4895 is still a first-round draft pick. Like IMR 4895, it's ideal for the .30–06. In my .411 Hawk, H4895 and H322 have been star performers. H4350 bridges the gap between H4895 and H4831.

Hodgdon manufactures and markets Pyrodex, a black powder substitute developed by entrepreneur Dan Pawlak in Washington State. Dan was later killed in a factory fire.

It's fine for light-bullet loads in magnum cases, and a top pick for heavy bullets in the .30–06. Varget is relatively new but so far has impressed many shooters with its consistency in the 7mm-08, .308 and other medium-bore short cases. BL-C2 has lost some of its shine to Varget, which burns at about the same rate and in some short-action cases outperforms it.

I bought 20 pounds of H335 once because, again, I'm a sucker for big tubs of powder. It's a good choice for light bullets in the .308 family of cartridges and for medium-size .22s. H380 is a bit slower and more useful for big game hunters. A charge of 51 grains will launch a 180-grain .30–06 bullet at 2700 fps while generating less than 50,000 copper units of pressure. H414 rivals H4895 in its service to the .30–06. H870 behaves like a ball powder version of H50BMG. It excels in the .264 Winchester Magnum and "overbore" cartridges with plenty of room for fuel and small exit ports. For heavy-bullet loads in the 7 STW, these two Hodgdon fuels rank among the best of few choices.

Brewster E. Hodgdon died in 1997. He lived modestly near the company he founded. A middle-class three-bedroom home was enough. He didn't travel a great deal or otherwise spend a lot on himself. He gave generously to the NRA Foundation and donated land to the Millcreek Gun Club. His interests included goose hunting and smallbore rifle competition. He taught his sons from the Bible and supported them in every way he could. When he turned the company over to them in 1976, it was with no strings.

Hodgdon Powder Company reflects Bruce's values and lifestyle. It has established itself as more than a seller of propellants, a source of Pyrodex. Hodgdon represents the kind of straight-dealing organization that Bruce worked hard to build. Its people are ready to answer questions because they are shooters and because they seem to believe that customer questions are important. You're treated as if Hodgdon thinks it can sell you 50,000 pounds of surplus 4895.

A Look Back at Bullets

In old western movies, the scowling men in black were occasionally made to "eat lead" as more becoming hombres under white hats gave 'em what for. A lot of unfortunates did die of lead poisoning on the frontier, and even now you hear of hunters "flinging lead."

But bullets weren't always made of lead. In the fourteenth century, when guns were still a new idea in warfare, hostilities commenced at very close range. Accuracy mattered less than noise and smoke. In big-mouthed cannons, a load of rocks or metal scrap flew far enough to count. Iron balls wrought such devastation that they were outlawed in central Europe. Even after 17-horse teams drew great bronze culverins to the battlefront, hand-to-hand combat still accounted for most of the casualties.

Lead became popular with shooters beginning in the fifteenth century. Lead's high density meant it had great inertia. Once launched, lead bullets held their speed against the push of the air, and they weren't easily stopped by obstacles like doors, saddles, and light armor. Lead's low melting point made for easy molding to various bore sizes. As rifles came into use, lead's softness at normal temperatures enabled a marksman to size it groove diameter, yet still load it from the muzzle with ease. Hollow-based lead bullets obturated (expanded) under the press of gas from the rear to "take" the rifling and seal the gas behind. Alloys to harden lead as bullet speeds increased were followed by gas checks (metal heels on bullet bases). Jacketed bullets combined the malleability of lead with cupro-nickel "skins" that kept the lead from melting into the rifling when bullets were driven faster than 1800 feet per second.

At Plymouth, lead was a precious commodity for the Pilgrims—for trading as well as shooting. Rifles built in shops on the American frontier were initially given small bores to conserve lead. Even when American riflemen discovered the utility of the patched ball, with the patch taking the rifling, nobody though seriously about replacing lead. Nothing killed like lead; nothing flew as accurately as lead; nothing was so easy to size as lead. Undersizing balls for use with a patch made sense then, and still does. A 50-caliber rifle may shoot accurately with a ball sized to .500, but after a few shots fouling makes loading from the front all but impossible with a tight ball. A .490 ball with a snug-fitting patch usually performs almost as well as tighter balls in a clean bore and may outperform them after a few shots.

During the 1830s, when percussion ignition began to replace flint and the short, heavy, big-bore plains rifle was edging out the elegant Kentucky on the American plains, conical bullets surged in popularity. At first, target shooters found the conical "picket" ball difficult to load and sensitive to loading technique. Alvin Clark invented the false muz-

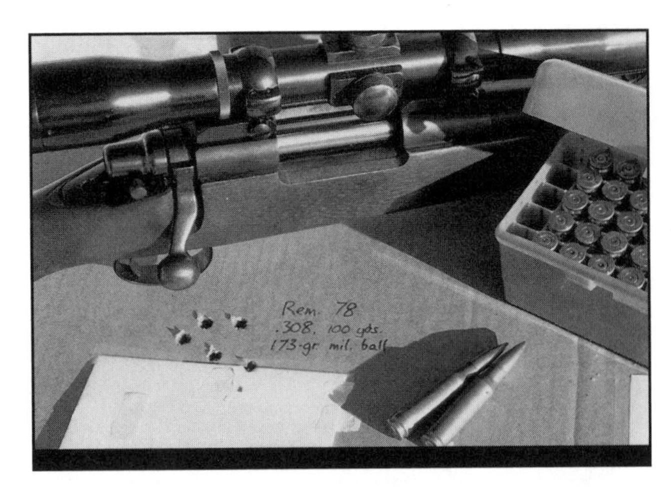

The first bullets in smokeless cartridges were full-jacketed bullets designed for military use. They're not humane or effective (or legal) for most big game hunting. But they make practice cheap!

zle in 1840, enabling riflemen to start a bullet square with the barrel proper. Paper-patched bullets and false muzzles soon became the choice in competition. But on the frontier patched balls remained popular. In fact, many frontiersmen chose smoothbore guns because they were so much faster than rifles to load and slower to foul.

A conical bullet in a smoothbore gun holds its velocity better than a ball because it has a higher sectional density (essentially, a ratio of weight to bore diameter). But its speed is wasted if it doesn't hit where it is aimed. A bullet shot from a smooth bore is likely to lose its point-forward orientation. It may even tumble. Once it starts to wobble, it loses the advantage of its sectional density because it is flying part of the time at an angle to the target, and it is using some of its energy in useless motion. Rifling prevents tumbling by stabilizing the bullet in flight, like a football thrown by a star quarterback or a top spinning on a counter. The more perfect the rotation of the football, the farther and straighter it flies. The top remains upright until its rate of rotation drops below a certain threshold. Then it wobbles and falls down.

Loading a groove-diameter bullet from the muzzle of a rifled barrel is hard work because you're engraving the bullet with the lands (the uncut portions of the surface left after rifling grooves have been cut into the metal) all the way. And a bullet's shank gives you much more resistance than the small bearing surface of a ball. Patches don't wrap evenly on bullet shanks. In 1826 Henri Gustave Delvigne of France attacked the problem of front-loading bullets with a new chamber design. His chamber had sharp corners that arrested and upset an undersize ball as it was rammed home. It worked. But the balls and bullets expanded to groove diameter in this process were also deformed. The central chamber peg de-

Hunting bullets are generally not the most accurate, but from super-accurate barrels, like the Krieger on this .375 by D'Arcy Echols, they can be. The author fired this 100-yard group with a 3x scope.

Hunting bullets must not only fly predictably; they must upset or mushroom predictably. This water jug shows the hydraulic effect of an expanding bullet.

vised by another Frenchman, Thouvenin, gave similar results. In 1834 a Captain Bernier of the British Brunswickers resurrected a century-old design to come up with a two-groove rifled bore and a ball with a belt that fit in the grooves. Originally proposed by a Spanish army officer, this arrangement required that the ball be oriented each time; still, pushing it home was easy. Shortly thereafter, the English gunmaker William Greener developed an oval ball with one flat end. A tapered hole from the flat end to the center of the ball held a metal peg whose round head conformed to the shape of the ball. Either end could be placed in the muzzle first. Upon firing, the peg drove into the undersize ball, expanding it to fit the grooves. Despite good accuracy, British Ordnance rejected the Bernier because it comprised two pieces.

About this time, other British gunmakers experimented with out-of-round bores that spun a bullet without engraving it. They didn't pass muster in military trials because the bullets had to be perfectly sized and oriented exactly. Production of out-of-round bullets also proved difficult. A fellow named Delvigne took another approach, figuring that powder gas could expand a hollow-base bullet to fit the grooves. He developed one in 1841. British Colonel Brunswick followed with a metal skirt soldered to the base of a ball. But these projectiles, like those upset by special chambers, often tipped in flight.

Bullet shape affects accuracy. But nose form is far less important than shank concentricity and a heel that's uniform. The author tore these bullet noses off with a shears, then fired the shanks into 3 m.o.a.!

British General John Jacobs abandoned the hollow bullet base for a four-finned bullet mated to four-groove rifling. Failing to win any government contract for his design, Jacobs persisted. He invented a heavy conical bullet for long shooting and a 32-bore double rifle that fired an exploding bullet. African explorer and elephant hunter Sir Samuel Baker chose instead a 4-bore Gibbs with standard rifling. Its 36-inch barrel had two grooves, pitched one turn in 36 inches. As was customary in that day, the grooves were broad, the lands narrow to facilitate loading. (Barrels built for patched balls had broad lands and narrow grooves to give more support to the undersize ball.) Baker's rifle used 16 drams of powder to heave a 4-ounce conical bullet. Recoil was at the nose-bleed level.

In 1847 French Captain Claude-Etienne Minie (pronounced me-*nay*) designed a bullet with an iron cup in its hollow base. Powder gas drove the cup forward, expanding the bullet into the rifling's grooves. Expansion was sometimes so violent, however, that the bullets were cut in two, leaving a ring of lead in the bore. Minie modified his design and tested it against balls from smoothbore muskets. At 100 yards the musketeers hit a 6x20-foot target 149 times out of 200 shots. Riflemen shooting the Minie bullet scored 189 hits. At 400 yards he recorded 9 hits for the muskets, 105 for the rifles. Later Minie would market a conical bullet to the British military for 20,000 pounds. William Greener protested and was awarded 1,000 pounds for the base-expansion idea. Minie's bullet, lubricated with mutton fat or beeswax, became issue ammunition for the Union Army's Enfields during the American Civil War. At first a wooden plug was placed in the bullet's hollow base to prompt expansion. This proved unnecessary.

Eager to find more effective ammunition for its troops, the British government hired Joseph Whitworth in 1854. A brilliant technician, Whitworth was also persuasive; he requisitioned and got an elaborate 500-yard range for his experiments The range was destroyed by a storm soon after it was finished, but another one was built, and Whitworth proceeded to try various rifling types and twist rates. The standard twist in military rifles then was 1–78 (one complete rotation in 78 inches of forward travel). Whitworth found the best all-around twist for short bullets to be 1–20. Skeptics thought the

Brush can make you miss. Even heavy 250-grain bullets from a .35 Whelen can deflect, or tip. One here entered the target sideways, making a classic keyhole.

sharp spin would retard the bullet, but a hexagonal bullet of Whitworth's own design gave more than twice the penetration of a standard ball from a slow-twist barrel. His smallbore hexagonal bullets flew flat and at long range drilled out groups a sixth the size of those shot with ordinary patched balls. Hexagonal bullets, however, were expensive to manufacture and slow to load. Whitworth's efforts gave other designers plenty of ideas. William Greener experimented with narrow-land rifling pitched 1–30. This relatively quick spin stabilized smallbore (.40 to .52) bullets out to 2000 yards. Gunmaker James Purdey built two rifles featuring Greener-style barrels in 1856. He called them "Express Train" rifles because of their great power. "Express" became a common descriptor of powerful British hunting cartridges.

Breechloading rifles eliminated the need to design an accurate bullet that could be loaded from the front. Bullets loaded from the rear could be made harder and longer and cast to full groove diameter. Hard bullets could be stabilized with shallower grooves and thus driven faster. Sharp rifling pitch could be used—even rifling that varied in pitch and depth from breech to muzzle. William Ellis Metford experimented with "gain twist" that allowed a bullet easy acceleration from the chamber but increased the rate of spin during its barrel travel. He used a 34-inch barrel that gave the bullet an exit spin of 1-in-17 but started it so gradually that the bullet turned over only once before reaching the muzzle. Metford favored the wide, shallow grooves that now work well with jacketed bullets. He also developed segmented rifling, with rounded lands and grooves. By the end of the century, gain twist and progressive grooving (deeper grooves at the breech than at the muzzle, as in England's Enfield and Martini-Henry rifles) were largely abandoned.

Charles Newton, American inventor, gunmaker, and wildcatter, reasoned that since a bullet bears on only one shoulder of a land, the other is unneces-

Ammunition firms now buy special bullets from other sources for their best-grade ammo.

sary. His "ratchet rifling" tested this concept. Like the Lancaster oval bore derived from Bernier's elliptical design, Newton's idea had merit. It has not, however, been embraced in the marketplace. Marlin, the U.S. gun firm that makes more .22 rimfire rifles than anyone else, uses multiple, narrow grooves in its "Micro-Groove" barrels. They deliver good accuracy, but probably no better than traditional Metford-style bores with fewer grooves. Metford rifling features wide groove bottoms the same radius as the bullet, plus flat-topped, square-shouldered lands. Groove number has varied since World War I, when U.S. soldiers carried two-groove Springfields. Now most rifle and pistol barrels (Marlin's Micro-Groove bores excepted) have four to six grooves. You'll see polygonal rifling on Heckler and Koch rifles.

Proper rifling pitch depends on the length and speed of the bullet. Long bullets require a faster pitch than short ones, slow bullets quicker spin than fast ones. Round balls benefit from rifling, because it prevents precession, a forward "rolling" motion. In 1879 British ballistician Sir Alfred George Greenhill announced a formula to derive proper spin for all bullets with a specific gravity of 10.9. The specific gravity of pure lead is 11.4, so most jacketed bullets qualify. Here is the Greenhill formula, first published in the *British Textbook of Small Arms,* 1929: Required twist in all calibers equals 150 divided by the length of the bullet in calibers.

This means a 180-grain .308 bullet an inch and a quarter long should be spun: 150/4=37.5. Remember, "37.5" is in calibers. To convert to inches, you multiply it by .308. Answer: just over 11 inches. Most 30-caliber barrels are rifled 1–10 or 1–12, so the formula says they have a useful pitch for that bullet. Most modern rifle barrels have the proper twist for common bullet weights. A perfect mathematical match is not necessary. Top accuracy can be hard to achieve with bullets at the extremes of a wide weight spectrum.

Jacketed bullets first appeared in the 1890s, to deal with the high velocities made possible by smokeless powders. They were of steel, with a cupro-nickel coating. Satisfactory in the .30–40 Krag, they fouled badly at the higher velocities of the .30–06. Tiny lumps of jacket adhering to the relatively cool steel near a rifle's muzzle tore at the jackets of other bullets, accelerating the process. Shooters fought the fouling with "ammonia dope" a solution of ½ ounce ammonia bicarbonate, 1 ounce ammonia sulfate, 6 ounces ammonia water, and 4 ounces tap water. Poured into a plugged barrel and allowed to "work" for 20 minutes, this brew was then flushed out with hot water. Drying and oiling followed. Shooters took special care with this operation, as ammonia spills on exposed metal could cause pitting.

Bullets must be the proper weight for the rifling twist. This .240 Hawk cartridge preferred light bullets (which it drove to 3900 fps!) in a rifle the author tested.

To reduce metal fouling, the U.S. Army issued "Mobilubricant" to soldiers just before World War I. Alas, pressures in the .30–06 began bouncing from 51,000 to around 58,000 psi. Coating the cartridge with Mobilubricant sent pressures even higher. This problem was exacerbated by tin-plated jackets. The tin could "cold solder" itself to the case neck. A lubricated case gave the neck no room to expand while it increased back-thrust on the bolt. Add a bullet stuck tight in the neck, and you could seize a bolt. One bullet recovered at a shooting range still wore the case neck! The Army soon dropped Mobilubricant and tin-plated bullets, choosing instead to incorporate tin in the jacket alloy. Later, cupro-nickel (60 percent copper, 40 percent nickel) became the jacket of choice. Gilding metal (90 percent copper, 10 percent zinc) was initially thought to be too soft for the high-speed 150-grain bullet in .30–06 service ammunition; but Western Cartridge Company's jacket of 90 percent copper, 8 percent zinc, and 2 percent tin worked well. It was called Lubaloy. In 1922 Western earned the honor of providing Palma Match ammunition with 180-grain Lubaloy-coated bullets. That year, experiments at Frankfort Arsenal showed that gilding metal could stand up to high velocities. It remains a popular jacket material for hunting and target bullets. Most bullet makers now favor jackets comprising 95 percent copper and 5 percent zinc (though Nosler stayed with a 90–10 alloy for its Partition bullets turned on screw machines before 1970). Softer, almost pure-copper jackets began showing up on bullets designed for deep penetration, like the Bitterroot Bonded Core that was one of the first so-called "controlled expansion bullets." A ductile jacket was less likely to break apart on heavy muscle and bone. The liability of pure copper is its tendency to foul barrels.

Bullet-Making

Lead bullets that fit the bores of muzzleloaders are fairly easy to make. Frontiersmen cast their own, and you can too, with a pot and a mold. But the high velocities achievable with smokeless powder generate lead-melting friction in the bore. In the 1890s bullet jackets became necessary to prevent leading and ensure accuracy.

The lead cores of big game bullets have a dash of antimony to make them harder. The usual ratio is 97.5 percent lead, 2.5 antimony. A little antimony makes a big difference. Six percent is about the limit in commercial bullets. Sierra uses three alloys for rifle bullets, with antimony proportions of 1.5, 3, and 6 percent. Some game bullets have pure lead cores. Thick copper jackets on the rifle bullets keep soft cores from disintegrating on impact. "Pure," incidentally, means unalloyed. Even pure bullet lead has traces of copper, zinc, nickel, arsenic, and aluminum. As little as .1 percent copper can cause hard spots. Core material is commonly cut from lead wire, extruded from bar stock in the appropriate diameter, then annealed to prevent expansion during forming. The Barnes X-Bullet and Jensen J-36 feature solid-copper (or copper alloy) construction; Speer's African Grand Slam has a tungsten core.

Bullet jackets are born two ways: by "cup and draw" and impact extrusion. Drawn jackets begin as wafers punched from sheet metal. Formed or drawn over a series of dies, they become progressively deeper cups that are eventually trimmed to length and stuffed with lead. The bullet is then shaped and finished off at the nose. Jackets given the impact extrusion treatment begin as sections of metal rods that are annealed and fed into a punch press that slams them into cup shape with 60 tons of force. Bullets like the Nosler Partition have cavities fore and aft and so must be punched twice. Nosler claims impact extrusion guarantees concentricity. Its old machine-turned jackets had a hole in the partition that varied in location and hardly contributed to fine accuracy.

Cannelures help the jacket grip the core but serve mainly as crimping grooves. Cannelures have gradually disappeared on rifle bullets. They've been retained on bullets for heavy-recoiling rounds like the .458 Winchester (to prevent bullet creep in the

Mike Brady of Glenrock, Wyoming, has developed a hunting bullet that drives deep and opens wide. Small shops like his have proliferated over the last 20 years.

magazine) and on pistol bullets (to keep short bullets in place and to smooth the cartridge profile for better feeding). Hornady, Winchester, and a few other makers routinely crimp bullets. Most cannelures are rolled on, but Nosler cuts the crimp in its 210-grain .338 bullet. Claims that crimping impairs accuracy have not been substantiated.

Jacket and core dimensions must be held to tight tolerances for utmost accuracy. Sierra, renowned for its match-winning target bullets, keeps jacket thickness within .0003 of a standard. The company limits bullet weight variation to .3 grain. Test lots of 168-grain 30-caliber match bullets that don't shoot into .250 inch at 100 yards can send the entire batch back into the production line. Demands on hunting bullets are less stringent, at least in terms of dimensional uniformity and accuracy. Rate of expansion and weight retention are more important in the design of big game bullets.

U.S. hunters are blessed with a plethora of excellent big game bullets, some specifically for pistols.

Properly bonding jacket to core is crucial to making—and marketing—big game bullets. Hunters have known for some time that bullets must stay in one piece to do the most damage. "Core-Lokt," "Trophy Bonded," "Bonded Core," and "Hot-Core" are trade names that play on this theme. Remington's Core-Lokt bullet has an inner belt (derived from the Peters Inner Belted bullet, which followed the Peters Belted design, with an exterior metal girdle). The Speer Hot-Core process reportedly gives a tight bond because the lead is warm enough to snuggle into every void in the jacket. The other companies don't tell exactly how their superior bonds are formed.

Bullet nose design has a lot to do with how expansion proceeds. Hollowpoint bullets are typically used for thin-skinned game. Small cavities and thicker jackets keep hollowpoints from breaking up, but they also make expansion less dependable. Among early hollowpoints, the Western Tool and Copper Works bullet with a tiny cavity had perhaps the best reputation on big game. Westley-Richards offered a bullet with a nose dimple covered by a metal cap that protected the nose and kept the jacket from rupturing too early. DWM's "strong-jacket" bullet had a long, narrow nose cavity lined with copper tubing and capped.

The sophisticate of early hollowpoints was Peters' Protected Point. Its jacket enveloped a flat-topped core, the front third wrapped in a gilding metal band and crowned with a pointed cone. On impact, the flattening nose drove the band down under the jacket. The band initiated and controlled the core's expansion, which split the jacket as it progressed. A Protected Point bullet required 51 operations and three hours to manufacture. Winchester's Silvertip is a cheaper rendition of this design. It lacks an inner band.

Remington put a hard peg in a hollow bullet nose to form its Bronze Point spitzer. Upon striking a target, this peg was driven back into the bullet to start expansion. A ballistic wonder, the Bronze Point opened pretty violently at close range. RWS has counterparts to this design: the TIG (Torpedo Ideal) and TUG (Torpedo Universal) bullets. The TIG's tail section has a funnel-shaped mouth into which the nose core fits like a plug. Expansion starts not only at the nose but at the core juncture. The rear core opens more reluctantly because it's harder. In TUG bullets the joint is reversed: A cavity in the nose accepts a conical protrusion from the rear section, which acts like a trailing bullet when the nose breaks apart in game. The pointed tail portion resists deformation, penetrating like the trunk of a Nosler Partition.

Bitterroot Bullet Company also uses unalloyed copper for its jackets, which in .338 can be .060-inch thick. The nose of a Bitterroot Bullet has a cavity to initiate expansion. These features make Bitterroots long for their weight, requiring deep seating and a careful eye to pressures bumped up by extended bearing surfaces.

Perhaps the best known of bullet-makers specializing in extra tough bullets is Barnes. This company started in 1939, predating John Nosler's by nine years. Barnes "Original" bullets, from 22 caliber to

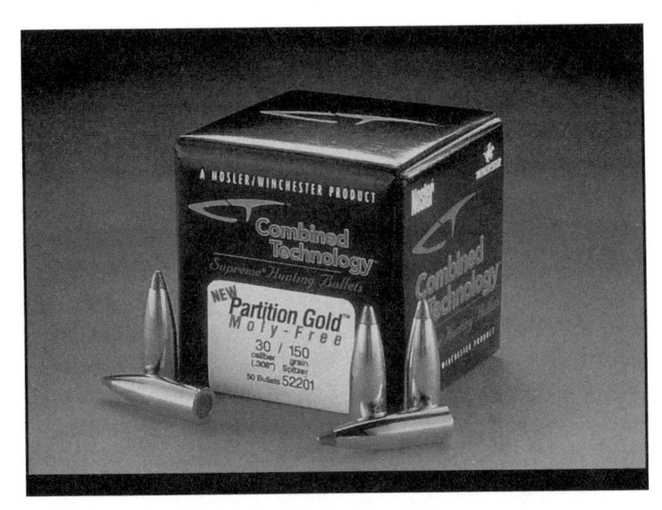

Nosler and Winchester got together to build and market bullets under the Combined Technology label.

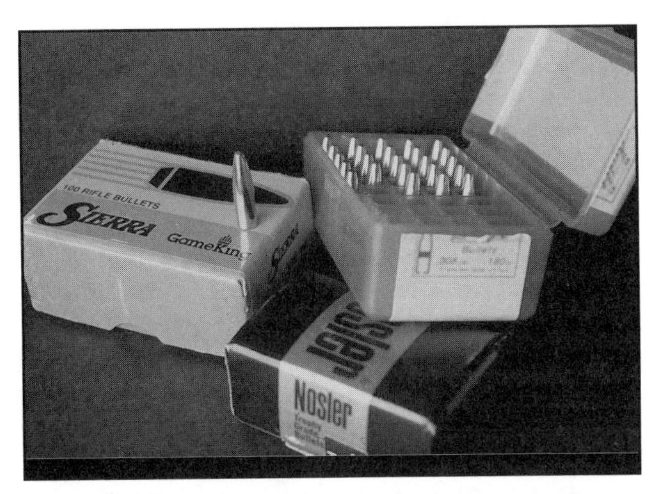

Hornady's XTP pistol bullet expands reliably at mid-range (center) and stays together up close (right).

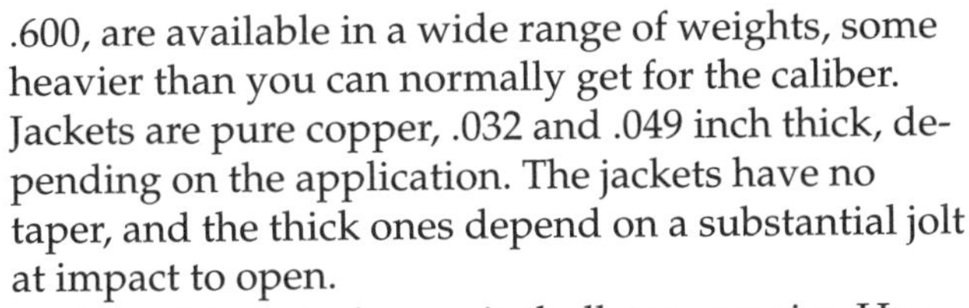

.600, are available in a wide range of weights, some heavier than you can normally get for the caliber. Jackets are pure copper, .032 and .049 inch thick, depending on the application. The jackets have no taper, and the thick ones depend on a substantial jolt at impact to open.

One of America's premier bullet companies, Hornady, has now been in the jacketed bullet business for half a century. The Grand Island, Nebraska firm began when Joyce Hornady and Vernon Speer, who had been working together, split to seek their own fortunes. By fashioning bullet jackets from spent rimfire casings, they'd circumvented war-time metal shortages in the early 1940s. Handloaders would, they hoped, sustain a market for jacketed hunting bullets.

Joyce Hornady was quick to spot value in the specialized machinery and tooling dormant in government arsenals after the war. The Hornady plant still uses Waterbury-Farrell transfer presses, manufactured as early as the 1920s and now updated with computerized controls. The rural town has grown, too, since the war, but Hornady is still run by people with Midwest smiles who make time for shooters with technical questions and hunting tales.

Mike Timmerman, Plant Supervisor, explains that every bullet run is tested at 100,000-unit intervals—rifle bullets for accuracy, pistol bullets for expansion. "We insist that pistol bullets not only shoot consistently but that they open reliably and penetrate. Wide velocity spreads make mushrooming more difficult to control in handgun bullets. Our .38s and .45s are tested at 750 fps and 1500, the .454 Casull bullets at 1000 and 1600. Those velocities work for muzzleloaders too. We use some pistol bullets—like the 300-grain .45 XTP (Extreme Terminal Performance)—as sabot projectiles in muzzleloaders. We sell our own sabot packets, and we make bullets and sabot hulls for Thompson-Center."

Mike shows me a rugged one-holer, "We shoot four five-shot groups with rifle bullets. The average must meet our standards."

"What standards?" I ask.

"Depends on the bullet. For .30s it's .600 at 100 yards, for .17s it's .400. The 6mms come in at .450, the .338s at .750. We have more stringent standards for match bullets: .350 for .22 match. We check 'em more often too: every 30,000 units. Our 30-caliber match bullets are tested at 200 yards. They have to shoot inside .800."

Dave Emary, chief ballistics engineer, says that jacket design largely determines expansion. "Nose cavity dimensions and bullet hardness matter too," he says. Hornady gets its lead in ingots, which are melted and formed into cylindrical blocks about the size of a roll of freezer paper (but lots heavier!). A massive press squirts these cylinders, cold, through dies to form the lead wire that's then cut into bullet lengths. Antimony content is specified at purchase, from 0 to 6 percent. More antimony means a harder bullet. Most bullet cores have 3 percent antimony.

Jacket cups are punched out of sheets at the factory. "Jacket concentricity is vital to accuracy," Dave adds. "We hold it to .003."

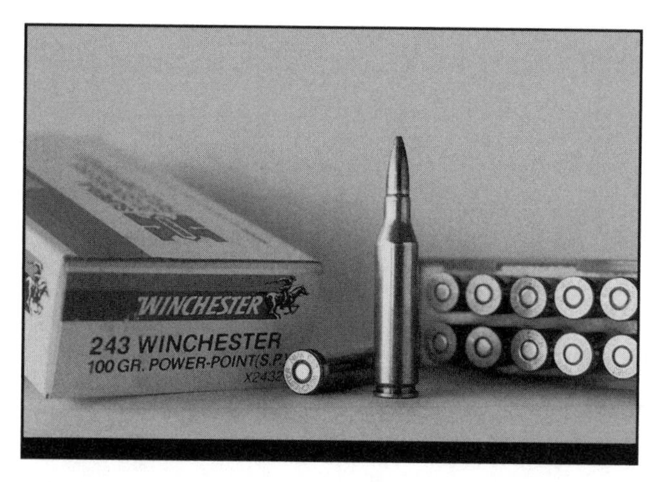

Winchester's Power Point is one of the author's favorite factory-loaded softpoints for deer-size game.

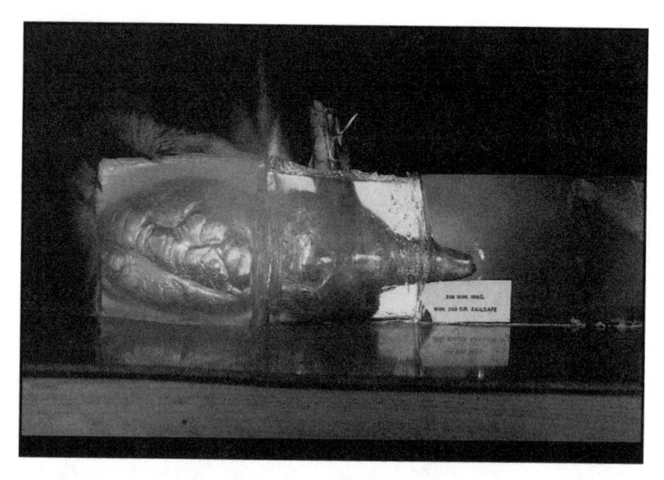

Winchester's .338 Fail Safe bullet was tested for expansion and penetration in ballistic gelatin—with the addition of an elk femur! Note the broken bone, the expanding cavity and the long bullet track.

Like many of Hornady's 140 employees, Dave Emary is an active shooter. His high-power rifle skills have earned him a coveted spot in the President's Hundred at Camp Perry. Before coming to Hornady, he worked with artillery, "shooting 5-pound bullets at 7000 fps and 12-pound bullets at 6500" from 90mm and 120m cannons. The expertise he acquired in the armed forces helps him design 25-grain 17-caliber spitzers and 500-grain .458 solids. But not all of the bullet line is in production all the time.

"Contrary to what many shooters think," says Dave, "we can't forever dedicate one machine to, say, 200-grain .338 bullets. We run enough to build up our stock, then switch to another bullet on that machine." Batch size and the duration of a run depend on market demand.

Hornady's Spire Point softpoints have been a mainstay of big game hunters for decades. Now the firm also makes A-Max (Advanced Match Accuracy) spitzers for competitive shooting. V-Max bullets, introduced in 1995, share the A-Max's plastic nose insert. They're designed with very thin jackets for explosive effect on small animals. The SST (Super Shock Tip) plastic-nose bullet for big game came out not long ago. "We got the SST designation from an old 1950s Hornady bullet board," explains Director of Sales Wayne Holt.

Some shooters might question the need for new softpoint bullet designs, given the broad selection available from Hornady and other makers. New product is as important in the bullet industry as in the automotive business, and customer expectations change. Bullet weight retention is a common measure of bullet performance in big game these days. So, though Hornady's softpoints have a lethal record, there's reason to build them to keep more of their original weight. While the R&D department cranks out new designs, Hornady production lines stay busy supplying handloaders and other companies that assemble ammunition. "We just finished a run of two million FMJ bullets for Federal," Wayne says. "We make the V-Max and boattail softpoints for Remington loads. In fact, we've manufactured bullets for every major ammunition firm."

But the Grand Island bullet maker is not about to change its focus to loaded ammo. Says Dave Emary: "We launched our Light Magnum line in 1995 after I saw at Olin-St. Marks how these super-powerful loads could be concocted. It was a good move, and we plan to expand the Light and Heavy Magnum stable." But shotshells and rimfire ammo are out of the question. There's no profit in either, claims Wayne Holt, adding that with market incentives, loaded centerfire rounds are as cheap now as they were 20 years ago, when you could buy a brand-new automobile for $4000.

Hornady is counting on its wide selection of high-quality bullets to keep orders coming. "We're developing new bullets all the time," says Lowell Hawthorne, a ballistician who's been with Hornady 18 years. He tests those prototype in the lab, below a rack of gleaming Hart and Schneider barrels screwed onto M70 and M700 receivers: the test bat-

Accuracy like this is recommendation enough for any bullet! Here it's the Nosler Ballistic Tip.

tery. There's a three-point free-recoiling machine rest filling the mouth of the 200-yard shooting tunnel, where ballistic coefficient is derived from starting and ending velocities. Some bullets, like .348s for Winchester's Model 71, are tested in stocked rifles from the shoulder. Dave shows me a cloverleaf group from a well-worn 71.

I ask about an instrument I've not seen before. "That's a Heise Gauge, a hydraulic device for calibrating chamber pressure," says Dave Emary, explaining that it has a transducer with a quartz crystal. Pressure registers through electrical discharge.

Dave points out that peak pressure as commonly measured in copper units (CUP) is not sufficient because it tells nothing about the pressure *curve.* "Time matters. You can boost velocity by extending the peak of the curve forward without making the curve higher." He says that a strain gauge falls short because it can't be calibrated to a standard in the manner of the Heise. Dave allows that handloaders get adequate indication of high pressure levels by measuring case web expansion. "But it's crucial that you mic the same spot, because a .001 bulge at the web can become .005 a touch farther forward. It's best to measure forward, because you get more reaction from the brass."

Typically, he says, you'll start to pierce primers at 70,000 CUP, and see blown cups at 80,000. Pressure that's high enough for most shooters to notice is already well above SAAMI spec. "We get higher speed from Light and Heavy magnum ammo by pushing the pressure curve forward so the peak occurs when the bullet is 3 inches out. The powder has about 4 percent more nitroglycerin than ordinary double-base powders. Surface deterrents slow the intitial burn so the bullet doesn't outrun the burn so soon."

The Light Magnum line includes popular rimless rounds, and the .303 British. "The .303 has a great record on big game," Dave says. "And we still get lots of orders for 174-grain FMJs from Australia." The Light Magnum load features a 150-grain Spire Point bullet at 2830 fps.

Hornady manufactures between 60 and 70 percent of its cartridge cases in house.

Niche markets definitely get attention at Hornady. Dave Emary points out with some frustration that the firm's .308 match bullets are fully the equal of Sierra's popular boat-tail. "Our bullet tips are also rugged, surviving even the violent cycling of M1-A rifles, which can bloody most bullet noses. But we haven't promoted target bullets so much as our game bullets." The official Palma Match bullet is still the 155-grain Sierra.

However, Hornady has a stellar reputation in Africa. Professional hunters I've talked with favor Hornady solids over most others for elephant culling. And the firm recently announced a new solid bullet for the .458. "Material for our old copper-coated steel jackets is no longer available in the quantities we can use," says Wayne Holt. "We'd have to buy enough for the next 50 years all at once. Now we'll use brass with a cap on the base so the core is enclosed." The new solids are stout enough to shoot through three-eighths steel plate. Hornady also markets .458 Heavy

Some controlled-expansion bullets with soft jackets leave metal fouling in the barrel. Vigorous brush work with Sweet's or Barnes CR-10 solvent can remove it.

Magnum ammunition that kicks a 500-grain bullet downrange at 2300 fps. At this speed, the bullet hits harder than one from a .470 Nitro Express, yielding 15 percent more energy than current .458 loads.

If the .458 hits hard, it lacks the reach of the .50 BMG. Surprisingly popular since someone found it would detonate dynamite at 1000 yards, this military round got a boost when Hornady announced its A-Max bullet for the big .50 in 1994. A sleek profile and long aluminum tip keep weight to the rear for best accuracy and a ballistic coefficient of over .600.

"We tried the aluminum tip on 162-grain 7mm bullets too," says Dave Emary. "They worked but proved too costly. We replaced them with a long plastic tip, but that gave erratic accuracy." The company then switched to small plastic tips (all Hornady red) in both match and hunting bullets, adjusting the nose contours when necessary. These produced small groups consistently, according to Dave, partly because the tips are easier to manufacture to uniform dimensions, partly because the bullets come out a bit shorter and are thus less finicky about rifling twist. Aluminum is now used only in A-Max .50s. "They're expensive, all right," Dave admits. "But alternative bullets are mostly hand-made and even more expensive."

Those half-inch missiles draw lots of attention but account for a small percentage of Hornady's sales. The traditional Spire Point and thin-jacketed V-Max dominate Hornady's output. The SX will remain a staple in the smaller bores.

Hornady bullets reflect Dave's experience at the 1000-yard line. He points out that bullet design is an exercise in compromise. "Though it would reduce drag in the barrel, lengthening the nose to shorten the shank gives the nose greater leverage if it swings off the bullet's axis in flight," he says. A long nose also mandates a more gradual curvature forward of the shank—which increases the odds for bullet misalignment in the throat, and for subsequent yaw. "A short transition is best, so the bullet is forced quickly into full contact with the rifling." Dave says he likes to keep the ratio of bullet bearing surface to diameter at 1.5 or higher, though nose type and other variables also affect that ratio.

He explains that the "secant ogive" nose profile of Hornady's traditional Spire Point bullets reduces their nose weight. "Tangent ogives generate a more rounded form," he explains. "We've used the secant ogive for years to minimize air friction. It does not necessarily mean a sharper junction between the conical and cylindrical sections of the bullet; however, it can appear that way." (Secant and tangent refer to the point from which an arc, whose radius is measured in calibers, scribes the outline of the bullet nose. The tangent measure is perpendicular to the bullet axis at the cylinder/cone juncture. Secant measurements are taken from behind that point.)

The blue Teflon-like coating on some Barnes bullets reduces bore friction, boosting velocity without unduly raising pressure.

Do Hornady bullets ever come back? "Well, yes," says Dave Emary. "But they shouldn't." He points to long shelves full of red boxes labeled seconds. "We cull our bullets severely, even for surface blemishes. The rigid testing schedule just doesn't allow substandard bullets or ammo out the door." He smiles. "Well, they *do* go out the door, I guess. Employees get the seconds."

Perhaps that's one reason the average shop tenure at Hornady is 12 years.

"What happens when a customer insists he's got bad bullets?"

Dave laughs. "One of those characters phoned me not long ago to say his bullets wouldn't stay closer than 3 inches. He seemed to know a little about shooting. Because he was also local, I asked him to bring his rifle to the plant, expecting I'd have to suggest major gun work. When he came in, the first thing I did was load up some of his bullets and take his rifle to the test tunnel. It shot into three-quarter inch. I showed him the target and gently told him that perhaps his shooting technique needed tuning. To his credit, he took it gracefully. Not all customers do. Shooting can be an ego thing."

Bullets for Tough Game

The elk kept pace with the retreat of morning shadows. Puffing from a sprint across a saddle, we bellied onto a ridge opposite. The hunter poked his rifle through the branches of a big sagebrush. I reached over the rifle to clear the sightline of twigs as he steadied the rifle in a fork.

"Can you see?"

"Yup."

I turned to the elk with my Swarovski 8x50s. "The cow in the middle is the biggest."

"Got 'er." His Browning BAR in .300 Winchester steadied. "How far?"

"Two fifty."

"Golly, that looks far."

"An honest two fifty *is* far. But hold center. Take your time. Gentle on that trigger. Keep your scope on 'er after the shot."

It was more advice than he needed. I couldn't help giving it, a litany I repeat often to myself.

BOOM! The cow humped and kicked at her abdomen as the other elk scattered. They stopped with their radar on. She alone struggled to keep her feet.

"Shoot again," I whispered. But the elk didn't cooperate. In and out of brush, seldom broadside and then not still, she presented a difficult target. Eventually the hunter fired. Then again. And again. I watched a bullet miss but couldn't see the other strikes. Once during the fusillade the elk fell; then she got up. At last she collapsed.

I expected to find a paunch wound and a superficial hit. But upon reaching the animal, I saw that the first bullet had penetrated both lungs, just a little above center. We found the 180-grain Sierra in far-side tissue. Fragments had cut tracks to the side of the bullet wound. Rupture must have been violent and immediate. There were two peripheral hits besides.

The next month I was climbing to timberline in Wyoming with my own cow tag in my pocket. On a steep, timbered ridge I glimpsed a patch of tawny hide above me. Easing to the ground, I found the elk in my scope. She was angling toward me. I waited until she turned to present her forward ribs. At the shot, she dropped behind a log, and the trees erupted with elk. They thundered across the narrow spine, plumes of snow marking their wake as they plunged into a canyon to my right. I stood and pocketed the empty, listening. Then I hiked to the log, certain I'd find the cow dead behind it.

She was not behind it. Nor was there any blood or hair in the snow channels gouged by the stampeding elk. It was as if the ground had gulped her down.

I scoured the site in widening circles. *Nada.* Then I soft-pedalled after the herd.

She lay on her back, dead, below the trail, where she'd been snagged by a gnarled pine. The 150-grain .270 Swift bullet had clipped the rear of her scapula, lanced the near lung, penetrated the paunch and lodged in the off-side ham. When I dug it out, the upset bullet weighed 141 grains.

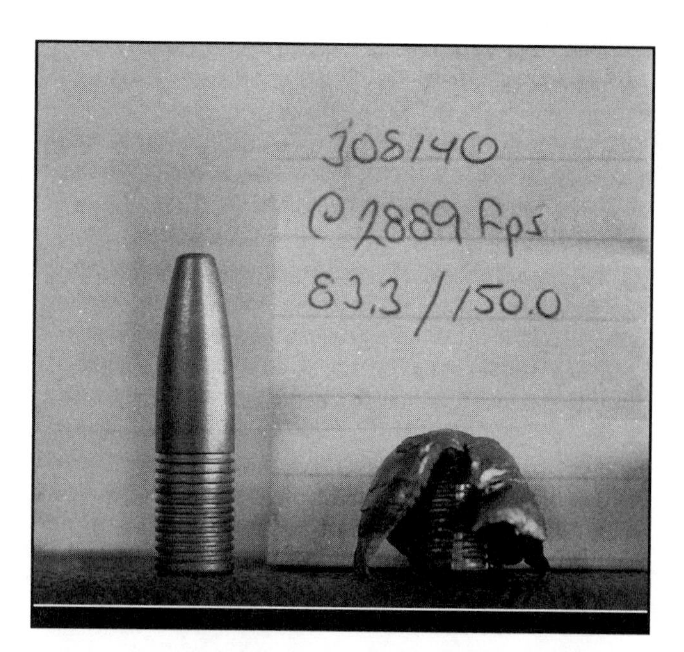

Mike Brady's bullets have no lead in the heel, only in the nose. The circumferential rings reduce bore friction. Expanded bullets were recovered from ballistic clay.

Which bullet worked better? In my view, both worked just fine. Why didn't either elk drop right away? Because elk prefer to stay on their feet after they're shot to get away from hunters who shoot them.

An expanding bullet releases energy as it decelerates. Flattening, the nose does not *slide* through as would the sleek ogive of a jacketed bullet failing to open. It *pushes*, boosting the deceleration rate and energy release. Its rotating nose petals tear tissue. Comparing channels made by a bullet that did not expand and one that opened to, say, double diameter, you'd find that the cross-sectional areas do not have a 1 : 2 relationship. The expanded bullet has destroyed much more peripheral tissue. Frontal area is not twice but roughly four times the frontal area of an unopened counterpart. Ragged edges of the petals on the still-rotating bullet, plus the hydraulic action of displaced tissue, add to the damage.

In my youth, hunters didn't weigh recovered bullets, or measure wound channels. They just bought ammo at the hardware store and shot it at game. If they shot well, the game died quickly most of the time. Most softnose bullets are quickly lethal if they're sent to the vitals.

John Nosler was one detail-minded hunter who decided he could build a better softnose bullet. Soon after World War II, he began making Partition bullets, mainly for himself. The bridge of jacket material separating the core into two parts ensured that even if the front broke up, the bullet's heel would continue on track like a full-jacket solid. John's design wasn't the first with internal structures to boost penetration. Before the Great Depression, American entrepreneur Charles Newton developed a wire-core bullet. During the 1940s the German H-Mantle bullet appeared. The last couple of decades have produced a tide of bullets built to drive deep and retain weight in heavy game. Here is a short list:

Barnes X. This solid copper hollowpoint is sleek in form but lightweight for its length. That means heavy X-Bullets must be seated deep to clear some magazine boxes and rifling. On the other hand, the X is built to drive deep without losing much (if any) weight. So you can use a lighter bullet and get fast flight, a taut trajectory, and deep penetration at the same time. X-Bullets don't offer the frontal area of many lead-core bullets after expansion, and pass-throughs are common. Reports on accuracy vary from very good to sub-par. You're smart to heed factory hints on seating depth. The XLC Bullet, with its thin shell of dry lubricant, adds velocity without boosting pressure. It also reduces fouling.

Barnes Original. You don't hear much about this bullet now, because the X-Bullet was designed to take its place. The Barnes Original is less aerodynamic than the X, but its lead-core design produces a broad mushroom. A very thick jacket controls upset. It burrows tenaciously through muscle and bone, and you'll find heavier weights in this line (195-grain 7mms, 250-grain .308s) than in most others. Accuracy is adequate for what the bullet was originally designed to do: kill big elk up close.

Nosler Partition. Developed by John Nosler to penetrate better than bullets he could buy, the Partition was initially distributed only to his hunting buddies. The wall of jacket material between front and rear core sections stops expansion cold and ensures that the bullet shank will penetrate. Often, Nosler Partitions will lose significant weight from the nose during penetration. But those fragments can destroy a lot of vital tissue, and the shank plows dependably forward. Partition jackets were once machined from solid stock; now they're impact-extruded, resulting in closer tolerances, better accuracy.

Speer Grand Slam. This dual-core bullet has a hard lead heel (5 percent antimony) held in place by an internal jacket lip. The nose core is softer, for quick upset and a broad mushroom. Grand Slam gilding metal jackets are 4 percent thicker than those on Speer's Hot-Cor bullets. During manufacture, both types receive their cores in molten form, to eliminate voids that cause jacket-core separation. The Grand Slam has an aerodynamic ogive behind a "protected" (flat) nose.

Swift A-Frame. This bullet features a bonded core plus a mid-section wall of jacket material to give deep penetration and keep the bullet nose from fragmenting. The A-Frame mushroom *looks* like a mushroom, and plows a broad wound channel. Commonly it retains more than 90 percent of its weight, even in the toughest game.

Swift Scirocco. First available in 1999, the Scirocco combines the sleek form and polymer tip of Nosler's Ballistic Tip bullet with the ductile copper jacket and bonded core of the Swift A-Frame. Result: flat flight and controlled expansion at high impact speeds. Weight retention averages a bit less than the A-Frame, but expect over 70 percent, even at strike velocities of over 3000 fps. The Scirocco opens at terminal speeds as low as 1440 fps.

Trophy Bonded Bear Claw. Jack Carter designed this lead-core bullet for high weight retention in heavy muscle and bone. The thick, ductile copper jacket includes a heel section that extends to near the mid-point of the bullet. Mushrooming yields a broad nose that rarely loses more than 10 percent of its weight in the toughest going. Like the Swift A-Frame and Speer Grand Slam, the Trophy Bonded Bear Claw has a sleek ogive and a flat bullet tip.

The author shot this bear with a 165-grain Trophy Bonded bullet (Federal High Energy ammo) and got 31 inches of penetration, front to back. The bullet retained 92 percent of its original weight.

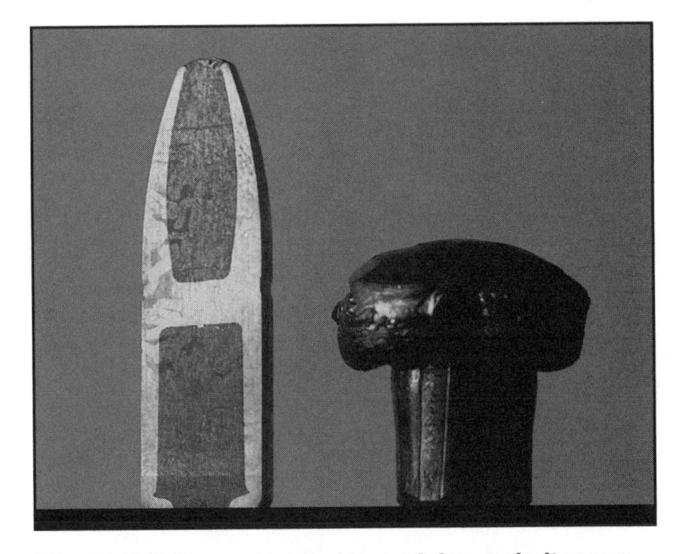

The Swift A-Frame has an internal dam to halt expansion, like the Nosler Partition. The Swift jacket is thicker, softer, and bonded to the core.

Winchester Fail Safe. A hollow nose of copper alloy is notched to give uniform four-petal upset. A steel cup keeps the lead heel core from ballooning upon impact. Weight retention typically approaches 100 percent, with penetration as deep as you'll want for the heaviest game. Pass-throughs are the rule. The Fail Safe's black jacket coating is not molybdenum disulfide (moly). Still, it does help reduce copper fouling.

Winchester Partition Gold. Developed under the "Combined Technology" banner that marked the recent collaboration between Winchester and Nosler, Partition Gold bullets resemble their Nosler namesake. Differences: The dam has been moved forward, so if the nose disintegrates there's less core lost. A steel heel cup prevents deformation of the rear jacket on impact. The Partition Gold has a black oxide finish, per Winchester's Fail Safe.

Off-shore bullet companies offer competitive products. The TIG, TUG, and H-Mantle bullets available to handloaders in Europe are loaded in RWS ammunition. The Old Western Scrounger imports RWS ammo (and can supply bullets and cartridges for many obsolete rifles). Australia's Woodleigh bullets, now offered by Federal in some loadings, have a solid reputation; core and jacket are as hard to separate as a pit bull and a burglar. Expansion tests in water (a severe medium) yield terminal weights that exceed 90 percent of initial weights. The petals open out wing-like, rather than collapsing to support a classic mushroom, as with Swift, Hawk and Trophy Bonded bullets. But despite their action in braking the bullet, and their exposure to rotational pressures, the petals of Woodleighs seldom break or lose the lead on their forward surfaces.

Do you need a controlled-expansion bullet? Well, probably not often. If you hunt where oblique shots at elk or moose are common, or if you prefer a shoulder shot on big bears, you'll want a cohesive bullet that plows a long furrow. But most big game is taken with bullets through the ribs. Even animals that weigh more than half a ton usually succumb quickly to an ordinary softpoint striking the forward ribcage. Some popular bullets that cost less (and sometimes shoot more accurately) than bullets built to retain weight are listed below. On deer-size game, they may also give you shorter blood trails, because explosive energy release in the vitals kills right away.

The effect of velocity on bullet performance can hardly be overstated. If you aren't pushing your little rockets at 3400 fps, you don't need bullets that hold up to impact speeds of 3200 fps. If you choose a stout bullet and shoot a deer with your .30–06 at 300 yards, the remaining velocity (say 2000 fps) may not give you the dramatic upset you want. I once asked a fellow at Nosler why there was a Partition-style 170-grain .30–30 bullet. He agreed with me that at .30–30 velocities, a Partition didn't make much sense, but added that retailers carrying Nosler bullets wanted to stock bullets for hunters using .30–30s, and they wanted to limit the number of suppliers they had to deal with. To earn their business, Nosler filled out its line.

Actually, the drift toward stouter bullets is paradoxical. As hunters buy more powerful scopes, and cartridges designed to give bullets flatter flight, they seem to be banking on long shots. The odds of having to shoot far and break both shoulders or thread an animal lengthwise are slim. At extended range, there's usually time to wait for a better angle and to place the bullet well. By the time even a high-velocity bullet has covered 300 steps, it is moving considerably slower than when it started (figure about 25 percent loss for a pointed bullet at normal launch speeds). You shouldn't worry about disintegration. These less costly bullets kill big game with no trouble most of the time:

Federal Hi-Shok Softpoint. The standard bullet for Federal's Classic line, Hi-Shok is available in round-nose, flat-nose, and spitzer form, in diameters .224 to .375. Like ordinary softpoint bullets loaded in PMC cartridges, the Hi-Shok is unremarkable in design. However, it is remarkably effective, a bargain in loaded ammunition.

Hornady Interlock. The reputation for fine accuracy that attended the company's flagship Spire

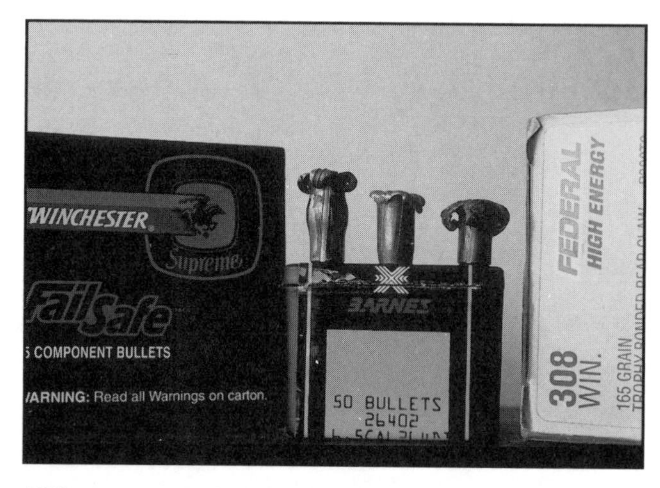

Different expansion characteristics of the Winchester Fail Safe, Barnes X and Trophy Bonded are plain here. All three are noted for deep penetration.

Point for decades carried over to the Interlock. Inner jacket belting holds the lead core in place during upset. The roundnose versions are dependable killers at moderate ranges. They're as accurate as their pointed brethren, and they pack more weight for any given length.

Hornady Super Shock Tip (SST). This bullet has a red polymer nose, after the style of Nosler Ballistic Tips, and the internal jacket belt of Hornady Interlock bullets. A tapered heel gives the Super Shock an edge at extreme range. Steve Hornady's first tests showed the SST a bit fragile for tough game, so it was beefed up to better handle high-speed impact. The inner belts help keep the core in place. Half-minute groups turn up at Hornady's test range.

Nosler Ballistic Tip. The polycarbonate tip of this bullet is color-coded by caliber and serves as a wedge to initiate expansion upon impact. Ballistic Tip bullets have an enviable reputation for accuracy, and their sleek profile ensures flat flight. They're noted for quick upset and are not recommended for big-boned game or oblique shots. However, .338 Ballistic Tip bullets have thick jackets for deeper penetration.

Remington Core-Lokt. In both roundnose and pointed form, this veteran may have killed more big game animals than any other softpoint. It has an internal lip to hold the core in place and is available in many weights. The Core-Lokt delivers a deadly combination of fast upset and deep penetration.

Remington Bronze Point. This decades-old design features a bronze nose peg to initiate violent expansion. Aerodynamic form and ready upset make the Bronze Point ideal for long shooting. It's not a top choice for oblique shots up close.

Sierra GameKing. Long renowned for superior accuracy and flat flight downrange, Sierra boat-tail bullets are quick openers. Expect lightning-like kills on deer from both softpoints and hunting-style hollowpoints. The 250-grain .338 and 300-grain .375s have extra-heavy jackets.

Speer Hot-Cor. The huge selection of these solid performers adds to their popularity. You'll find weights not available elsewhere, and even .366 spitzers for 9.3x62 rifles, common in Europe and Africa. Their traditional softnose construction and a sleek profile make Speer Hot-Cors ideal for long shooting at deer-size game.

Winchester Power Point. This softpoint might well be considered an archetype. It has no special features, save the nose notches on its tapered jacket. They ensure quick expansion. Still, penetration is often to the far side of moose-size game. The Power Point has recently been moly-coated for stepped-up loads in Winchester Power Point Plus ammo.

Winchester Silvertip. For years the company's heavy-game bullet, Silvertip got a more fragile nose cap in the 1960s. This bullet now opens more quickly—sometimes even quicker than the Power Point. Expect sure kills with rib shots. Weight retention doesn't match that of bullets built to penetrate.

Winchester Ballistic Silvertip. This is another Combined Technology product, a cosmetic variant of the Nosler Ballistic Tip. Its gray polymer nose and black jacket coating distinguish it. Fine accuracy, quick upset, and flat flight are hallmarks.

If you buy loaded ammunition, you benefit from the collaboration of bullet firms with the ammo makers. Federal offers Barnes X-Bullets, Nosler Partitions, Sierra GameKings, and Trophy Bonded Bear Claws in its Premium line of rifle ammunition. PMC catalogs the Barnes X and Sierra GameKing, while Remington loads Nosler Ballistic Tips and Partitions, plus Swift Sciroccos and A-Frames—besides its own Core-Lokt and Bronze Point bullets. In the Winchester line you'll find not only the company's Fail Safe, Power Point and Silvertip bullets, but Ballistic Silvertips and Partition Golds.

These are the most prominent commercial offerings, though keen hunter interest in big game bullets continues to drive a cottage industry. Blue Mountain Bullets and Elkhorn Bullets, both Oregon

Controlled-expansion bullets like the Swift A-Frame make the .270 more effective on elk-size game.

companies, specialize in bullets for elk. Warren Jensen of Blackfoot, Idaho, recently improved on his sleek J-26 spitzer by replacing the polycarbonate tip with one of soft copper and making other, less visible changes. The new bullet, the J-36, is designed expressly for long shooting at elk-size game but has proven its integrity at close range too, where high impact velocity tears weaker bullets apart. Mike Brady of Glenrock, Wyoming, has a controlled-expansion bullet with a ribbed heel that reduces bore friction. Northern Precision has focused on big-bore bullets but has now added new bullets in smaller diameters, custom-made in almost any weight. Hawk bullets, designed in central Wyoming by Bob Fulton, have a rounded form but fly surprisingly flat. The ogive, or forward curve of the bullet body, has a greater effect on trajectory than does tip profile. Hawk and other semi-spitzers keep up with their pointed counterparts at normal hunting ranges. Hawk bullets also feature thick copper-tubing jackets; so do the Kodiak bullets made in Alaska. These apparently pass the bear test.

What a bullet looks like after it comes out of an animal is no indication of its lethality. I once watched a hunter shoot a big bull elk at 250 yards with a .30–06. The elk took the bullet a bit far back, but the shot angle kept it on course into the off lung. Within five seconds that bull was on the ground. By the time we reached him he was dead. The 180-grain Hornady was pretty well minced.

"Not much left of that slug," said the hunter, inspecting it.

"No," I said. "But this is a 370-point bull, and that tattered bullet gave him to you. Be charitable." After all, a set of worn tires may not look like much either, but if they've carried you 50,000 miles, have you a right to expect more?

Finding ammo that shoots well in your rifle inspires confidence and leads to more accurate shot placement afield. That's more important than seating a bullet designed to thread six railroad ties, a stack of New York phone books and a 1950 John Deere engine block without losing more than 3 percent of its weight. Before you settle on a bullet, buy several kinds, splitting boxes with friends to trim cost. You don't need 50 shots to tell you a given bullet is inaccurate or acts like a grenade inside wet newspapers. Keep your mind open to different bullet weights. When thin-jacketed softnose bullets were the only option, a heavy bullet made sense for tough game, because a lot of bullet weight would be lost right after entry. But modern controlled-expansion bullets needn't be so heavy. One of my friends at Winchester prefers a 150-grain Fail Safe in his .300 Winchester Magnum. He has killed many, many animals of all sizes with that combination. In fact, the option of using lighter bullets, with higher velocity and flatter arcs, might push you toward a controlled-expansion bullet even if you don't expect quartering shots.

The Swift Scirocco was designed for ballistic efficiency and deep penetration.

Whether you handload or shoot factory rounds, there are more than enough bullet choices to keep you busy testing during the off season. More shooting means that whatever bullet you choose is more apt to land in the right place and kill right away.

PART II

Ballistics to Measure

The Violence Inside the Gun

When the primer spits fire into the powder charge and burning commences, gases form, increasing pressure inside the case and (because pressure produces heat), accelerating the burn. On a graph, you get a pressure peak that rises steeply after a short horizontal line that represents the time delay between primer detonation and powder ignition. After the peak, which usually happens within a millisecond (1/1000 second) after the powder starts to burn, the pressure curve comes back down. This slope is not so steep as the bullet moves forward, increasing the volume behind it. The faster the powder, the steeper the slope on both sides of the peak. The area under the pressure/time curve translates to bullet velocity. Between 2 and 3 milliseconds after the striker hits the primer, pressure has dropped to zero. The bullet is on its way.

A 180-grain bullet from a .300 Weatherby Magnum exits the muzzle of a 26-inch barrel about 1¼ milliseconds after it starts to move. Here's a summary of what happens (data adapted from a pressure/time curve in the excellent text *Any Shot You Want,* a reloading manual by Art Alphin's A-Square company):

time (seconds)	*pressure (psi)*	*velocity (fps)*	*distance (inches)*
0	0	0	0
0.0001	12000	60	0.02
0.0003	36000	500	0.60
0.0005	60000 (near peak)	1400	2.80
0.0007	42000	2350	7.40
0.0009	24000	2970	13.80
0.0011	6000	3250	21.30
0.0013	100	3300	26.00

A few things are noteworthy here. First, peak pressure comes when the bullet has moved only about 3 inches, even with the slow-burning fuels appropriate for a .300 Magnum. The pressure drops off fast, too, losing 90 percent of its vigor in the next 18 inches of barrel. But the bullet continues to accelerate as pressure behind it diminishes. Between 14 and 21 inches, pressure loss totals 18,000 psi. But the bullet speed increases 300 fps! With very little pressure remaining behind it at the muzzle, the bullet is still speeding up. The value of a long barrel is clear, even if nearly all of it is used to control the tail of the pressure/time curve.

A pressure/distance curve differs from a pressure/time curve in slope, but it has the same general shape: steeper at the start than at the finish. The area under a pressure/distance curve represents the amount of energy available from the bullet. However, the energy generated is not all available down-

Pressures in excess of 50,000 pounds per square inch are trying to tear your rifle apart under your cheek each time you fire.

Break-action single-shot rifles used to be limited to low-pressure rounds, but new steels and designs have made them stronger. This custom rifle on a T/C Encore action is chambered in 7 STW.

range. A lot of it is lost in thermal (heat) transmission, expansion of the case into the chamber wall, bullet/rifling friction, even the bullet's rotation. Plotting a load's pressure/distance curve is important for designers of gas-operated autoloading rifles because these rifles must tap the gas at some point in the bullet's travel. Too much pressure and the slamming can damage rifle parts. Too little and bolt travel is insufficient to clear the fired case.

How can pressure be measured? Well, in the mid-1800s, Alfred Nobel and an American named Rodman were wondering about that. They apparently came up with solutions to the problem at the same time. Rodman's, the crusher system, is still in use. It's a factory procedure not easily or safely performed in a home shop.

A small cylindrical piston is slid into a hole in the barrel of a test gun, and a copper or lead pellet inserted snugly between the top of the piston and a stationary anvil. When the gun is fired, the piston pushes against the pellet or crusher, shortening it. The difference in lengths of the crusher, before and after firing, is then converted mathematically to a pressure range, in units of CUP or LUP (copper units of pressure or lead units of pressure). Copper crushers are generally either .146 in diameter and .400 long to start with, or .225 in diameter and .500 long. The choice depends on application. Copper crushers work best for centerfire rifles and handguns that generate substantial pressures. Lead crushers (.325 x .500) typically register the low-pressure loads in rimfire guns and shotguns (though small-diameter copper crushers can be used too). Crushers are calibrated in a test press. Subsequent measurements and the pressures that produced them yield "Tarage Tables" that enable technicians to calculate the pressure in a rifle chamber tested with a similar crusher.

Crushers get hammered by peak pressure, but they don't register peak pressure accurately because the flow of the copper is slower than the change of pressure inside the chamber. Also, the moving piston must be brought to a halt, which skews a reading in the opposite direction. There's no balancing out here, just conflicting forces. *Copper units of pressure and lead units of pressure are not the same; nor can they be interchanged with another common unit of pressure, pounds per square inch (PSI).* CUP value can be the same as the PSI value; for example, SAAMI data lists 28,000 as maximum average pressure for the .45–70. Both units apply. But maximum average pressure for the .243 is 52,000 CUP and 60,000 PSI. Most other cartridges show similar discrepancies. Sadly, there's no formula to convert CUP to PSI or vice versa.

A more modern device for pressure measurement than the crusher is the piezoelectric gauge. It registers via transducer an electric charge produced when a certain type of crystal is crushed. The pressure applied to the crystal yields a proportional

Managing gas pressure becomes critical in gas-operated autoloading rifles like this Remington.

transducer reading in pounds per square inch. Conformal transducers are installed in the barrel, just like crusher pistons, and become part of the barrel. External transducers can be mounted on the barrel, then taken off for replacement or recalibration. Unlike crushers, piezoelectric gauges need periodic calibration checks. The calibration machine is sophisticated and very expensive.

One pressure tester that's become popular among shooters is the strain gauge. Developed for the consumer market by chronograph guru Ken Oehler, it's essentially a piece of wire that you glue to the outside of your chamber wall. When you fire the rifle, the chamber expands, and this wire stretches. A measure of the wire can be compared then with a previous measure to get a stretch reading that translates into pressure. Readings do not equate with readings from the crusher or piezoelectric gauge. The way to use a strain gauge is to read pressures from factory loads and maximum recommended loads and determine the relative pressures of hunting loads by comparing values.

For years handloaders without pressure gauges have monitored their loads by examining fired cases, paying particular attention to ejector-hole marks on the head and flattened primers. Both indicate high pressure. A sticky bolt lift means you're near maximum. Extruded primers may or may not be caused by excessive pressure. The most reliable way to gauge pressure: Use a micrometer to measure case expansion just in front of the extractor groove. Speer ballisticians recommend that you allow no more than .0005 inch expansion after loading once-fired cases.

Headspace

When you fire a cartridge with a quick blow from the striker, all the important events follow right away. Because the brass case is ductile, it expands outward under gas pressure, ironing itself to the chamber walls. The bullet blows forward and the head of the case blows back against the bolt face. During this mayhem, the chamber keeps everything in place. In order for cartridges to fit the chamber snugly but not so tightly as to cause problems in the field, and because cartridges vary slightly in dimensions, the chamber must be a little bit bigger than the case. If it is too big, however, the brass is stretched unduly, shortening case life. And then there's headspace, the distance from the face of the locked bolt to a datum line in the chamber that arrests the forward movement of the cartridge.

Gas pressure popping the bullet like a cork is also thrusting rearward: For every action there is an equal and opposite reaction. What complicates things is the taper of the case wall, from thick at the web to very thin at the shoulder. The web itself is a solid partition of brass around the flash-hole. As the front of the case collapses into the chamber wall, the rear section stays close to its original diameter, slightly smaller than the chamber—and, thus, mobile.

Now, if the case is held tight to the bolt face, everything is OK. The shoulder blows forward, the forward body and neck outward to contact the chamber; and the case head simply absorbs the rearward thrust of gas without moving because it is sup-

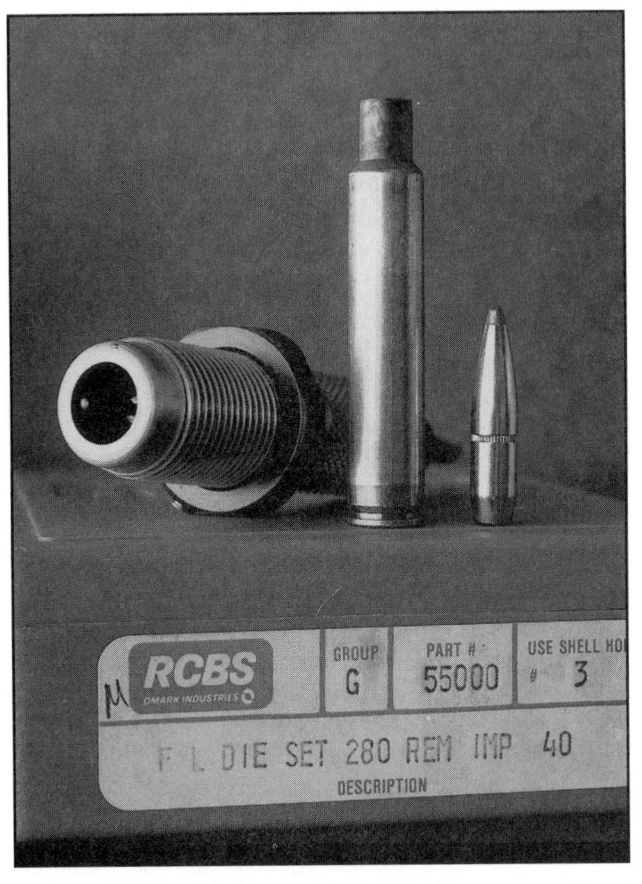

The .280 Improved has a sharper shoulder than the .280 Remington but the same headspace measure to the datum line on the shoulder.

ported by the bolt face. But if there's too great a distance between the bolt face and the point in the chamber that stops the forward motion of a cartridge, you have a condition of excess headspace. Again, a blow from the striker pushes the case forward until it contacts the datum point in the chamber. Case expansion sticks it there. The case head isn't stuck because it doesn't expand as much, so it moves rearward to meet the bolt face. But now the bolt face isn't close enough to support the case head, so the case stretches under the pressure of building powder gases. The case wall just forward of the web can be stretched to the point of cracking or even separation.

A cracked case caused by excess headspace is dangerous because it spills powder gas into the chamber. This hot, high-pressure gas instantly seeks release, speeding down the tiniest corridors at velocities that can exceed bullet speed. It jets along the bolt race, through the striker hole, down into the magazine well. It can find your eye faster than you can blink. To avoid case ruptures, you must know something about headspace.

Headspace is measured from the boltface to the mouth of a straight rimless case like the .45 ACP, whose mouth stops the case from going farther forward. In a belted magnum, the stop is the leading edge of the belt. On a .30–30 case it's the front of the rim. The datum line lies on the shoulder of a rimless bottlenecked cartridge like the .270 or on the rebated .284 case. The term "headspace" originated when all cartridges had rims, so the measurement was made only at the head. Now, of course, rimless rounds are the rule, and headspace as a dimension includes all measures of bolt face to cartridge stop.

Semi-rimmed cartridges theoretically headspace on the rim, but sometimes (as in the case of the .38 Super Automatic) the rim protrusion is insufficient given the action tolerances needed for sure function. The case mouth then serves as a secondary stop. The semi-rimmed .220 Swift has a more substantial lip; but most handloaders prefer to neck-size only, so after a first firing, the case actually headspaces on the shoulder. More on that later.

Headspace is measured with "go" and "no go" gauges. The "go" gauge is typically .004 to .006 shorter than the "no go" gauge for rimless and belted cartridges. The bolt should close on a "go" gauge but not on a "no go" gauge. Theoretically, if the bolt closes on a "no go" gauge, the barrel should be set back a thread and rechambered to achieve proper headspace. However, many chambers that accept "no go" gauges are still safe to shoot. The "field" gauge, rarely seen now, has been used to check these (mostly military) chambers. It's roughly .002 longer than a "no go" gauge.

Minimum and maximum headspace measurements for chambers are not the same as the corresponding minimum and maximum case dimensions. For example, a .30–06 chamber should measure between 1.940 and 1.946, bolt face to shoulder datum line. A .30–06 cartridge usually falls between 1.934 and 1.940. Case gauges machined to close tolerances perform the same check on cartridges that headspace gauges do in chambers. An obvious difference: case gauges are female.

Now, case gauges don't measure headspace, because that is really a chamber dimension. They simply indicate whether a cartridge will work in a chamber that's correctly bored and fitted. Thus, crimping a .45 ACP to put the front .003 to .005 into the bullet's cannelure or moving the shoulder of an undersized .280 Remington case forward by firing it in a generous chamber doesn't change headspace. Headspace is a steel-to-steel measure in the gun. Altering case dimensions *does* change the relationship of cartridge to chamber. Reducing the head-to-datum line length of a hull can result in a *condition* of excess headspace, even if the firearm checks out perfectly.

Not long ago I was sizing cases for a wildcat 6mm cartridge, the .240 Hawk. The custom sizing die had been made to reduce the neck diameter of the .30–06 case without changing the shoulder or datum line. I set up the die initially to full-length resize, so I'd be starting with cartridges that would all fit the chamber. After one firing, I would neck-size only to get a tighter fit and prolong case life.

My first shot blew gas out the crevises of the Remington 700 action. The case showed a circumferential crack forward of the belt. Because the loads were not stiff, and because the bolt lift did not indicate high pressure, I fired another round. Same result. The third shot (which I should not have fired, as something was obviously amiss) nearly separated the case. Escaping gas blew the extractor.

Back at the bench, I measured the sized cases against the fired cases. The sized .240 hulls were shorter by nearly .1 inch. I screwed the die into the press but left it one-eighth inch short of contacting

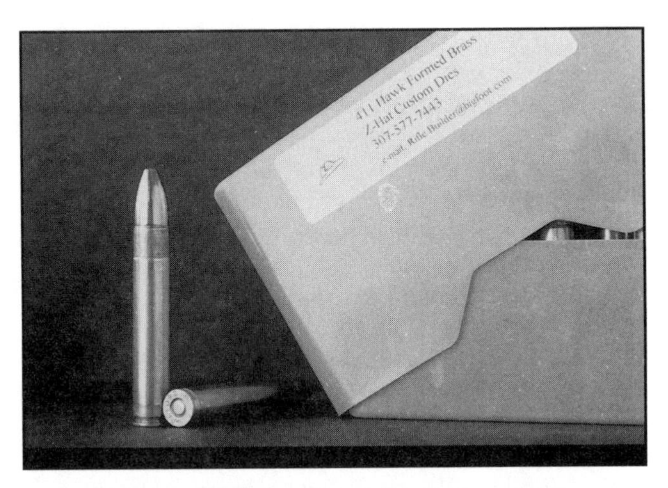

The .411 Hawk is based on a .30–06 case with the shoulder blown forward. The large-diameter neck doesn't give the shoulder much purchase, just enough to ensure against creep and excess headspace.

the shell-holder. I ran a case into the die, then tried chambering it in the rifle. It wouldn't go. I ran the die down a thread and sized the case once again. No go. Lowering the die incrementally and trying the case each time, I finally closed the bolt. It was a snug fit. I looked at the relationship of die and shell-holder. There was nearly .1 inch gap—the measure of the difference between fired and unfired cases. Unlike most commercial dies, this one had not been manufactured to full-length size when it was flush with the shell-holder. Screwing it down that far, I had made the case .1 shorter than the rifle's chamber. When I fired, the striker drove the case forward .1 inch. The front of the case expanded into the chamber to grip it, and the rear of the case backed up .1 inch against the bolt, pulling the brass apart just ahead of the web.

Because neither rifles nor cartridges can be manufactured to zero tolerances, you can't expect skin-tight fit of an unfired cartridge in the chamber. If you could, you wouldn't want it in a hunting rifle. The least bit of dust, water, snow or residual oil might prevent easy bolt closure and a missed opportunity. Changes in temperature can also affect cartridge fit. In a revolver, the breech face cannot be in close contact with the case head without tying up the cylinder. A tight fit of straight cases in autoloading pistols means that any jump at all in case length prevents the slide from coming into battery.

If headspace can legitimately vary .006, and the corresponding case dimension another .006, it's possible to load a round .012 shorter than the chamber will allow, bolt face to stop. Full-length sizing dies, when set so the shell-holder presses tight at the end of the stroke, should not bring cases below minimum dimensions. However, dies vary too, and in a small die you can compress a hull so much as to create a condition of excess headspace if the chamber is cut big.

Full-length resizing compresses a cartridge case; firing stretches it. Like the repeated bending of the tab on a can of soda pop, "working" brass can make it fail. To prolong case life, you're better off neck-sizing only, so the brass moves little upon firing. Because a hull shrinks after firing to allow extraction and subsequent chambering, there's no need to further reduce its dimensions *unless you plan to use the ammunition in another rifle that has a slightly smaller chamber.* The only other reason to full-length size (or to use small-base dies that squeeze cases down even further) is to accommodate autoloading or lever- or slide-action rifles that have little camming power. Some hunters full-length size the cases they'll use on the hunt to ensure easy chambering if they must fire quick follow-up shots.

Neck sizing is a particularly good practice with belted cases, because chambers for these hulls are often cut generously up front. The critical dimension, after all, is the distance from bolt face to belt face (.220 to .224, "go" to "no go"). If you full-length size magnums, you may be shortening the *head-to-shoulder* span considerably each time—which means the case stretches considerably at each firing. Eventually (sometimes quite soon), you'll notice a white ring forming around the case just ahead of the belt. If you insert a straightened paper clip with a small "L" bend at the end and feel around the inside of the case, you'll detect a slight indentation forward of the web, where the ring is. The white ring signals a thinning of the case and shows where the case will separate if you keep full-length sizing the case. As on a rimless case in a chamber with excess headspace, the brass is being shoved forward by the gas pressure against the shoulder. Though the magnum's belt keeps web and head from moving very far forward when the striker hits, the front of the case can still move considerably in a chamber of generous dimensions. This brass movement forward has been called flow by some handloaders. That's

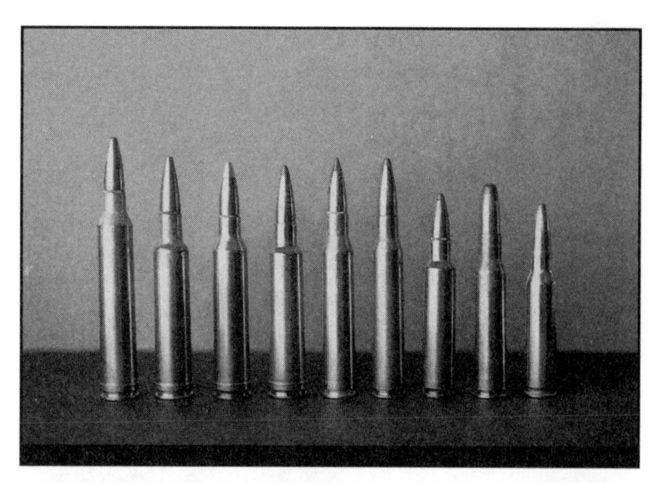

This lineup of 7mm cartridges shows rimmed (7–30 Waters, right), rimless (7x57, 7x64 and .280, 2nd, 4th and 5th from right), rebated (.284, 3rd from right) and belted (7x61, 7mm Rem., 7mm Wby. and 7 STW (left end, from right). Headspacing for rimless and rebated cases is on the shoulder.

not quite what happens. At any rate, the brass *does* get thin like a stretched rubber band just ahead of the belt. Cases separations of belted magnums may thus occur even when the headspace measurement is perfect. Prevent excessive case stretch by neck-sizing after the first firing.

Rechambering a rifle from one belted round to another (say, reaming a 7mm Remington Magnum to 7mm STW) can produce excess headspace if the reamer is allowed to eat away at the belt recess. The leading edge of the web will then be unsupported and could blow out as the head backs into the bolt face.

Rechambering rifles to Improved cartridges does not change headspace measurement. The reduced body taper and steeper shoulder angle of such rounds provide greater case capacity, but the datum point on an altered shoulder should remain the original distance from the bolt face. That's why you can fire factory ammunition in an Improved chamber safely. Expect velocity to be lower because some of the gas energy is absorbed blowing the standard case outward to fit the new chamber and is dissipated in the larger hull. Accuracy can range from poor to very good.

True wildcats that require case forming in dies must sometimes be given a false shoulder to serve as the chamber stop before firing full-power loads. For example, the Gibbs wildcat rounds pioneered in Idaho by Rocky Gibbs in the 1950s, and the Hawk line more recently developed by Bob Fulton and Fred Zeglin in Wyoming, are both based on the .30–06 with the shoulder moved forward. To fashion a .338 Hawk case, you might neck a .30–06 up to .375, then neck that hull down to .338 in a Hawk die. The result would be a case with two shoulders. The rear one would disappear upon firing.

An alternative method would be to neck the .30–06 to .338, then seat a bullet to contact the lands as the bolt is closed. Hard against the lands, the bullet keeps the head of the case tight to the bolt face. Both techniques are best used with reduced charges of powder. Handloaders who prefer to make false shoulders commonly use a small amount of fast-burning powder behind a case filler of corn meal and a wax cap.

Often shooters who don't understand headspace blame it for all sorts of problems not related to headspace. The first centerfire rifle I owned was a surplus Short Magazine Lee Enfield. The rimmed .303 British cartridges fed like Vaselined sausages, but roughly one in four separated upon firing. Now, several things could have been wrong. There could have been too deep a cut for the rim. The chamber could have been scored by gas from a ruptured case, leaving all subsequent hulls unsupported at that point. Because headspacing on SMLE rifles can be changed by switching bolt heads, the bolt itself was even suspect. The cases in this instance were of new commercial brass so were above suspicion.

I didn't know enough about rifles then to figure out what was wrong, but I did know that repeat shots at whitetails were improbable if I had to wait for the barrel to cool and use a rod with a tight brush to extract the fronts of broken cases. I was loathe to part with the rifle, because for months I'd ogled it in the Klein's magazine ads and eventually paid $15 for it. My elders, mostly farmers with 97 Winchester pump guns and Krags as old as my .303, didn't know what to tell me. In desperation I took it to an ill-tempered gunsmith, who peered into the black breech, examined one of my two-piece cases and declared I had a "swelled chamber." I didn't know how chambers could swell, but he convinced me to sell him the rifle. My second SMLE cost twice as much, and I plowed an additional $7.50 into it with a Herter's walnut stock blank that took weeks to fit. Somewhere I found $10 for a set of Williams sights. The cases all extracted, and some deer bit the dust.

Years later I noticed queer marks on hulls kicked out of a Model 70 Winchester in .338 Magnum. They were white lines, but not in the usual place. I knew enough by then to neck-size only. Inspection of the chamber revealed a gas cut made sometime before the pre-64 rifle came my way. Responding to my call for suggestions, Winchester advised rebarreling. That didn't make sense unless I could locate and afford a vintage barrel. So I threaded the flash-hole of a fired .338 case and inserted a quarter-inch rod. Then I chucked the rod in a variable-speed drill and secured the barreled action in a vise. After smearing the case body (but not the belt!) with J-B's abrasive paste, I carefully fed the case via the rod into the chamber. The drill, turning slowly, polished out the chamber without removing so much metal as to materially change its dimensions. It dulled the edges of the gas cut and reduced its depth. The marks on fired cases all but vanished.

Recently a friend had a rifle chambered for the wildcat .338–08. No matter how clean he kept the chamber and cartridges, or how potent his handloads, or whether he full-length sized or neck sized, cases would stick. There were no flat primers or ejector marks on the empties, and he even had trouble extracting loaded rounds. The claw had plenty of pull, but not quite enough to dislodge the fired brass. We hunted in Africa, he toting the rifle and either his tracker or me trotting behind with a cleaning rod so we could punch out the empties. The cartridge, with both Barnes X and Nosler Partition bullets, performed well indeed. But neither of us could finger the cause of the sticky cases. Back home, my amigo took his problem to a gunsmith, who found the chamber to be out of round. Careful touch-up with a new .308 reamer made everything right.

Headspace is a length measurement. It has nothing to do with diameter. Chambers that are egg-shaped or belled by unprincipled cleaning, or whose diameters are larger or smaller than normal may cause problems, but they're not headspace problems. Headspace hinges on reamer dimensions as well as on the machinist's eye and expertise. After long use, reamers cut slightly smaller chambers than when new. New reamers or those used aggressively can bore oversize chambers. The natural elasticity of cartridge brass allows for some variation in chambers. While you won't get great accuracy from a loose cartridge fit, you won't necessarily be imperiled when you pull the trigger.

With each firing of your centerfire rifle, some compression of the locking lugs and lug seats occurs. The elasticity of the steel keeps headspace essentially the same. But many firings with heavy loads can diminish that elasticity. A permanent increase in headspace results. A rifle with hard lugs and soft seats and generous headspace can eventually develop so much headspace that a field gauge can be chambered. At this point the rifle is unsafe.

Headspace is a chronic problem with early lever-action rifles built for rimmed rounds. Shooters seldom fret much about it, partly because these cartridges, some originally loaded with black powder, are kept to relatively low pressures. Riflemen who handload for old lever guns wisely keep charges mild for the sake of both case and mechanism. Also, loose, rear-locking actions are intrinsically difficult to hold to the tolerances of our modern front-locking bolt rifles. Some handloaders with ailing saddle guns prolong case life and make shooting safer by peening the rims of cases before firing. Careful peening from the side, with frequent testing in the action, will boost rim thickness to match the chamber cut. It's tedious but, I'm told, effective.

There's nothing mysterious about headspace. It's simply a measurement. Proper headspace is a close but easy fit between cartridge and chamber stop.

Figuring Bullet Flight

To the first ballisticians, bullet flight was no mean mystery. Bullets flew so fast you could not see them. In 1537 Trataglia, an Italian scientist, wrote a ballistics book and described the bullet's path as an arc. (Though this seems quite obvious to us, it was not so then. The arrow's visible arc notwithstanding, some people envisioned a bullet flying almost straight until its energy was spent, then dropping abruptly to earth.) Trataglia also experimented to determine the launch angle that would give a bullet its greatest range, and found this angle to be near 45 degrees. This is much steeper than the angle you'd choose to give your .270 bullet its greatest reach. But Trataglia's conclusion was valid. Reason: At the very low velocities of his day, gravity has a far greater effect than air resistance on bullet flight. Modern high-speed bullets, in contrast, are influenced less by gravity than by drag.

A hundred years after Trataglia studied bullet flight, Galileo published his descriptions of and explanation for a bullet's arc. Dropping cannonballs from the Leaning Tower of Pisa was only one of many experiments he undertook as a ballistics consultant for the arsenal at Venice. Galileo concluded that the acceleration of a falling body was a constant and that, as a result, bullet trajectories must be parabolic. This was a great discovery. However, Galileo concurred with Trataglia in assigning 45 degrees as the launch angle for maximum range. His experiments did not take into account the variable of drag, because, again, in those days projectiles were very heavy and very slow. Compared to the acceleration of gravity, drag on a cannonball dropped from a window was inconsequential!

Another century went by. Then, in 1740, Englishman Benjamin Robins invented a device to measure bullet speed. He called it a ballistic pendulum. The pendulum had a heavy wooden bob, which Robins

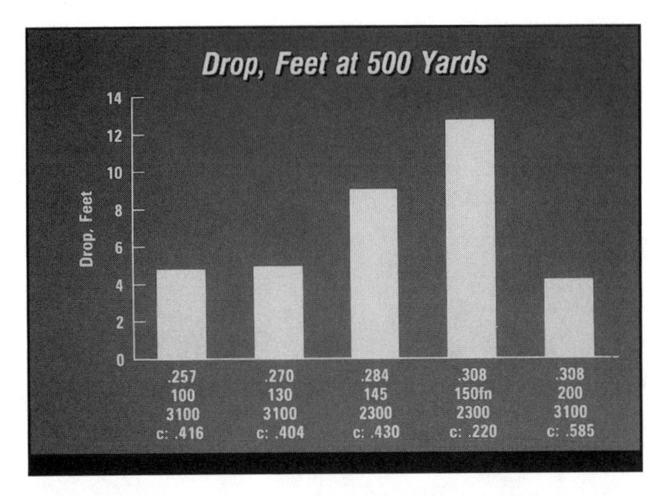

Bullet drop depends on starting velocity and ballistic coefficient, C, as shown by these examples. At right, a very aerodynamic match bullet at high velocity shows almost as much drop as ordinary hunting bullets with Cs of .400. Reduce velocity or C significantly, though, and drop increases.

weighed. Then he weighed the bullet to be fired. The pendulum was next brought to rest, hanging vertically, and the bullet was fired into it. By noting the height of the swing, Robins was able to calculate impact velocity. He made a series of measurements with 75-caliber musket balls, recording velocities between 1400 and 1700 feet per second. People of the day were skeptical—how could anything travel that fast? When Robins increased the firing distance, his results drew even more criticism. To account for the readings at the greater range, air resistance had to be 85 times as strong as the force of gravity! Now *that* was a hard concept to sell. Of course, it was true.

Robins was not alone in his beliefs. Sir Isaac Newton, who died only 15 years before Robins began his experiments, had come up with several important observations in the fields of physics and mathematics. One of them was the universal law of gravitation, which specifies that the force of gravity varies with altitude. Newton's fundamental laws of mechanics, and his development of calculus (at the same time Leibnitz worked on calculus in Germany) were crucial pieces in the complex puzzle of ballistics. Newton showed that drag increases with the density of the air and the cross-sectional area of a projectile. He also demonstrated a relationship between drag and the *square* of the projectile's speed. Because he had no way to measure the speed of musket balls, he could not know that drag increases dramatically when projectiles approach the speed of sound (1120 fps).

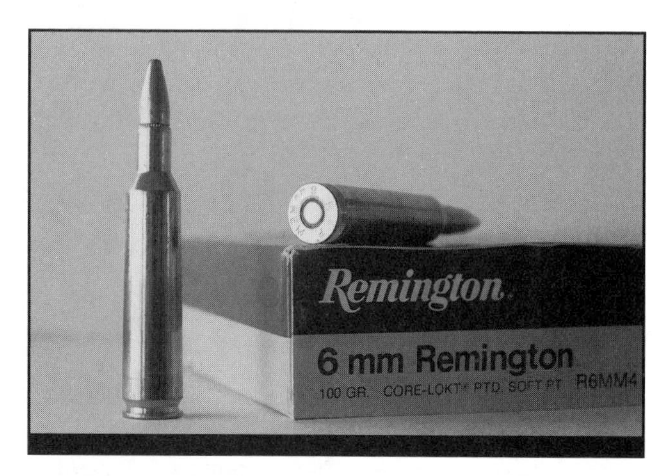

A 100-grain 6mm Remington bullet shoots flat because of its high velocity and high C. Reduce bullet weight, and you get faster flight, but at long range drag will pull the lighter bullet back.

Beginning in the mid-1700s, ballisticians in Europe worked hard to describe, mathematically and precisely, the flight of high-speed bullets. But accurate measurements of drag eluded them until the late 1800s, when chronographs were invented in England and Germany. By this time, conical bullets were in widespread use, and cartridge firearms were supplanting muzzleloaders.

Around the middle of the nineteenth century, ballisticians hit upon the idea of a "standard bullet" that could be used to develop benchmark values for drag and other variables of bullets in flight. The flight characteristics of other bullets might then be computed using a constant factor that defined their relationship with the standard bullet. Thus was born

Long shooting taxes both bullet and marksman. A flat-shooting bullet won't hold the rifle still.

the concept of a ballistic coefficient, usually expressed as "C." In mathematical form, C = drag deceleration of the standard bullet / drag deceleration of the actual bullet. Though this relationship is exactly true only when both bullets are of the same shape and density, it holds well enough when the bullets are of similar form. By 1865, ballisticians around the world were hard at work determining drag characteristics of standard bullets. The best-known studies were conducted by Krupp in Germany (1881) and by the Gavre Commission in France (1873–1898). Meanwhile, around 1880, an Italian ballistician named Siacci discovered a short alternative to the laborious calculus in determining a bullet's trajectory. He found that by using calculus to plot the path of a standard bullet, he could calculate with simple algebra the path of any other bullet with a known C. This shortcut proved a boon to every shooter and ballistician who followed.

The Gavre Commission's tests were perhaps the most extensive, including velocities up to 6000 fps and bringing in data from other countries. But variations in atmospheric conditions led to inaccuracies in the Gavre drag figures. The Krupp studies proved more fruitful. The Krupp standard bullet was a flat-based conical, 3 calibers long, with a 2-caliber ogive. Shortly after the Krupp data were published, a Russian colonel named Mayevski developed a mathematical model that showed the drag deceleration of this bullet. U.S. Army Colonel James Ingalls used Mayevski's work as the basis for his Ingalls Tables, first published in 1893 and revised in 1917. The Ingalls Tables show the ballistic behavior of a standard Krupp bullet—which, it turns out, is quite similar to most modern bullets used in big game rifles.

Still, the actual C for other bullets had to be determined before the standard bullet flight data could be applied in determining *their* flight paths. One way of determining C is with firing tests, comparing the ballistic properties of the test bullet with those of the standard bullet. Another method, developed by Wallace Coxe and Edgar Beurgless at DuPont in the 1930s, is essentially a matching exercise. The test bullet is compared to bullet profiles on a chart of many profiles, and when a match is found, the chart provides a number that can be used to quickly calculate C. This method has been widely used, though ballisticians at Sierra Bullets report that C values so determined typically vary up to 10 percent from the C values from firing tests.

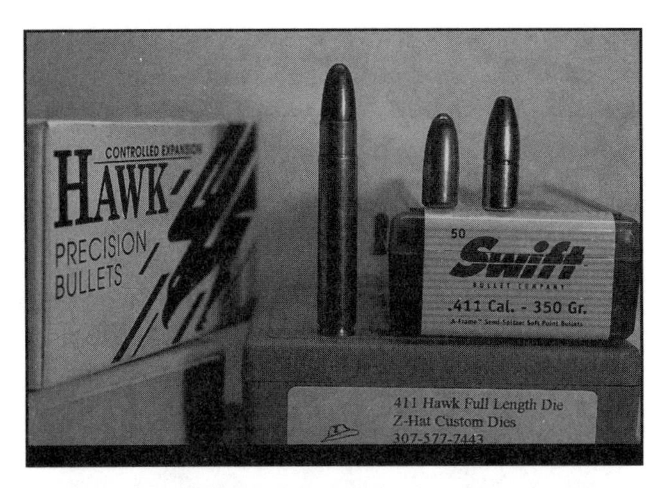

The long .411 bullet won't fly as fast as the shorter, lighter one at normal ranges, but its greater weight can boost penetration and tissue damage. Added weight offsets velocity loss in calculations of energy.

Development of the analog computer during the 1940s made ballistics calculations easier, while the weapons used in the second world war demanded more detailed plottings of trajectories. The digital computer later increased speed and accuracy in figuring complex trajectories, such as those of bombs dropped at high altitude. In 1965 Winchester-Western published ballistics tables that grouped small arms bullets into four families, each differing from the others in shape and each having its own drag function. "G1" drag applied to all bullets not described in the other categories. "G5" drag was for sleek boat-tail bullets that set up little air resistance and friction. "G6" drag was for flat-base spitzer (pointed) bullets with full metal jackets. "GL" drag applied to hollow-point bullets. All drag functions except the GL had already been determined by the U.S. Army Ballistic Research Laboratories at Aberdeen Proving Ground in Maryland. These shapes covered almost all rifle bullets in common use by hunters. G1 drag closely matches that found in the Ingalls Tables (the "G" descriptor, incidentally, was chosen in honor of the Gavre Commission).

It's important to note that ballistic coefficients are useful in comparing the flight charactistics of bullets *only when those bullets are driven at the same speed.* Because drag increases with speed, you can't expect shooting tests of the same bullet at different speeds to yield the same C. For the same reason,

The .338 Nosler Ballistic Tip and Jensen J-26 both deliver flat flight and adequate penetration for quartering shots on elk.

tests of different bullets at different speeds are invalid for determining C.

Few shooters and hunters have the equipment and facilities for finding C in firing tests. A simple math formula is an alternative: C=w/id2, where w is the bullet weight in pounds, d is bullet diameter in inches and i is the form factor. The higher the ballistic coefficient, the flatter the bullet will fly and the better it will conserve speed and energy. The standard bullet has a C of 1.000. High C values for hunting bullets range from .200 to .600, roughly, with most bullets falling in the .300 to .500 range. A flatnose .45–70 bullet has a C of around .200, as does the standard 170-grain flatnose .30–30 bullet. A flatbase spitzer bullet of medium weight for the caliber, such as the 100-grain .25, 140-grain 7mm and 165-grain .30, show ballistic coefficients of .350 to .400. Adding weight or a tapered heel can boost C to .450 or so. Only the most streamlined hunting bullets approach .500 in C. Above that value you'll find heavy boattail bullets for target shooting at long range.

The effect of C on bullet drop is less than its effect on drag (which by increasing flight time also contributes to drop). Sierra tables show that a bullet with a ballistic coefficient of .600, driven at 3000 fps, drops about 58 inches at 500 yards. One with a ballistic coefficient of .400 (clocking the same speed) drops 65 inches. That's a relatively small difference, given the 33 percent drop in C value and the 500 yards of bullet travel. The difference in remaining energy is more striking, though similar in terms of percentage: 2256 foot-pounds for the bullet with a C of .600, and 1929 foot-pounds for the bullet with a C of .400. Again, 500 yard is quite a reach, probably double the distance even experienced hunters can kill big game with certainty under field conditions.

If you cut velocity by 500 fps, to 2500 fps, both bullets drop much farther over 500 yards. The bullet with a C of .600 drops 85 inches—27 inches more than it did at a starting velocity of 3000 fps. The bullet with a C of .400 drops 96 inches—31 inches more than it did with the faster start. Remaining energy: 1835 and 1551 foot-pounds. It seems *a reduction in velocity only half as great (by percentage) as a reduction in ballistic coefficient* has a greater effect on bullet drop and remaining energy at long range.

A couple of other things to remember: A change in C has a larger effect on remaing velocities at high muzzle velocities than at low ones, because as bullets speed up, drag increases as a percentage of the forces impeding bullet flight. Secondly, a change in C has a greater effect on drop at lower velocities than at high ones, because at low speeds, gravity takes a more active hand than drag in depressing a bullet's arc.

Ballistic coefficients change markedly near the speed of sound.

Chronographs and Bullet Speed

Most hunters talk about bullet flight in terms of speed as measured in feet per second. The first instruments to measure bullet speed were mechanical—the Robins pendulum of the eighteenth century. The movement of the pendulum when it was struck by a bullet gave Mr. Robins the information he needed to calculate speed, because speed (with mass) is a component of momentum. We no longer measure velocity with pendulums, but with electric eyes that register the passage of a bullet's shadow as the bullet travels a known distance. The device is a chronograph.

Chronographs are a century old, but it wasn't that long ago that Texan Dr. Ken Oehler did shooters an enormous service by designing and building the first chronographs meant for consumer use. Before the Oehler instruments, chronographs were found only in the laboratories and shooting tunnels of ammunition companies. They were fixed in place. They were very costly. Consequently, shooters could only take the catalog ballistics charts at face value, and hope their handloads were producing the pressures and bullet speeds listed in the loading manuals. Now every serious shooter I know has a chronograph. Portable and easy to use, some are less expensive than an ordinary rifle scope. A chronograph is vital to understanding what your bullets are doing downrange.

The author sets up Oehler Sky Screens 4 feet apart. A bullet flying over the electric eyes registers its shadow, and based on the time it took to travel 4 feet, the chronograph computes its speed.

Portable chronographs have "screens"—the electric eyes—set up a short distance apart on a bar or even directly on the electronic box that gives you a velocity reading. The chronograph measures the time (hence, "chrono") between the bullet's passage

Chronographing needn't be done from a bench, but it's handy to check the accuracy of new loads as you determine their speed.

over the first screen and its passage over the second. Just as you could drive a car a quarter mile and compute the number of miles per hour your car would travel at that speed, so you can time bullet flight for a short distance and extrapolate. Some chronographs allow you to adjust the distance between screens. The greater the distance, the more accurate the reading. The chronograph must be precisely calibrated for the span between screens, or it won't read your velocities correctly. Also, you must shoot squarely through the opening above the screens, as an angled shot, either horizontally or vertically, effectively increases the distance the bullet travels during its timed flight.

You'll get plenty of information from a chronograph: bullet speed for each shot, of course, but also average bullet speed for a string of several shots. The average, or mean velocity, is the sum of all recorded velocities divided by the number of shots you fired. The instrument will also tell you the extreme spread (ES)—that is, the range of velocities, slowest to fastest. ES is useful because you want a load that gives you uniform results. Great variation in ES often shows up at the target as poor or mediocre accuracy. It's unrealistic to expect all shots to stay within 5 fps of each other; on the other hand, I've seen more than 100 fps variation between the slowest and fastest readings. If a hunting load delivers good accuracy and the speed that I want from my bullet, and ES stays within 25 fps, I'm generally satisfied. Some cases and powders can be expected to do better; some combinations work hard to meet that standard.

The other kind of data you get from a chronograph is standard deviation. This is a statistical term that probably originated in the late 1890s. American statistician Karl Pearson is generally credited with the formula. Without getting into mathematics, which had me in a headlock from grammar school through my PhD program, I'll summarize standard deviation as the positive square root of the variance. Now, what's variance? It's simply the sum of the squares of the deviations from the mean of your chronograph readings, divided by a number that's one less than the number of times you shot. A high standard deviation indicates a lot of spread in your data—that is, a great deal of variability among your readings. A low SD means that most of your velocity readings were clustered close to your mean.

This chronograph shows the bullet just fired clipped along at over 3500 fps.

There's more to milk from standard deviation. With it, you can construct a bell curve that shows how your velocities grouped around the mean, and, for any given speed range, the percentage of shots likely to fall within that range. Once in awhile you'll get a velocity reading you don't believe. It may be the chronograph didn't register the bullet's passage properly. Or that the load was somehow defective. A rule of thumb you might consider is to throw out any reading more than 2½ times the SD from the mean.

Bullet velocity varies not only with the powder type and charge and bullet weight; it's influenced by chamber and barrel dimensions, throat shape, and length and bore finish. A tight chamber reduces the amount of energy lost in case expansion. So does a tight throat. But a long throat that allows the bullet to move before engaging the rifling and permits long seating of the bullet to increase powder space enables a handloader to "heat up" his loads, boosting velocity. That long throat is generally thought to be less than desirable for accuracy; but on hunting rifles it's not a liability. Roy Weatherby made good use of long throats and ambitious Norma loadings to give his magnum cartridges lots of pep.

Bore dimensions and rifling type affect pressures and bullet speed. Some barrels seem to "shoot faster" than other barrels that by all appearances should give the same performance. Bore finish is pretty much invisible, but you can feel it. The consensus of barrelmakers is that smooth lands and grooves boost accuracy. "But you can get a bore too smooth," said one. "If you make it glass-smooth, you increase friction." Increased friction means higher pressures, which can increase speed; however, as with the throat, if your ammunition already performs at the top of the chart, higher pressure is not what you want. More velocity *without* higher pressure is the goal.

Barrel length also makes a difference as regards bullet speed. How much difference per inch depends on the original and finished lengths, the original velocity, the cartridge, and powder and bullet. The only way to tell for sure how much velocity you'd gain with a long barrel or lose with a short barrel is to chop it shorter an inch at a time. Tests conducted this way show great variation. One, conducted by A-Square with a .300 Winchester pressure barrel, measured velocities at one-inch increments from 28 down to 16 inches. Here are some of the velocities, with loads of 70.5 grains IMR 4350 and a 150-grain Nosler Ballistic Tip, and 78.0 grains RL-22 and a 180-grain Sierra Spitzer:

Three registry point on this Oehler chronograph make the final read more accurate.

barrel length	*velocity, fps, 150-grain*	*velocity, fps, 180-grain*	*velocity loss, fps, 150/180*
28 inches	3346	3134	
26 inches	3268	3089	78/45
24 inches	3211	3016	57/73
22 inches	3167	2966	44/50
20 inches	3108	2930	59/36
18 inches	3014	2874	94/56
16 inches	2903	2748	111/126

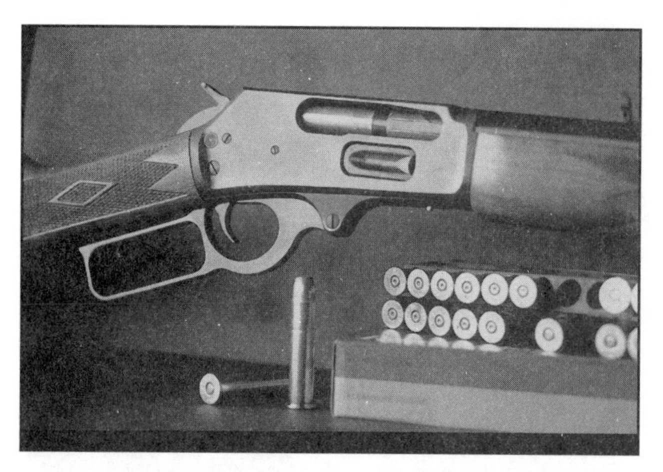

The .45–70 bullet is not quick, but its weight makes it lethal at woods ranges. A steep trajectory makes shot placement difficult beyond 200 yards.

Velocity loss per inch of barrel length varies from a low of 22 fps to a high of 56 fps for the 150-grain bullet, and 18 fps to 63 fps for the 180-grain bullet. That range could be expected to vary with other .300 Winchester Magnum loads. Certainly it would differ for other cartridges with faster-burning powders. Note that rate of velocity loss increases substantially as the barrel is lopped to less than 20 inches in length. Logic tells us that we're cutting into the descending sector of the pressure curve, before it flattens out. A lot of pressure is being released to the atmosphere instead of staying in harness behind the bullet. There's not much change in the rate of velocity loss between 20 and 26 inches, and variability in the data prevents any conclusions as to differences in rate of loss between 150- and 180-grain bullets.

Based on the figures presented here, you lose relatively little speed and energy chopping a barrel from 24 to 22 inches. For some reason, the loss is greater when barrel length drops from 26 to 24 inches—but again, there's too much variation in the data to establish a smooth trend line. Rule of thumb: Unless you cut barrels short enough to invade the "working" part of the pressure curve, estimate 30 fps per inch of barrel for most hunting loads. If the barrel is cut very short, the bullet exits when the volume of expanding gas has a greater effect on raising pressure than the increasing volume behind the bullet has in lowering pressure. And you lose a lot of speed with small reductions in barrel length.

Changing components can also affect pressures and velocities. Substituting magnum primers for standard primers, for example, can boost pressure by several percentage points. You can expect a smaller change in velocity, however. At full-power levels, increasing pressure (by whatever means)

The flat flight and mild recoil of the .25–06 make precise shooting easier at long range. Though a 100-grain bullet seems light for elk, some hunters use it with great effect.

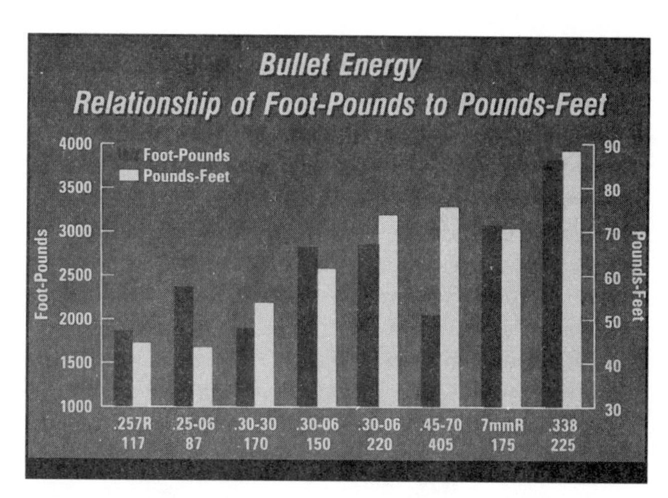

There are several ways to measure energy. The common unit of foot-pounds gives fast bullets an edge over heavy ones because in the equation for kinetic energy, velocity is squared. Pounds-feet, an option proposed by shooters like Elmer Keith, gives the advantage to heavy bullets. Here's how the measures compare when applied to popular hunting cartridges.

Early efforts to plot bullet flight were made with slow projectiles more affected by gravity than by drag. When muzzleloaders gave way to high-velocity centerfire rifles, maximum ordinates changed.

kicks velocity up a little—but not proportionately. Similar changes occur when you substitute cases with smaller powder chambers (thicker walls or webs). Pressure goes up; velocity follows but not at the same rate. Switching bullets, you may see no significant change in velocity. However, if the new bullet has a longer shank and generates more friction in the bore, or is seated closer to the lands on take-off, or is very slightly larger in diameter, or has a "sticky" jacket, pressures can rise. Remington's choice of Nosler Partition bullets for the first .300 Ultra Mag loads no doubt disappointed the folks at Swift, who were already supplying Partition-style A-Frame bullets to Remington. The rationale: A-Frame jackets were softer, resulting in more bore friction and higher pressures. To get 180-grain bullets going as fast as they wanted them too, Remington engineers had to use the Noslers.

Bullets on Leashes

There's lots of movement to a bullet leaving a rifle. Not only forward movement, but rotational movement and precession. No bullet flies straight. Barrels are pretty straight, but once a bullet leaves, it is no longer held in line with the bore's axis. Gravity pulls it to earth. Wind pushes it. A dam of air counters the thrust of powder gas and wraps the bullet in drag. Other forces keep the bullet on a leash all the way to the target. Some work for you, some against you. Bullets begin losing their battle with air as soon as they clear the muzzle. But like a football, a bullet is most likely to travel in a straight line when spinning around its longitudinal axis. If it remains nose-first, it also travels farther, because drag is kept to a minimum.

Long, streamlined bullets like these Barnes X-Bullets retain velocity and energy very well at long range. But gravity still pulls them to earth at an accelerating rate.

Rifling twist is the rate at which a bullet is spun, expressed as the distance it travels while making one complete revolution. A 1-in-14 twist means the bullet turns over one time for every 14 inches of forward travel. If all bullets were the same weight and shape and traveled at the same speed, one rate of twist would work for all. That's not the case, however. The proper rate of twist varies. A patched round ball in a muzzleloader requires a very slow twist, while a long bullet in small-bore rifle cartridge needs a quick twist to stabilize. The proper rate of spin for the muzzleloader might be 1-in-66 for a round ball—and 1-in-32 for a conical bullet. Some muzzleloaders are rifled 1-in-48 as a compromise. Rifles designed for long .264 rifle bullets may have twist rates as short as 1-in-9. In 1879 Briton Sir Alfred George Greenhill came up with a formula that works for most bullets most of the time: The required twist, in calibers, is 150 divided by the length of the bullet in calibers. So if you have a 180-grain 30-caliber bullet 1.35 inches long, you first divide 1.35 by .30 to get the length in calibers (4.5). Then you divide 150 by 4.5 and get a fraction over 33. That's in calibers, so to bring it into inches of linear measure, you multiply it by .30. The final number is very close to 10, which is a useful rate of spin for most popular 30-caliber hunting cartridges, from the .308 to the .300 Weatherby Magnum. The formula is valid for most jacketed lead bullets (specific gravity 10.9).

A .308 bullet spun one turn in 12 inches at a muzzle velocity of 3000 fps reaches an animal 400 yards away in about half a second. (Average velocity is a shade under 2500 fps.) In that half second, the bullet spins 1500 times. Compare that turnover to the rps's of an automobile engine carrying you down the highway at 60 mph. Maybe the engine tachometer registers 3000 rpm. That's only 500 rotations per second, 250 per half-second. The bullet is turning six times faster. Give it a one-in-nine twist, and you boost rotational speed by 25 percent, to 3750 rotations *per second.* Unlike a bullet's forward speed, which slows under the influence of drag, rotational velocity remains essentially constant. It affects the reaction of bullets to twigs and other obstructions in flight, and the expansion of softnose bullets in game. Expansion inside game can affect bullet rotation. Jacket petals spread far to the side add drag to both forward and rotational movement. So besides reducing penetration, they slow the rotation, reducing stability. The bullet is more easily moved off course. Petals also act as lever arms to tip the bullet. They do inflict some additional damage because of their rotation—damage that is evident in high-speed film footage of bullets driving through ballistic gelatin. Contrary to what some shooters think, however, there's limited "buzz-saw" effect in bullet wounds. The .308 bullet turning once every 12 inches of travel barely gets through one revolution in penetrating the chest cavity of a big deer.

Precessional movement is the rotation of the bullet's nose about the bullet's axis. It is undesirable but virtually unpreventable. A bullet seldom leaves the muzzle perfectly. Any tipping of the bullet—due to a damaged muzzle, nicked bullet base, or lack of concentricity in the bullet's jacket—can put the nose into its own orbit around the bullet's axis. Like a top that "goes to sleep" after you give it a hard spin, the bullet may rotate more smoothly, with less precession, after covering some distance. That's why sometimes you can get smaller groups (in terms of minutes of angle) at long range. A rifle that shoots into 1½ inches at 100 yards might keep all the bullets inside 3 inches at 300 yards. Penetration is deeper if the bullet is not precessing significantly. That's partly why sometimes slow bullets have, in tests, driven deeper than the same bullets driven fast.

It's prudent, once in a while, to think beyond the muzzle and consider the bullet not as a rocket but as a fragment driven yonder by an explosion. This mindless sliver of lead, spinning madly, must rip through powerful forces that we know collectively as air. Air is not a void. It is like water. It offers resistance to penetration. If you dip your hand in the water, you feel a little resistance but not much. When you swim, you feel much more resistance. A belly-flop from the 7-meter board will show you more resistance still. Bullets trace paths unique to themselves and subject to conditions. Looking through a high-power target scope as you launch a .22 match bullet downrange, you'll see the path of this slow projectile as a hook, the steep trajectory fore-shortened in the lens. You might be surprised that the image of the bullet's track, as brief as it is, shows you distinctly that a bullet striking center is out of line for a center hit for almost its entire flight.

Thinking of bullets as curve balls thrown by a pitcher can help you put them in the vitals of big game. It's important that you know the range and shoot to that yardage. This may seem elementary. Still, many hunters make poor shots because they think in straight lines.

A BB is hardly more accurate than a stone from a slingshot. But long ago my Daisy lever-action hit

Doping wind is usually easier when you have a second opinion. A spotter can often see bullet strikes the shooter can't, because of recoil.

small targets at ranges to 50 yards. The reason: I shot it so much that I learned how the steel pellets reacted to air and gravity. Their lazy arcs came to me in my sleep. Intuitively the coarse stamped-steel sight would find some corner of space above and to the side of what I wanted to hit, and in due time the BB would land there. BBs behaved just like the 25-cent arrows I launched from my 18-pound Shakespeare bow into passing flocks of starlings. From these crude tools I learned that aiming is simply an attempt to arrange a collision at some point in space.

Not being able to see bullets is a handicap to learning. That's why machine guns and anti-aircraft cannons are fed to spew tracer bullets at regular intervals. Indeed, if you doubt the influence of wind and gravity, it's a good idea to watch footage of aerial and naval combat shot during the second world war. Drag, drift and deceleration all show up with great clarity in the tracer paths. Gunners trained to shoot at Axis airplanes were often started with BB guns that made trajectory and flight time easy to see close up.

As soon as a bullet leaves the muzzle of your rifle, it starts to drop at an accelerating rate of 32.16 feet per second *per second.* But bullets seldom stay aloft for an entire second. A 160-grain 7mm magnum bullet started at 3150 fps will strike a deer or elk 250 yards away in about a quarter of a second, given deceleration that brings average velocity to 3000 fps (1000 yards per second). During that quarter-second the bullet drops 3 feet (not 8 feet because gravity pulls it faster and farther the last quarter second than the first). If your line of sight were parallel and tangent to the line of bore, you'd hit 3 feet low.

A slower bullet drops exactly the same distance during the same time. It just doesn't cover as much ground during that time. Say your 165-grain .300 Savage bullet clocks an average 2400 fps (800 yards per second) over its first 200 yards. After a quarter-second of travel it will pass that 200 yard mark. So instead of striking 3 feet low at 250 yards, as with the 300 magnum, you're 3 feet low at 200. Bullet speed doesn't affect gravity; it determines the dimensions of the bullet's parabolic arc.

Crude illustrations of trajectories show them as rainbows. As Galileo discovered, they are not. Because air is pushing against the bullet's nose, sucking on its tail and clawing at its shank, the bullet slows down as soon as there's no more thrust on its base. The slower it goes, the less ground it covers per unit of time, while gravity brings a constant load to bear. The arc gets steeper downrange. If you dropped a bullet from your fingers at the same time a bullet was fired horizontally from a rifle the same distance above the ground, the two bullets would come to earth at very nearly the same time—assuming, of course, that the fired bullet didn't meet an obstacle or come across uneven ground first.

Scopes like this Pentax Lightseeker have made accurate aiming easier at extreme range.

Cartridges like the .30–378 Weatherby and 7mm Remington Ultra Mag are deadly at long range because the velocity that gives them the impact of a wrecking ball also flattens trajectory. A moose that's 450 yards away will escape your bullet if you're shooting a .30–06 and underestimate the range as 350 (easier to do than you might think!) A .30–378 bullet flies flatter and will still strike the vitals, albeit several inches low.

A lot of shooters have been bamboozled into thinking a bullet rises above line of bore during its flight. It does not, ever. The misunderstanding results from illustrations of trajectory that aren't carefully drawn. *The sightline is not parallel to the boreline,* but at a slight converging angle. The sightline dips below the boreline and the bullet's arc. Sightline never meets boreline again, because both are straight. They cross and forever get farther apart. A bullet will hit above sightline at midrange, but only because the sightline has been angled down through the trajectory. If the sightline were parallel with the bore, it would never touch the bullet's arc, which starts dropping away from it as soon as the bullet leaves the muzzle.

Sighting in or zeroing a rifle is simply adjusting the sights so you are looking where the bullet hits. You cannot change the impact of the bullet relative to the bore (except by changing loads or bedding). You *can* change your line of sight so it intersects the trajectory at any ranges you choose.

Zero range is not point-blank range. Shooting point-blank is shooting with no compensation for a bullet's trajectory. Hunting big game, you can ignore small differences between point of aim and point of impact. If you're willing to ignore a 2½-inch deviation, a 200-yard zero will give your .30–06 a point-blank range of 230 yards or so, depending on the load. Maximum point-blank range is the distance at which a bullet falls so far below line of sight that you must compensate for gravity's pull by elevating your rifle.

Another force that moves your bullet away from sightline is wind. Wind is gravity on another plane. As the bullet slows, a constant wind has greater effect. But, unlike gravity, wind force and direction are *not* constant. Either or both can vary between rifle and target. In competition, I've often fired through wind that straightened the target flags one way while swinging mechanical wind vanes the opposite direction on the line. Limp flags at either end or midrange were no consolation. While a wind change at the muzzle has more effect on impact point than does an equivalent shift downrange, the bullet is more easily moved the farther it gets from the muzzle because there's more time downrange (per unit of distance traveled) for the wind to work its mayhem. A slight pickup or letoff or fishtail can flip your bullet wild after you press the last ounce from that trigger. A "full value" or 3- to 9-o'clock wind has the greatest effect on strike point.

The best way to learn to shoot well in the wind is to shoot often in the wind and study its effect on your bullets. But you'll get a good start as a wind-doper with a ballistics program from Ken Oehler or the people at Sierra Bullets. Punch in the velocity and ballistic coefficient of your bullet, and the Oehler Ballistic Explorer will plot a bullet track for you, telling you about energy, point-blank range, and wind drift beside showing bullet drop. The Sierra program is fully as useful. There's even a way to calculate ballistic coefficient from simple flight data you collect at the range.

Another thing that affects bullet behavior en route is temperature. Warm air is thinner than cold air, so your bullet meets less resistance on a warm day—just as an airplane gets less lift at any given speed on a warm day. But you won't see much difference in bullet impact due to the effect of *air temperature on bullet flight.* Figure no more than half a minute of elevation for every *100 degrees* change in temperature. Of course, bullet flight is also affected by breech pressure: the higher the pressure, the faster the bullet. A hot day can raise the temperature of the powder, generating higher pressure. A cold day can make the cartridge perform sluggishly. Tests run by Art Alphin (A-Square) with .30–06 ammunition showed that at 40 degrees a charge of 51 grains RL-15 generated 54,600 psi to push a 180-grain Nosler Ballistic Tip at 2675 fps. The same rifle and load registered 59,900 psi and 2739 fps when the temperature was 120 degrees. It's important to note that the temperature of the ammunition is what's important. A cartridge kept in a warm pocket and fired soon after loading in a rifle on a cold morning will perform as if the chamber temperature were warmer than the ambient temperature. Cartridges left on a hot dashboard in a safari vehicle can get much warmer than the rifle and cause higher pressures than the ambient temperature would indicate. The effect of changes in temperature on ammunition performance varies with the load. Some powders are more sensitive than others, and if you're approaching the safe pressure limit, you may get a dangerous spike in pressure with a relatively modest increase in temperature. Rule of thumb: figure a velocity change of 3 fps for every degree of temperature change.

As you might suspect, altitude also affects bullet flight. The higher you go, the thinner the air and the less resistance it offers to the bullet. The effect is greatest with flat-nose bullets. According to T.D.

The late Jack Slack of Leupold was an excellent rifleman and shot often at long range. But he knew also that big game animals most often were found in cover, where accurate shooting had to be quick.

Smith, a former fighter pilot who has used his military experience to develop long-range shooting aids for small arms, a .30–30 bullet hits about 9 inches higher at 300 yards when fired at 10,000 feet than when fired at sea level. In contrast, a ballistically more efficient 175-grain bullet fired from a 7mm Remington Magnum will strike less than 3 inches higher in the thinner air. The longer the yardage, the greater the difference in strike-point due to changes in air resistance. If you're a hunter, you'll say "Aha! But the higher you climb, the colder the air!" That's so. Temperature and elevation changes can thus cancel each other out. In the mountains, air resistance can be greater (because of the cold) and less (because of the elevation) than air resistance at sea level. Truly, few animals if any are lost because changes in temperature and elevation change bullet impact. On a hunt, I ignore these variables. Shooting competitively at 1000 yards, you may notice their effect when you check your zero on the sighter target.

Another thing to ignore is the Coriolis effect. The earth's rotation causes any projectile to drift slightly right in the northern hemisphere, slightly left in the southern hemisphere. Get on a merry-go-round moving clockwise and toss a ball to someone else on the merry-go-round. The ball seems to curve to the left. It doesn't, of course; only your frame of reference gives you that illusion. Coriolis acceleration for a rifle bullet can be expressed by this mathematical equation: Y=2wVsin(latitude), where w is the earth's rate of rotation (.0000729 degrees per second) and V is the bullet's average speed in feet per second. For a bullet moving 2800 fps at a latitude of 45 degrees north, the Coriolis acceleration comes out to .30 fps/second, or roughly 1 percent of the acceleration of gravity. At 350 yards, you'll get about half an inch of displacement. That's not enough to notice, because your hunting rifle probably won't shoot half-inch groups at 100 yards, let alone 350. And if you can hold inside a half-inch circle at 350 yards, you need to be selling your secrets to people like me!

Now, at long ranges, the Coriolis effect can be significant. Air Force F-16 fighter cannons are not wired for Coriolis correction, says T.D. Smith. "But the on-board computer *is* programmed for a 6-inch correction at 5000 yards with either the Mark 82 or Mark 84 dive bomb." Longer bombing ranges bring a higher correction value. If you're shooting at extreme range, figure a 1-inch correction for each second of bullet flight time. It takes a 170-grain flatnose .30–30 bullet a second to cover 500 yards. The bullets most hunters use for big game these days spend much less time en route to the target.

A rotating bullet generates other forces that cause drift from boreline. One of these is gyroscopic effect, which amounts to about twice the Coriolis drift—and is thus negligible in most hunting situations. Spun by a right-twist bore, a bullet moves slightly right. A left-hand rifling twist pushes the bullet left. The reason: torque within the bullet throws the bullet nose slightly in the direction of twist. Because of bullet rotation (right-hand for most rifles) a wind from the left commonly pushes bullets to 4 o'clock, one from the right to 10 o'clock.

Zero

A rifle shoots where you aim it only if you've first aligned the sight with the bullet's path. The factory can roughly align iron sights, but the final zeroing is up to you. Factory technicians can't know which load you're going to use, or at what range you want to zero; and they may not look at the sights the way you do. You may be able to hit the side of a washing machine at 50 yards without touching factory-installed open sights. But if you want bullets to land in a smaller target, you'll likely have to adjust. Some rear open sights have a crude step elevator to shift the point of impact vertically. Moving point of impact to the side is a job for a drift punch. Incidentally, you move the rear sight in the direction you want the bullet to go.

When you mount a scope on the rifle, you'll have to zero from scratch.

Zeroing, or sighting in, is a pretty simple job. But a lot of hunters don't mind the details, and they wind up with a rifle that doesn't hit point of aim. Result: unexplained misses and crippling hits. Here's how to zero so your bullets land where you look.

Zeroing begins with bore sighting: centering a target in the bore, then, without moving the rifle, adjusting the sight until it too is "spot on." Shooting follows, to refine the zero.

Before you start, though, it's a good idea to consult a loading manual or a book like this to find out what kind of flight path your bullet will take. Ballistics charts show you the arc of your bullet so you can determine the most useful zero range and proper holdover for longer shots.

Your line of sight is straight, and it will contact the bullet's arc only in one or two places. It's possible to adjust the sight so it comes tangent to the bullet's arc, but a better tack is to thread the line of sight *through* the arc. Two intersections give you a zero range at the second crossing. Imagine a parabolic arc, the path of a decelerating bullet that drops ever more steeply at long range. Now imagine a straight line angling into the arc from above and behind but almost parallel to its starting path. The line will cut the arc, travel beneath it, then cut it again as the bullet plunges. That's your sightline. Between intersections, you want the sightline to stay close enough to the arc that you can ignore the gap.

If you're a target shooter, you must adjust your sight to zero for every yardage on the course. As a hunter, that's impractical because shots can come quickly, and you'll seldom know the exact yardage. So before zeroing, specify a maximum gap that you'll tolerate between line of sight and the bullet's arc at mid-range (between those crossing points). For big game hunting, a reasonable maximum is 3 inches. If your bullet hits 3 inches above point of aim when you're shooting at a deer's chest, you'll still kill the deer. And, of course, that much error occurs only at one distance, a little over halfway to the second intersection of your sightline and the bullet's arc.

By the same logic, you should be able to "hold center" on big game animals to the range at which your bullet drops 3 inches *below* sightline. This is maximum point-blank range. You determine maximum point-blank range when you zero your rifle. Fast, flat-shooting bullets have a longer point-blank range than slow bullets that drop quickly, but the actual yardage is also determined by your zero and, by extension, how much gap you'll allow between trajectory and sightline at midrange. Varmint shooters get less latitude here than do big game hunters. A prairie dog or ground squirrel is easy to miss if your bullet hits 3 inches high. Better to zero for a shorter range to keep the sightline close to the bullet's trajectory, or take shots within a narrow block of yardages—say, 175 to 225 yards if you're zeroed for 200. Of course, you can compensate by holding over or under too.

Depending on the bullet's starting velocity and ballistic coefficient, and the kind of shooting you expect, zero range can be short or long. Zero ranges have changed over the last century. The .30–30's blunt bullets have a ballistic coefficient (C) of about .260. In other words, their shape and sectional density (ratio of weight to diameter squared) is such that they don't cleave the air easily. They are not aerodynamic, compared to long, pointed bullets with a C of over .400. Also, they're slow. When the .30–30 appeared in 1895, its velocity of just under 2000 fps was fast. Now most hunting bullets fly over 2700 fps, and many over 3000. Scopes enable us to aim precisely at game so far off as to be barely visible with iron sights, and modern bolt rifles help us hit it.

Zeroing is best done from a bench, to minimize human error.

The first intersection of line of sight with bullet arc is roughly the same for a .30–30 and a .300 magnum: between 30 and 40 yards from the muzzle. But the second crossing, the zero range, is much different. The best zero range for a .30–30 carbine may, in fact, have less to do with the limited range of the cartridge than the more limited range at which you can shoot accurately with iron sights, or the even more limited distance you can see in the cover you generally hunt. With a 150-yard zero, Remington tables put 170-grain .30–30 Core-Lokt bullets 2 inches high at midrange. A 200-yard zero lifts the bullet nearly 5 inches above sightline at midrange. While the 150-yard zero is a good choice, a 100-yard zero may be more practical, especially if you hunt in dense timber and most of your shots are very close.

Most hunters these days use cartridges more potent than a .30–30. Say you have a .30–06 and want to shoot 180-grain bullets at 2700 fps. (Chronograph those factory loads; they commonly fall short of listed velocities.) In Nosler's reloading manual, trajectory data for the 180-grain Partition bullet is based on a 200-yard zero. At 100 yards, the bullet hits 2 inches above line of sight, and at 300 yards it's 8.5 inches low. The top of the arc comes a little beyond the 100-yard point (remember, the arc is parabolic), but still the bullet stays well within the 3-inch maximum "lift" that we specified. So with this load, you could zero your rifle a bit beyond 200 yards, increasing maximum point-blank range. If you were willing to live with a 3.5-inch lift and drop, you could zero at 220 or 230 yards and stretch point-blank range to nearly 300.

Say you have a .300 Weatherby or .300 Remington Ultra Mag, whose 180-grain bullets race out the muzzle at 3250 fps. Their high speed makes a 250-yard zero practical. According to the Barnes manual, a 180-grain X-Bullet stays within 3 vertical inches of line of sight to 300 yards.

Sight height affects zero because it determines the angle of the sightline to the bullet's arc. A flat line of sight that first crosses the bullet's path at 35 yards will cross it again sooner (closer) than a line of sight that's steep by virtue of high scope mounts. Keep that in mind when you're comparing points of impact with the data in ballistics tables, some of which specify a 1.5-inch gap between bore center and sightline at the muzzle.

It's important to use the same hold every time, to pad the forend well, and **never** ***to rest the barrel.***

To make things easy, I zero almost all my big game rifles at 200 yards. That's a bit short for 7mm and .30 magnums, if a 3-inch midrange gap is acceptable. But I'd as soon have less lift, and a point-blank range of 250 yards is long enough for me. I rarely shoot at game farther than that; when I do, I can shade a little high. A top-of-the-back hold with my .270 will kill deer to 350 yards—far enough to make me giddy. I zero my .30–30 Marlin and .356 Winchester at 150 yards, my .45–70 at 100, along with 12-gauge sabot shotgun slugs.

Before you start a zeroing routine, make sure the scope is mounted firmly, the base screws tight and the rings secured to the base. Before you snug up the rings, make certain they are aligned. Don't use your scope for this! Dovetail rings are best turned into alignment with a 1-inch dowel. When the scope drops easily into the belly of the rings, slip the tops of the rings over the tube, but don't snug them. Shoulder the rifle to see that the reticle is square with the world and you have the proper eye relief. You should see a full field of view when your face rests naturally on the comb. I like the scope just a little farther forward than most shooters, for two reasons: First, when I cheek the rifle quickly, I want the field to open up as I thrust my head forward. I don't want to waste any time pulling it back to see more through the scope. Often I'll have to make a fast shot as with a scattergun, my face well forward on the comb. Secondly, I want some room between the ocular bell and my eye should I have to shoot uphill or from the sit hunkered over the rifle. I've been bitten many times during recoil by scopes set too far to the rear for a stock-crawler like me. My rule of thumb is to start with the ocular lens directly over the rear guard screw on a bolt rifle, then move the scope back and forth incrementally to fine-tune its position.

After you lock the scope in place by cinching the ring screws (alternately, as you would tighten the wheel on an automobile hub), remove the bolt from your rifle and set the rifle in a cleaning cradle or on sandbags. You don't have to be at the range, only somewhere that gives you a long view toward a small, distant target. (I often use a padded chair in my living room and look through the window to a rock on the far side of the Columbia River.) Remove the bolt and line up the target in the center of the bore. Brace the rifle so it does not move. Adjust your

Standard graduations on the adjustments of scopes built for the U.S. are quarter-minute: one click moves point of impact ¼ inch at 100 yards. In Europe the value is typically about ⅓ minute.

scope so the reticle centers the target. It's that easy! The rifle is now bore-sighted and should plant bullets close to your point of aim at 35 yards. Bore-sighting saves time and ammo at the range. Lever, slide, and autoloading rifles, and muzzleloaders, require a collimator, an optical device you attach to the muzzle with a bore spud. Collimators show you a grid that takes the place of a distant target. If you shoot with iron sights, zeroing requires live fire from the start. Recent laser instruments that fit in the muzzle and project a beam on a wall can help shooters with irons.

No matter what the collimator says, or how carefully you align the sights with a laser beam or the image of a distant object in the bore, you still must shoot to get zeroed! Shoot all second-hand rifles, no matter how carefully you think the previous owner zeroed the rifle, or how he says it shoots. Before heading for the range, check guard screws and scope screws to ensure that they're tight. Then make a list of what you'll need at the range. Here's one:

- Sandbags and adjustable shooting rest
- Spotting scope and bench tripod
- Targets, with pasters or tape to cover bullet holes
- Stapler and staples, or tape, to attach targets
- Trouble-shooting kit, with screwdrivers to fit every rifle and scope mount screw
- Rifle cleaning kit
- Old sweatshirts to pad your elbows on the bench
- Recoil shield or towel for your shoulder
- Shooting glasses and hearing protection (use muffs *and* plugs under a roof)
- Notepad and pens to record sight changes, general observations

In your haste, don't forget the rifle and ammo. I once ran off without the rifle's bolt. If you've no range nearby, you may have to bring your own bench. Maybe that's an old sleeping bag to cover the hood of your pickup. Or maybe it's a portable bench, like the clever fold-up polycarbonate device by Storehorse.

Try to pick a still day for zeroing. You don't want to fight the wind for control of the rifle, and you don't want to zero for a right or left drift. You can chronograph loads while zeroing, and test them for accuracy. But it's best if you first zero with one load, then focus on other things. If you must change zero for a different load later, it won't take long.

My range isn't developed. It has a bench but no target boards. I substitute large cardboard boxes, with white paper squares affixed. These white squares are my favorite targets. I size them according to the sight. For iron sights at 100 yards, they're a foot to the side, or bigger! For a 4X scope, I'll use a 6-inch square, for a 6X scope a 4-inch square. If you're shooting a varmint rifle with a 16X or 20X scope, you'll want a much smaller block. The white paper shows up plainly against a brown box, and holes are easy to see no matter where the bullets land. Black lines and target faces hide little bullet holes. I prefer square to round targets because

It's important to mark your first shot from a cold barrel. That's the one most critical on the hunt.

they're easy to quarter with a crosswire. With a dot or a front bead, the corners of the square remain visible, clearly indicating if the sight is off-center. I also like these targets because they are as cheap as typing paper.

Your first shots should be at 35 yards, whether or not you've bore-sighted. No sense wasting even one bullet off the backer, or hitting a target frame. If the bullet hits within a couple of inches of center, take the target to 100 yards and start zeroing in earnest. If it is off the mark at 35, adjust until you get a bullet at point of aim, then move. Zeroing for 200 yards, your goal at 100 is to print a 3-shot group just above point of aim. Adjust the sight until that happens. When you get beyond rough adjustments, fire at least two shots before making sight corrections. If those holes are more than an inch apart, fire another shot. Small groups tell you with greater certainty where to move the sight.

Group size is partly a function of rifle and load, but mainly it's a function of how still you hold the rifle and how well you execute the shot. Even with a benchrest, it's easy to make a bad shot. In fact, a bench can give you a false sense of stability, prompting fast, sloppy shooting.

An adjustable rest helps you make a good shot because with it you can "dial in" the exact position you want the rifle, bringing its natural point of aim to the bullseye. You can relax behind it, your body only a recoil brace. If you must *hold* the rifle on target, you introduce muscle tension, pulse and nerve tremors that can kick your bullet off course. Sandbags are OK; you'll just use more time getting them to hold the rifle where you want it. Whatever the base of your rest, make sure the rifle contact is well padded and in the same place, shot to shot.

The rifle is best supported just behind the forend swivel and just ahead of the stock's toe. Protect forward sandbags from the swivel stud on recoil by wadding a washcloth in front of the bag. Never zero a rifle with the barrel touching a rest. The barrel will vibrate away from the rest and throw the shot wide. Unless you're shooting a rifle of very heavy recoil and must hold it down lest it jump off the rest, keep your left hand off the forend. Use it instead to pinch the sandbags or beanbags that support the toe of the stock. A little hand pressure here can shift the rifle just enough to bring the sight to the exact center of the target. Your right hand should keep steady but light pressure on the grip as you pull the trigger straight back.

Remember those towels? Use 'em! I've heard other shooters chuckle when I've swathed my shoulder in extra padding. "Can't take it?" Well, yes, I *can* take it. But my purpose behind the rifle is not to see how much recoil I can absorb. It's to shoot well so I get a precise zero. I won't shoot well if I flinch. Flinching is commonly blamed on heavy recoil. It also follows *repeated* thumps by rifles of *mild* recoil. Zeroing, you want to take the human element out of the shot. To flinch is human.

Excepting external knobs on target scopes, you move a scope's aiming axis by turning the dials on

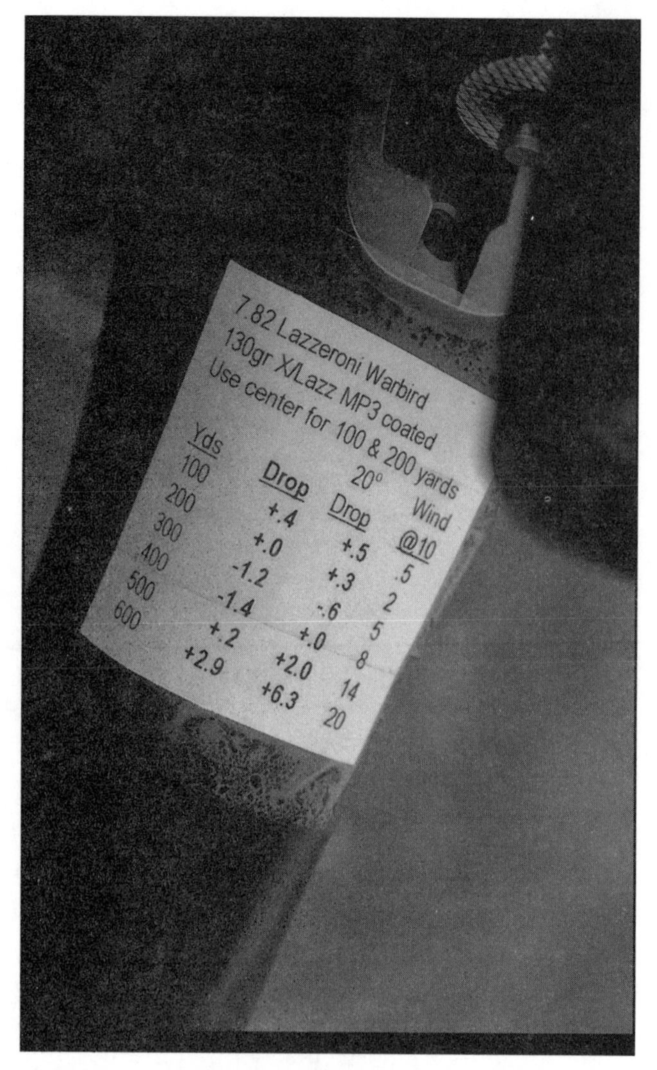

Typing up drop and drift data and attaching the notes to your stock can help you hit later.

the turret with a coin or screwdriver. Newer dials have raised ribs or knurled knobs for finger adjustment. Dial "clicks" or graduations are specified in inches of movement at 100 yards. Some shooters say "minutes of angle." A minute of angle is actually 1.047 inches at 100 yards, but it's commonly rounded to an inch. At 200 yards a minute is 2 inches, at 300 it's 3 inches and so on. Scope adjustments with quarter-minute clicks move point of impact an inch every four clicks at 100 yards, 2 inches every four clicks at 200. A target scope may have adjustment graduations as fine as 1/8 minute, but most hunting scopes feature half- or quarter-minute clicks.

When you get a group slightly to 12 o'clock at 100 yards, move the box or backer to 200 yards and change the target, doubling the size of the square or bullseye. Now fire another 3-shot group. It should be close to center. Adjust until it's there. If you have room, move the box or backer to 300 and 400 yards, using appropriate targets, and note bullet drop.

Next, get off the bench and shoot one 200-yard group each from the positions you most often use when hunting. I shoot a lot from the sit with a tight sling. I've found that the sling pulls the rifle down and left, so my shots don't hit where they would if the rifle were benched. One .30 magnum put a sitting group *9 inches* below the centered group I'd fired from the bench. A barrel-mounted sling swivel exacerbates this problem, but a stud on the forend is no guarantee that groups will stay together. A forend that applies lots of pressure to the barrel can release that pressure when pulled by a sling. A big change in point of impact may result.

Keep the barrel cool. I take no more than 10 shots before setting the rifle aside, bolt open, to bring the bore temperature down. If I have to take more than 30 shots, or if the groups open up, I clean the bore. Having two or three rifles at the range is a good idea. The second and third keep you occupied while the first cools. If you must twiddle your thumbs, you won't let the barrel cool enough.

Unless you visit a backwoods range like mine, you'll likely have to share the line with others. A

Check zero from hunting positions. It may change when you get off the bench and snug that sling.

couple of courtesy tips: Stay as far from others as possible, especially if your rifle has a brake. The blast from a braked rifle is beyond disconcerting. It's downright annoying. If you must talk to another shooter, walk behind and wait quietly until he or she has an unloaded rifle. Don't interrupt a shot; don't yell across the line. Obey all firing line commands instantly, no matter how inconvenient. If there's no range officer, ask individual shooters for permission for a cease-fire, then wait until all rifles are empty before calling a cease-fire. Leave your rifle on the bench, action open, whenever someone is downrange. When everyone has returned from the targets, wait until someone declares that the line is hot before loading up. Or ask if the line is ready before calling a hot line yourself.

Offhand shooting gives a rifle no firm support. It won't bounce up at the shot. Your bullet may hit low as a consequence. Check by shooting.

Post enough targets to occupy your rifles for at least half an hour. Frequent cease-fires waste time. At long ranges, especially, set targets on adjacent frames, if they're vacant. The walk may be good for your health, but it also raises your pulse and delays your shooting after you're back on the line. If someone calls a cease-fire too frequently, offer to loan targets or set up another frame for the shooter. If he or she doesn't have a spotting scope, offer to look at that target periodically.

After zeroing, thoroughly clean your rifle and run a lightly oiled patch through the bore. If you have time to let the barrel get stone-cold, shoot later that day at 200 yards to check point of impact. Pay attention to the first and second shots: where they land and how close they are. Those are the shots that count when you're hunting. Save that target. Visit the range at least once more before the hunt to see if the zero has shifted, again firing from a cold barrel at 200 yards. Composite first-shot groups from different days should form a knot no wider than 4 inches. Remember that wind affects point of impact; compensate for it.

Careful zeroing can teach you a lot about your rifle. Seeing those bullets chew out the center of a 200-yard target can boost your confidence. Learning where bullets hit at various ranges and from various shooting positions can tell you where to hold when you see game. It's a good idea to check zero from time to time, even if the rifle has been babied. Changing moisture and temperature can make wood stocks walk.

If you don't zero carefully, or check zero often, you may be hunting for a shot you can't make.

How Not to Miss

After a close election, a newsman asked an unseated incumbent why he'd lost. The reply: "My opponent got more votes." Before you write this off as a wisecrack, consider how basic votes are to any election. Analysts on the trail of the obtuse too often overlook fundamentals. So it is with shooters.

Ask any rifleman why he shot a 9 instead of a 10, or missed an elk. He'll tell you of tricky winds or problems with his rifle or ammunition, about deceptive terrain that skewed range estimation or a hidden branch that caught his bullet. A humble marksman might claim responsibility, conceding that he misread conditions or tried a shot beyond his ability to make. But even he will have neglected to name the cause of all misses.

We miss because the bullet is pointed in the wrong direction when we launch it. There is no other way to miss. We can't help but hit when the bullet heads in the right direction, just as a politician can't help but win when he gets most of the votes.

A long time ago I was shooting a regional smallbore match in the Midwest. The competitor to my right was Johnny Moschkau, an unflappable old man who consistently shot tiny groups. The morning sun soon flooded the targets with heavy mirage. With a bit of luck and lots of "sighter" rounds, I managed a ragged but still perfect 200 with my first 20 shots. Moschkau, I noticed, had dropped a point. This made me feel very good indeed.

During the next stage of the 1600-point match, I kept one eye on Johnny's target, the other on the mirage. I shaded a few into the X-ring, listening for pauses in rifle fire on the windward end of the line to warn me of impending let-ups or reversals. Moschkau apparently wasn't aware of all that. Still as a corpse on the mat, his left eye pressed to the spotting scope and his right in the sight-cup, he'd not moved a muscle except to finger the bolt up-back-down, up-back-down, with a mechanical drop of his hand to the loading block somewhere in the middle. But then Johnny stopped.

I looked at his target. Twelve holes. With the same count in my paper, I kept firing. Sighter shots had eaten much of my time. Johnny would run out of clock and eat some cartridges if he didn't keep shooting.

Suddenly the wind died, and my next bullet looped for a 9. It was my second. Heart thumping, I put three Xs in the sighter, and lost two more points for score before finishing with a minute left. I swung my scope to Moschkau's target: 16 holes. Still he lay like a beached crocodile, waiting for the condition he wanted. Then, with fewer than 25 seconds remaining, Moschkau's rifle spat a bullet. X. The efficient bolt manipulation and loading was the same as before, but the .22 cracked again almost without pause. X. Ten seconds. Bang. Now five. Bang. X. X.

"Cease fire!" The range officer called it before Moschkau had ejected his last case. Johnny un-

A target at caribou camp confirmed that travel hadn't affected this hunter's zero. Good news!

latched his sling, rolled over and asked how I'd weathered the bumps. Not too well, I said. He told me there were still lots of Xs left out there for people who aimed carefully and didn't let a shot go until everything looked good.

I must report that Johnny Moschkau whipped me soundly at this event. His cool, focused attention to the target and to shooting fundamentals proved more effective than my attempts to compensate for every vagary of the wind. My sidelong glances at other targets broke my concentration. I knew how to shoot an X, but dividing my attention netted me sloppy 10s. And 9s. That's what Johnny said, anyway. "Hitting is easier when you keep bullets on a tight rein."

He was right, of course. Choosing your shots wisely and executing them carefully helps you hit. But that's not all there is to know.

After you've established your zero, you must trust it, putting the intersection of the crosswire right where you want to hit. One of the most common failings of hunters is thinking when they should be shooting. The reticle goes where you want the bullet to strike. Compensation for range and wind is seldom necessary: first, because hunters typically overestimate range and secondly, because wind has little appreciable effect on hunting bullets at the ranges most game is killed.

Another hunter and I once trailed a herd of elk along the side of a large coulee. Just at dusk, we peeked over a rise and spotted them only 150 yards away. The hunter fired, dropping a bull. But the animal got up and staggered off. The man shot his rifle dry, then reloaded. None of the bullets hit. Meanwhile, the elk went down again. We ran forward and finished it up close. The bull's front legs had been shattered by the first shot; this fellow later admitted he'd aimed low every time. Why? He couldn't say.

I've had a great deal of experience aiming where I shouldn't. A couple of years ago I drew a bead on a bull elk that gave me one shot at about 280 yards. My 160-grain handloaded bullet would land about 4 inches low, I figured—no call for holdover. Perversely, I aimed at the spine. The elk died quickly, but the hit was high. Another time, I surprised a

Steve Hornady dropped this fine Shiras moose with three quick shots at 300 yards.

group of elk slipping through thin timber on their way to bed. The animals were walking fast, but at 60 yards there was no need to lead the bull. Maybe I confused the speed of my pulse with that of the elk—anyway, my bullet whizzed in front of its chest. Obligingly, this elk gave me a second chance after I tracked it a short distance in crunchy snow.

Another bull was not so generous, and when my Nosler clipped a wad of hair from his brisket, he throttled up. Dutifully I took the trail. It was easy to follow, but as bloodless as cabbage.

My willingness to aim anywhere but where I want to hit derives from my youth, when BBs cost a nickel a pack and I amused myself by attempting impossibly long shots with an air gun. The lazy coppery arc of a BB is more like that of an arrow than a bullet. At ranges beyond 20 feet, I had to elevate. The fickle ball sidestepped at the mere suggestion of a breeze. Aiming at the mark was the surest way to miss it.

Later, shooting .22 rifles in competition at 50 feet, I had no worries about gravity or wind. But my pulse and quivering muscles wouldn't allow the sights to settle. A slow squeeze on the trigger was OK for shooting from a bench or solid position, but offhand the only way I could nip an occasional 10 was to grab the last ounce as the muzzle strafed center. This abominable technique was known by those of us who practiced it as a controlled jerk. Like any other jerk, it moved the rifle; so we had to correct not only for rifle movement before the jerk, but for the barrel's leap or dip when our adolescent fingers bumped the striker into freefall.

While I still have trouble shooting offhand, my bullets don't stray quite as far as they used to. The reason: I learned, after many misses, that to hit I had to trust my rifle and hold it still.

Trust is more than adopting a center hold within point-blank range. It means you jettison every excuse having to do with rifle, sight, and ammunition. If your equipment isn't good enough, buy better. In competitive circles, amateurs remain so until they get rifles that shoot tight enough to win every match. They can then forget about equipment and build their skills to that standard. On the other hand, hunting rifles needn't shoot as well or cost nearly as much as target rifles. A big game rifle that shoots about an inch and a half at 100 yards should earn your confidence. A rifle that manages no better than 2 minutes of angle won't cause you to miss. At

Always look for a rest; pad it with your hand.

some point, you'll have to accept your rifle and ammunition as adequate and hold yourself accountable for your shooting. Practice, not excuses, makes you confident and competent. How would you like to ride in a Boeing jet behind a pilot who flew just once a year or, as he settled into the cockpit, told the passengers that someday he'd like to be assigned a *good* airplane?

Some rifles are choosy about ammunition, but their preference is easy enough to find. If you have a rifle that doesn't like *any* brand you feed it, sell it. You didn't marry it. The same goes for scopes. While you can wager your mother-in-law's good graces on the reliability of modern scopes, the one you bought must satisfy you. If it doesn't, get rid of it. One of my rifles, a Winchester 70 in .300 Holland, has shot poorly since I mounted a certain scope on it several years ago. I've been tardy in yanking that sight, but it *will* be shelved. The scope may have nothing to do with the wide groups, even if they shrink under a replacement. That doesn't matter. What counts is my confidence in the instrument.

This smacks of the superstitious. In fact, it is a practical shortcut to better shooting. Some riflemen make a hobby of their equipment, just as automobile enthusiasts change cams and wax valve covers. These obsessions are harmless. If top-flight marksmanship is your goal, however, you'll do well to get beyond the equipment hurdle as soon as possible so you can more productively invest your time in learning to hold a rifle and control its trigger.

Crossed sticks have been used by many hunters to steady their rifles. Plant the sticks well forward and pull them toward you.

Holding a rifle still without a rest is so hard that hunters commonly refuse to try it in public, shooting only from the bench. While resting a rifle while zeroing is essential, once you've obtained a zero, you've no reason to shoot from a bench. Shooting at big game, you won't have time to build one. Not that you shouldn't look for a stump or a rock to steady your rifle. In fact, smart riflemen always take advantage of natural supports. You can also carry shooting sticks or install a bipod on your rifle. Both help in open country but are a hindrance in thick cover. A one-inch sling with a shooting loop gives you more versatility and doubles as a carrying device. When you shoot prone, sitting or kneeling, a tight sling transfers rifle weight from your shaky left arm to your more firmly-anchored left shoulder and pulls the rifle snugly into your left hand and right shoulder. The tension deadens wobbles.

But a sling is of marginal utility offhand because your ground contact is so limited and your center of gravity so high. You sway like a tall tree, and there's no rest for your left elbow. Sling or no sling, your left biceps and triceps tire, leaving the muzzle free to dance to pulse, muscle twitches, and wind. As you mount the rifle, the reticle obeys your hand; but as you try to steady it on target, it gyrates with increasing vigor and randomness. There's no way to keep it centered. Your job is to reduce the amplitude of oscillations while taking slack from the trigger.

The way to become a better shot is to shoot a lot, practicing only good form so that it becomes habit.

Unorthodox field positions are fine if they help you steady the rifle. Hitting is what counts.

Shooting a lot demands more time than most hunters want to commit, and with centerfire rifles can become expensive. There's no substitute for the time, but you can save money by dry-firing in your living room at a thumbtack on the wall. Dry-firing won't hurt most centerfire rifles. And it won't make you flinch.

An understudy rifle can help with live fire. The J. Stevens Arms & Tool Company developed, in 1887, the .22 long rifle cartridge, and it has been the best training aid for riflemen since. Match-quality .22 rifles and ammunition are expensive, but not necessary for practice. Taping tire weights to a cheap .22 to give it the balance of your hunting rifle lets you practice without developing a flinch or draining your IRA.

I once watched a hunter miss an elk about 160 yards distant. His rifle was resting across a spotting scope on a tripod. The bull stood obligingly, waiting. When dirt flew from the hillside, both elk and hunter showed some surprise. In the firestorm that ensued the elk expired and the hunter said he didn't know how he could possibly miss a target the size of a California beach towel.

"Over the top by a foot," he conceded. "Can't fathom that."

I fathomed it because I'd been a dispassionate observer. The man laid the rifle across the scope without padding it with his hand. He shook. He shot too quickly. He yanked the trigger. "Some shots just can't be figured out," I said generously. "Mystery shots. I've had 'em too."

Indeed, I've committed every blunder possible with a rifle. The most bizarre have remained mystery shots, either because no one witnessed them or observers lacked the stomach to point out my failings. But a few of the many errant bullets in my past have, verily, been ambushed in mid-flight.

There was no other explanation for the lack of blood where a bull elk once filled my scope field less than 50 yards away. My crosswire had been steady on his rib. But a thorough search showed nary a clipped hair, no shattered twigs. The hoof-gashes looked normal enough, and I followed them as far as I could. Save for the empty case, there was no evidence of a shot at all.

"Sometimes ya jest miss." The old logger shrugged. Sometimes he knew more than he said, a habit I found irritating. If I was willing to listen, he of vast experience was surely obliged to expound.

"How many elk have you missed?" I asked.

"Couple."

"Two? Is that all?" After 45 years shooting elk?

"Ya want me t'say more?" He rubbed his chin and looked at the rising moon, a white sliver over the Cascades. "I s'ppose. . . ." Pause. "OK. Mebbe three."

He probably hadn't missed three, just couldn't remember. I told him so.

"Ya could be right," he said. He was sitting on a splitting block, his gimpy leg stretched.

I waited. The moon struck shadows in the orchard and glinted from the faded green top of his Volkswagen, parked tight against the blackberries. In his moss-mottled mobile home a dim light glowed.

"Get quiet."

He fished out a folding knife and eased the point under a fingernail blackened by chainsaw oil.

"What?"

He shrugged again. "If you want to hit, you gotta get quiet."

"You mean at the shot?"

He nodded. "Before the shot. You gotta control that bullet. Steer it, don't just toss it out there."

An understudy rifle like this .22 can help you refine shooting technique without having to endure recoil.

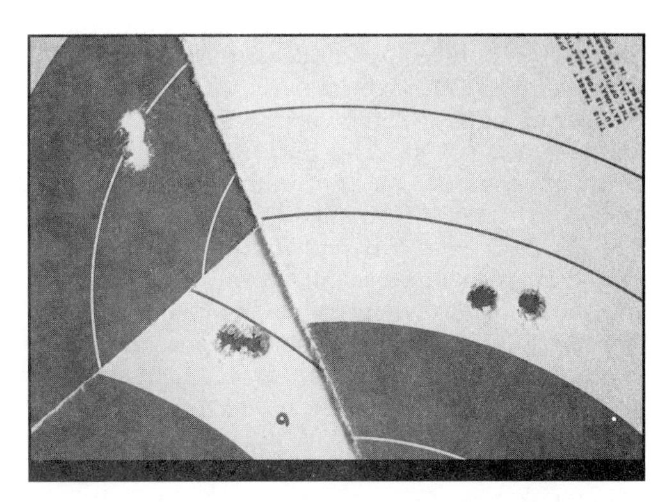

Two-shot groups don't count in bench matches, but in the hunting field, those first two shots matter a great deal. They should hit close together right where you want them, like the author's Freudenberg .30–338.

I remembered what he said because what he said always panned out as the truth. If he said he had missed two elk in five decades, he had. Most of those years, he'd shot more than one elk, some legally.

I got quiet on my next bull. Too quiet. I decided not to shoot. But that was OK because I had for the first time stepped outside myself to talk to my bullet, one on one. I told it to stay put. Yes, Coach, it said. The episode seemed weird at the time.

The next time I got quiet, and talked to my bullet, I let it loose. The deer dropped right away.

"It worked," I told the logger.

"Won't always. You'll get some mystery shots. Real ones."

Again, he was right. Since then I've kept track of mystery shots. About as often as August snow, I've had a bullet vanish in flight or veer to hit far from the intersection of my crosswire. My file of mystery shots grew faster when I tapped the experiences of other hunters. I got stories of ghost bullets and even specific animals that could not be killed. I came to believe that most of these hunters genuinely did not know why they missed shots that seemed impossible to miss. But probably most of them missed for the same reason: They weren't in touch with their bullet.

It's easy to get out of touch, to think that by pulling the trigger you're starting a chain reaction that after a violent explosion puts a hole in the vitals. Actually, a bullet is something you release. Control the trigger, and it doesn't matter how much you wobble. But loosing a bullet as the crosswire breaks hard left and accelerates into a dive is a bit like tossing a golf ball into a soup can from a galloping horse. You never get the release quite right.

Trigger control is the key to eliminating mystery shots. The trigger is your link to the bullet. As you take up that last ounce of pull, it's time to get quiet and get in touch with that bullet. You should know when the rifle fires where the bullet will hit. Calling shots helps improve trigger control. Concentrating on a squeeze to the exclusion of watching the reticle can leave you wondering where the reticle was when the sear broke free.

Familiarity with his .270 brought this hunter a great bull elk.

The best way to control the trigger, according to many expert marksmen, is with an intermittent squeeze, applying pressure when the reticle is close to where you want it, holding the pressure when your wobbles carry it away, then applying more pressure as it drifts back. The detonation should come as a surprise, but you'll see where the crosswire jumps because you're regulating finger pressure according to its movements. Shooters adept at calling their shots can explain most misses. And they soon figure out how not to miss.

Getting the Drift

If Wyoming wasn't anchored to Utah, Idaho, and Montana, it would probably blow away. I've hunted in all sorts of wind, but a Wyoming wind always gets my attention. It has a breadth and depth that makes the earth itself hunker down. A Wyoming wind bends thick trees and shaves the edges off big rocks. It certainly has its way with little things like bullets.

I thought about that as I bellied into the short sage on the ridge and peeked north through my 4X Lyman Challenger. The buck was still there. Wind-tears blurring my aim, I snugged the sling and rested the horizontal wire on the antelope's back. Then I pulled the rifle to the left, against the wind. How much? The wind seemed strong enough to throw a little spitzer into South Dakota. I gave it 8 inches and let my lungs collapse. The shot came a few seconds later, and the buck dashed off. Dust had spurted high and was almost instantly taken by the wind. I bolted in another round. Almost 100 yards farther out now, the pronghorn slowed, then stopped. A few inches higher and double the windage, I guessed. The Remington 722 bounced once more, and the 90-grain Remington softpoint sped away. This time when the buck ran, it was only for a few yards. Then he stumbled and fell and lay still.

I'd held on the animal's nose. Gravity had pulled my 6mm bullet about 18 inches low, and wind had moved it almost as far. The result: a heart shot.

No matter how accurate your rifle or how solid your position, you won't hit center if you ignore wind. Wind speed and angle both affect bullet path. Imagine being in the center of a huge clock face, with 12 o'clock straight out in front and 6 o'clock directly behind. Forget about wind coming from between 11 and 1 o'clock, and between 5 and 7 o'clock. Wind that angles toward you from 10 or 2, or flanks you from 4 or 8, must be reckoned with if you're shooting beyond point-blank range and the breeze is clocking over 15 mph. A "full-value" wind from 3 or 9 o'clock gives you the most trouble because, like gravity acting on a bullet fired horizontally, it is pushing at right angles to the bullet path.

A bullet's vulnerability to wind depends on wind speed and angle, plus bullet velocity and ballistic coefficient (C), a numerical sum of shape, weight, and diameter. Bullets of similar C show about the same wind drift. For example, consider a 130-grain .270 bullet, a 140-grain 7mm, a 165-grain .308, and a 210-grain .338. If they're all Partition spitzers from Nosler, with essentially the same nose shape, they'll be close in C value—from about .390 to .440. Launched at a common 3000 fps, all these bullets will drift about 6 inches at 200 yards in a 20-mph crosswind. Drop C to .289 with a 150-grain .308 protected-point bullet, and drift goes up 50 percent, to 9 inches. A wind far from the muzzle moves a bullet more than does the same wind close up, because downrange the bullet is traveling slower. On the other hand, wind at the firing point can change flight *angle* of the bullet, whose altered course then multiplies the wind's effect over distance.

Like gravity, wind forces a bullet off course. The higher the wind speed and the longer the bullet must fly, the greater the effect. But wind is not as predictable as gravity. Here's how a 10-mph wind (lower curve) and a 20-mph wind move a wind-resistant 30-caliber match bullet driven at magnum velocity.

Incidentally, rifling twist can put a vertical component into wind drift. With an ordinary right-hand twist, you'll get a lift to 10 o'clock from a 3-o'clock breeze, while a wind from the left will drive a bullet down to 4 o'clock. You probably won't notice such subtleties shooting at game, but a combination of little things working together sometimes puts your bullet where you don't expect it.

A wind from 12 o'clock or 6 o'clock has essentially no effect unless it is very strong, and then the result may not be what you think. A bullet fired at a distant target across level ground is actually launched slightly nose-up *and remains nose-up.* Surface exposure due to the bullet's in-flight attitude affects shot displacement. Unlike an arrow, a bullet is not heavy at the front and does not "porpoise." A bullet moving at 3000 fps won't appreciably slow down or speed up with a breeze on nose or tail. Even a stout 20-mph wind is moving only 30 fps.

Wind didn't matter much to me when I started hunting deer. Michigan's woodlots made for short shooting, and the trees throttled winds that could build up steam across the corn stubble. I moved west in 1972 and soon learned about wind.

It wasn't on a hunt. My first lessons in wind drift came on the target range, where I competed in smallbore prone matches. I'd shot indoors for years. Shooting where the wind could reach my bullet, I felt as though I'd been plucked from a swimming pool and dropped into the North Atlantic. Indoors, medals went to shooters who held the rifle still and executed their shots well. Outdoors, you had to hold, execute, *and* dope the wind. If you ignored the wind, you lost.

At first, I thought wind flags unnecessary. I dismissed as gadgets the ubiquitous "windicators"—small, wind-driven fans with tails that swung atop stems on ball bearings. Then I found myself watching the flags and listening to the hum of the windicators. Wind speed and velocity were affecting my shots. A light breeze could shove the bullet across a couple of scoring rings. If I zeroed during a predominant wind condition and the wind suddenly picked up or slacked off, my bullets would stray. A reversal would hurl my shot far out of center.

Sometimes the windicators wouldn't tip me off. They'd hum lazily with nary a flip of their tail, while my bullets jumped in and out of the 10-ring. The flags at 50 and 100 yards told me that downrange, the wind was capricious. Sometimes wind at the target would be opposite that at the line. I'd see flags in full flap at 100 yards, and limp at 50. Occasionally the windicators would spin furiously to the left, while the mid-range flag lifted to the right, and the 100-yard flag kicked out left again. A bullet sent through that gauntlet would fly a zig-zag course. Shooting when you got mixed signals was pretty risky. I noticed that when the wind was visibly contrary or undecided, the spatter of shots at the line would die out as shooters waited for more favorable conditions.

Favorable didn't necessarily mean still. It is possible to shoot very well in a stiff breeze, as long as you're zeroed for that condition or "shade" (compensate) for it properly. Savvy shooters make notes about the wind on a range so they learn its idiosyncracies. The Spokane range, for example, is built on a river bank. Wind typically angles across the firing line from 7 or 8 o'clock but then bounces off the bank and crosses the target line from 4 o'clock. If you minded only wind at the line, you'd make a big mistake. Zeroing for predominate drift gives competitors more shooting time without sight changes during a match. They can then afford to hold their fire during letoffs, pickups, and reversals, or at least reduce the number of shots they must fire in those conditions. They note differences between morning and afternoon drift, and they watch flags on both ends of the line, because wind conditions can differ between firing points.

"Windicators" on a shooting line tell competitors about wind speed and direction. Both variables can differ at the target—and en route!

Long shooting at Coues deer calls for a rifle like this David Miller Marksman in .300 Weatherby, and a powerful scope like its 6.5–20x Leupold. You must also be able to estimate range and dope wind.

When you're hunting big game, you're not using a .22 rimfire, and you're not shooting at X-rings the size of a dime. At modest ranges, bullets from most centerfire rifles can drive through pretty strong winds without significant deflection. For example, a 170-grain flat-nose .30–30 bullet drifts less than 2 inches in a 10-mph sidewind at 100 yards. A 25-mph wind, which is strong enough to sway trees, pushes that .30–30 bullet only about 4 inches off course. Bullets from most other big game cartridges buck the wind better. So at woods ranges, you really needn't think about wind. Remember too that deflection is generally figured for wind at right angles to the bullet's path. Even wind that howls through the trees and picks up small dogs and trash can lids can have little effect on your bullet if the angle is acute.

But distance makes a lot of difference, because bullets slow down. Just as the trajectory of a bullet becomes steeper the farther it gets from the muzzle, so wind deflection becomes greater at long range. You might think of drift as horizontal trajectory. A constant wind is, in effect, very much like gravity. Bullets scribe a parabolic arc under the pressure of wind for the same reasons their trajectory is parabolic. Double the wind speed, and you double the drift. Halve the wind speed, and you halve the drift. Reduce the wind's angle from 90 degrees, and you reduce drift proportionately.

Change the shot distance, however, and the drift may surprise you. For example, a 130-grain .270 bullet launched at 3000 fps drifts only about three-quarter inch at 100 yards in a 10-mph wind. But at 200 yards, it is 3 inches off course—four times as far as it was at 100. At 300 yards it drifts 7 inches, at 400, 13. Why?

Well, there's little drift at 100 yards because the bullet gets there fast: in one-tenth second. There's not much drop from boreline at 100 yards either. Drift and drop increase significantly between 100 and 200 yards. For most long-range cartridges, it's useful to think of wind and gravity as beginning their work at 100 yards. At that point, bullet deceleration has enabled both to change the bullet's course just enough to notice. It may be of academic interest that drift quadruples in the second hundred yards; the important thing to remember is that it increases with distance but not at a constant ratio. In fact, wind drift for the .270 bullet at 500 yards is about 60 percent greater than at 400. A handy rule of thumb is to assume an inch of drift at 100 yards, and double that at 200; triple the 200 drift at 300, and double the 300 drift at 400. Let's see how that works for a 180-grain .30–06 bullet at 2700 fps:

Drift for .30–06 bullet (180 grains) in 10-mph right-angle wind

	Actual drift (inches)	*Rule of thumb drift (inches)*
100 yards	0.7	1
200 yards	2.9	2
300 yards	7.0	6
400 yards	12.9	12

In this case, the estimate is within an inch of actual drift as far as most hunters can hit the vitals of big game. That's certainly close enough. Nobody I know can hold within an inch at 400 yards under field conditions, and darn few rifles will shoot even half that tight. Now, if you're shooting a .30–30 with a flat-nose bullet, the rule of thumb fails beyond 100 yards because the bullet is wind-sensitive and decelerating quickly. The rule works well enough for high-octane bullets like a 140-grain spitzer at 3300 fps from the 7 STW. Out to 300 yards, actual and estimated drift are close; but at 400 yards the STW bullet stays about 2 inches closer to line of sight than does the .30–06 bullet.

You might think that the problem with the .30–30 bullet has to do with its blunt nose. A bullet the shape of a soup can is not very well adapted for flight. There's a lot of air pressure on the nose, a high rate of deceleration. But lightweight spitzers—say, 70-grain .243s—also have low ballistic coefficients. Their low sectional density (ratio of a bullet's weight to the square of its diameter) acts like a blunt nose to reduce ballistic coefficient. In other words, a streamlined bullet that is short for its diameter can be as inefficient cleaving air as a bullet that is longer but has a blunt nose. A 70-grain .243 Nosler has a ballistic coefficient of .252; the .30–30's is .268. Wind drift for the .243 bullet at 100, 200, 300, and 400 yards is 1.0, 4.3, 10.3, and 19.7 inches respectively. That compares with 1.7, 7.6, 18.1, and 34.4 for the 170-grain .30–30 bullet. It's significantly more deflection than we expect from most big game bullets. Velocity, by the way, is not at issue until it drops well below normal impact speeds. While the .30–30 is sluggish at 2200 fps, the 70-grain .243s give their dismal performance in wind after leaving the muzzle at 3400 fps! At 300 yards, the .30–30 is far beyond the range at which it is designed to kill big game.

Fast bullets buck wind better than slow ones. Here are some comparisons:

Deflection in a 10-mph, right-angle wind

	range (yards)	*0*	*100*	*200*	*300*	*400*	*500*
.30–30, 150-grain	velocity (fps)	2390					1040
	drift (inches)		2	8	21	39	65
.30–06, 150-grain	velocity (fps)	2910					1620
	drift (inches)		1	4	10	19	31
.300 Win., 150-grain	velocity (fps)	3290					1810
	drift (inches)		1	4	9	17	28

A couple things are operating here. First, at 500 yards the .300 Winchester bullet drops 3 inches less than the same bullet from a .30–06. Not much difference, considering the 380 fps disparity in starting velocities. Note, though, that the 500-yard velocity spread is less than 200 fps. In other words, the .300 bullet slowed at a greater rate. Moving faster, it met stiffer air resistance, just as you feel the wind on your face grow stronger as you pedal faster on a bicycle. So though the .300 drifted less than the .30–06, its velocity advantage was moderated by a greater rate of deceleration. It showed more "lag."

The .30–30 launched a bullet of the same weight but with the blunt nose mandatory for rifles with tube magazines. Though its starting velocity is roughly 80 percent as high as that of the .30–06, this bullet drops *more than twice as far* at 500 yards. Why? The .30–30 bullet has a low ballistic coefficient and decelerates much more rapidly. Its 500-yard velocity is only 40 percent of its launch speed. The '06 and .300 retain about 60 percent of their initial velocity at that range.

Deflection in a 10-mph, right-angle wind

	range (yards)	*0*	*100*	*200*	*300*	*400*	*500*
.30–06, 110-grain	velocity (fps)	3330					1240
	drift (inches)		1	6	15	30	52
.308 Win., 180-grain	velocity (fps)	2620					1210
	drift (inches)		1	6	15	29	49

These bullets leave the gate 700 fps apart but at 500 yards are clocking about the same speed. Drift is the same out to 300 yards, as the greater weight of the .308 bullet offsets the velocity edge of the .30–06 bullet. But beyond 400 yards the .308 bullet stays on course better. It is catching up to the 110-grain spitzer, whose great rate of deceleration continues to make it vulnerable to wind. Without a speed advantage, it cannot match the .308 in resisting deflection.

Does greater bullet weight always mean better performance in wind? Not always.

Deflection in a 10-mph, right-angle wind

	range (yards)	*0*	*100*	*200*	*300*	*400*	*500*
.223 Rem., 55-grain	velocity (fps)	3240					1270
	drift (inches)		1	6	15	29	50

This bullet is half the weight of the 110-grain .30–06 but is a close match in velocity at the muzzle and downrange. Drift is about the same. One more comparison:

Deflection in a 10-mph, right-angle wind

	range (yards)	*0*	*100*	*200*	*300*	*400*	*500*
.300 Wby., 180-grain	velocity (fps)	3250					1990
	drift (inches)		1	3	8	14	23
.375 H&H, 300-grain	velocity (fps)	2530					1130
	drift (inches)		2	7	17	33	56
.458 Win., 500-grain	velocity (fps)	2040					1160
	drift (inches)		2	6	15	28	45

The 180-grain bullet from a .300 Weatherby, driven as fast as a 150-grain spitzer from the .300 Winchester (first block of figures), drifts 5 inches less at 500 yards, a function of the additional weight and 176 fps less lag at that range. Given the same bullet shape and diameter, you reduce deceleration as you add weight. At the same time, heavier bullets do resist the force of a crosswind better than do light ones, despite their greater surface area.

The effect of deceleration is confirmed by the greater drift of the .375 bullet compared to the .458. Terminal velocities at 500 yards are nearly identical, but the .458 loses only about 800 fps en route, while the .375 drops 1400 fps. (This .375 solid, incidentally, is one of the least aerodynamic of many .375 bullets available). So here the faster bullet drifts nearly a foot farther than the slower bullet. The .375's ballistic coefficient is only slightly lower than that of the .458 bullet. The great weight of the .458 bullet increases inertia, which extends maximum range and reduces wind drift.

If you want a bullet that performs well in wind, choose one that cheats gravity at long range, and holds its speed and energy well. That is, choose a pointed bullet just a tad heavier than average: say, a 100-grain bullet in .243, a 100- or 120-grain in .257, a 140-grain in .264. Pick a 140- or 150-grain .270 bullet, a 150- or 160-grain 7mm bullet, a 180-grain .30, a 225-grain .338. Heavier bullets, like super-streamlined Sierra 250-grain .338 boat-tails, can be the best

You may find moose in thickets, but this giant fell at nearly 400 yards. Wind did not interfere.

choice in big cases. In hulls of modest capacity, they keep starting speed so low as to impair overall performance at normal hunting ranges. A 200-grain bullet, for example, may be a fine choice in a .300 Weatherby. It is by most standards too long and heavy for use in the .308 Winchester case. In fact, a 165-grain bullet in the .308 might be your top pick. It shoots about an inch flatter than a 180 at 400 yards, and drifts about an inch and a half farther—negligible disparities.

Don't pay too much attention to the shape of the bullet tip. Winchester ballistician Alan Corzine says that the first .1 inch of the nose can be flat, round, or pointed without affecting trajectory or drift. The ogive—the leading curve of the bullet between tip and shank—*does* matter. You'll hear arguments for "tangent" ogives and "secant" ogives. These are engineering terms having to do with the placement of the center of the circle of which the ogive forms a segment. That center determine the segment's profile. A round-nose bullet with a sweeping ogive may fly along nearly the same track as a pointed bullet. Similar trajectories mean the bullets lose velocity at about the same rate, and that they'll respond similarly to wind.

Boat-tail bullets become an asset only at very long ranges—or in gale-force winds. A 30-mph wind that shoves a flat-base 7mm bullet 17 inches at 350 yards moves a similar boat-tail bullet 15½ inches. The lesser drift afforded by a tapered heel at higher wind speeds is academic, when you consider how hard it is to estimate drift in a wind that strong, or to hold a crosswire within a couple of inches at 350 yards. Also, the *percentage* difference in wind deflection between the two bullets is about the same for a 10-mph wind and a 30-mph wind.

Wind, Mirage, and Luck

Probably the longest shot I'll ever take at game happened long ago. I was at ridgetop in Oregon's Wallowa Mountains when I spied a fine mule deer buck far below. The sun behind me, I had a grand view, and the buck couldn't see me squirm onto a rock, padding the rifle with my jacket. As I thumbed the safety on the Model 70, it occurred to me that the sun was making this deer seem a lot closer than it was. Gauging distance with the center section of my crosswire, I came up with something around 450 yards. It was really too far, especially in the squirrelly wind. But a precipice in front of me made a stalk untenable, and I was full of vinegar and keen to kill the deer. I held a foot into the wind and 2 feet high—off the animal, as it was facing in my direction. The bullet hit between its legs. My second shot was high enough but to the right, the victim of a wind gust. Giving myself one last try, I held even with the top of the deer's head and about 18 inches left. The buck sprinted across the slope and piled up, heart-shot.

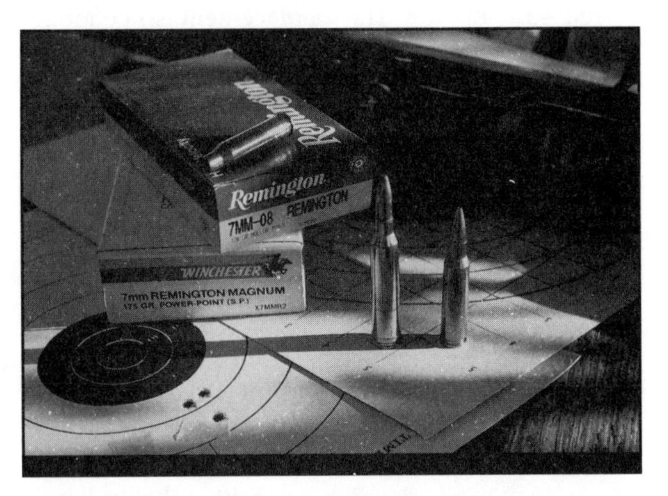

Though faster bullets are a bit less wind-sensitive, a rifle that recoils less may help you shoot better. Remember, wind only steps in occasionally. The rifle kicks every time you fire.

I wouldn't shoot at that buck now. It's irresponsible to fire when you know a lethal hit is unlikely. Any time the wind is blowing, I remind myself how little I know about conditions downrange. If the range is under 250 yards, and the wind less than 15 mph, the shot is all but guaranteed—assuming I can hold the rifle still. I'll shade for elevation and windage. If my estimates are off by as much as 20 percent, the bullet will still land in the vitals because drop and drift are both modest. Between 250 and 400 yards, range estimation becomes more important. However, if you're good at it or cheat with a laser rangefinder, you'll get within 10 percent of the yardage and make a fatal hit. Estimating wind that far away is harder, and you must consider not only prevailing wind and wind at the target, but *net* drift, which may be affected by reversals, pickups and letoffs at various points along the bullet's path. Remember: unlike gravity, wind is not a constant. You

A trophy mule deer buck. After watching it at long range in treeless terrain for two days, the author still hunted near where the animal was bedded and jumped it from a coulee.

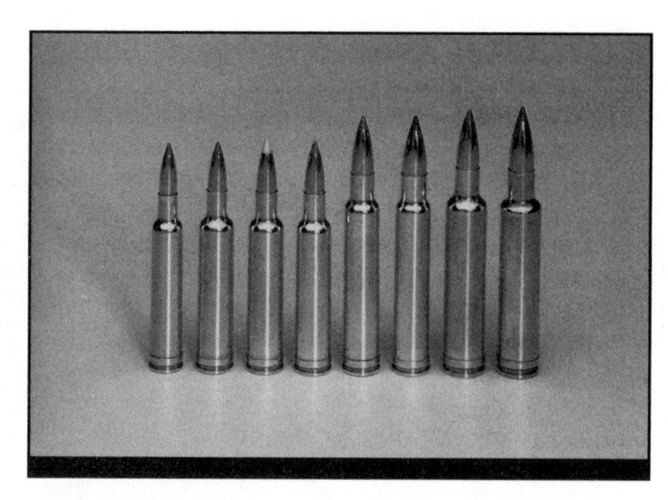

Their ability to cheat wind and gravity helps these Weatherby cartridges sell. The Nosler Ballistic Tip bullets shown have a sleek profile, stretching effective range even further.

might as well consider each shot unique as regards wind.

Reading wind is an acquired skill. When I started minding windicators and range flags, I thought myself pretty sophisticated. But still bullets strayed, even when I called the shot well. Dick Nelson, a fine rifleman who also helped Boeing engineer the first moon vehicle, took me aside one day. "You'll have to read the mirage. Do that, and those bullets will hop through the 10-ring like trained pigs." I had never seen trained pigs, but the part about the 10-ring got my attention. Here's what Dick and other savvy smallbore shooters taught me about mirage:

Mirage is a visual distortion caused by heat waves rising from the earth's surface. If you don't see it, it isn't there.

Mirage does not move bullets; its dance shows you wind that does.

You can't see mirage at all distances at once. You'll either see the strongest mirage or the mirage at the range for which your scope is focused. To get the most information about wind that most strongly affects their bullets, match shooters typically focus their spotting scopes to read mirage a little short of the targets.

Mirage that's really bumpy and moving slowly indicates a light breeze. Mirage that's flat and fast indicates stronger breeze. When mirage disappears suddenly with no change in light conditions, it's often because the wind has picked up. Mirage that boils vertically shows you a still condition—but beware! A boil commonly precedes a reversal in wind direction. Many competitive shooters zero for a light prevailing breeze, then hold their fire during boils and reversals, shading and shooting during pickups and letoffs.

Mirage can make you shoot at a target that isn't there, by "floating" the target in the direction air currents are moving. The displacement isn't enough to worry about when you're shooting at game, though it can cost you points on the small targets in a rifle match.

In the field, you may seldom see mirage. Fall hunting seasons bring cold weather, which all but cancels mirage. To read wind, you'll have to rely on coarser signs: nodding trees and grass, the leaves and snow, and mist that yield to wind. Remember that wind at the target is as important as wind at the muzzle. In fact it can be *more* important, because the bullet is moving slower, and over any given distance wind has more time to work its mischief. Wind at the muzzle has the advantage of leverage. That is, it can start a bullet on a new course, and distance will magnify the displacement of that bullet from boreline. If it strikes 2 inches to the left at 100 yards, it will not come back to center even if conditions are dead calm at 200 yards and beyond. In fact, the bullet will move ever farther from boreline at long range, if at a reduced rate, because wind at the muzzle has established an angle between the bullet's path and boreline.

Not long ago I spent half a day shooting at prairie dogs with a .223. I don't shoot a lot of prairie dogs, because it seems to me that shooting and hunting are different. When I shoot a lot, it's at paper. But this day, a couple of friends and I bellied to within 200 yards of a small dog town and battled the wind with our little 50-grain spitzers. It was tough, especially after the 200-yard sod poodles got wise and we were forced to stretch the rifles to 250, then 300. A stiff breeze from 8 o'clock carried the bullets from 3 to 6 inches, depending on distance and our timing. Shooting between gusts, we sometimes over-corrected. In spotting for one another, we learned more than if we'd been alone, trying to see bullet strikes during recoil.

Pronghorns are seldom hard to find, but the wind in pronghorn country can blow your bullet far off course. Solutions: learn to dope wind, and get closer!

Late in the morning, we moved to another rise so the wind came from 7 o'clock. Immediately we saw wind deflection shrink. It confirmed an old target-shooting rule: Unless it is *very* strong, ignore wind from between 11 and 1 o'clock, and from between 5 and 7 o'clock. It's hard, when you feel wind on your face or neck, to remember that a bullet moving 3000 fps is encountering tremendous wind resistance even in still conditions. It is generating its own headwind—a *2000-mph* gale. What difference do you think a 20-mph headwind or tailwind will have on this bullet's flight?

Next time you hunt where the wind blows, remember that it's just like gravity, an invisible force pulling your bullet off track. But wind is not as predictable as gravity, and to hit where you want to hit, you may have to aim where you don't want the bullet to go.

Getting the Angles Right

You probably hold a rifle with the sights on top and the trigger on the bottom, just as you drive a car with the shiny side up. But unlike automobiles, rifles aren't connected to the ground. Rifles can be tipped as easily as you tip your hand. They can be fired at a tilt or even upside down. While most shooters keep the sights on top, many do not. Those who tip their rifles are said to be canting.

"A cant isn't bad," a shooting coach told me long ago, "so long as you do it the same each time." Doing it the same each time presupposes that you know you're doing it in the first place. Riflemen who don't think about cant either tip the rifle at pretty much the same angle out of habit, or they allow the angle of the sights to change slightly with each shot.

It's quite easy to spot a cant if you're coaching the shooter, just as it's easy to see the tilt of a truck loaded too heavily on one side if you're driving behind it. In the truck's cab, you might not be able to tell; and when you're looking through the sights you often can't tell either.

A shooter used to canting his rifle will likely mount a scope so the crosswire is tilted off the vertical axis of the rifle. You've probably thrown a friend's rifle to your shoulder and had to consciously rotate the stock so the vertical wire would appear to be straight up and down.

"Good gravy," you say to your amigo. "How can you shoot with the reticle falling over like that?"

Shooting uphill or down, you'll hit a little higher than if shooting horizontally the same distance.

"Here, lemme see," he replies, grabbing the rifle and aiming it. After a pause: "Looks square to me. You're prob'ly just tippin' it."

"Ha! Not likely!" you shriek. "That scope's as crooked as your brand of poker."

"You callin' me crooked? Just because you can't hold a rifle. . . ."

"Now wait just one minute, you. . . ."

Discussions about canting can get animated in a hurry. About the only way to settle them, short of spiking the venison stew with cigar tobacco, is to lay the rifle in a rest or across sandbags on a bench and square up the stock by nudging the butt so a plumb line or carpenter's level shows it to be vertical. Now, without moving the rifle, look into the scope. The crosswires should appear to be vertical and horizontal. If they are not, you simply loosen the scope rings and twist the tube.

That is, unless you want to leave the reticle as is.

A canted reticle will not cause a miss. In fact, you can rotate the scope so the crosswire looks like an "X" and use it as effectively as before. A small disadvantage is that you won't have a vertical wire to help you hold off for wind deflection or show you the line of bullet drop at long range. You won't have a horizontal wire to help you lead running animals.

A canted *rifle,* however, is another story. No matter how the reticle appears, if the rifle is tipped, you'll have problems hitting beyond zero range because the bullet path is not going to fall along the vertical wire or directly below the intersection. If your sightline is directly over the bore, a long shot requires you only to hold high. If the sightline is forced by a canted rifle to the *side* of a vertical plane through the bore, you'll not only have to hold high, but to one side.

Here's the reason: Given that your scope is mounted directly above the bore, your line of sight crosses the bullet path twice. The first crossing happens at about 35 yards; the second is at zero range—say 200 yards. If the rifle is rotated so the scope falls to the side of the barrel, the sightline crosses only once, because gravity sucks the bullet straight down, while the line of sight has a horizontal component. Whether the scope is on top of the rifle or a bit to the side, the line of sight will converge with the bullet path and slice through it, then angle away. If the scope is on top of the rifle, gravity pulls the bullet in an arc back into and through the straight line of sight. If the scope is not on top, the bullet path still dips below the horizontal plane; but the sightline doesn't follow it down. The rifle shoots to the side of where you look.

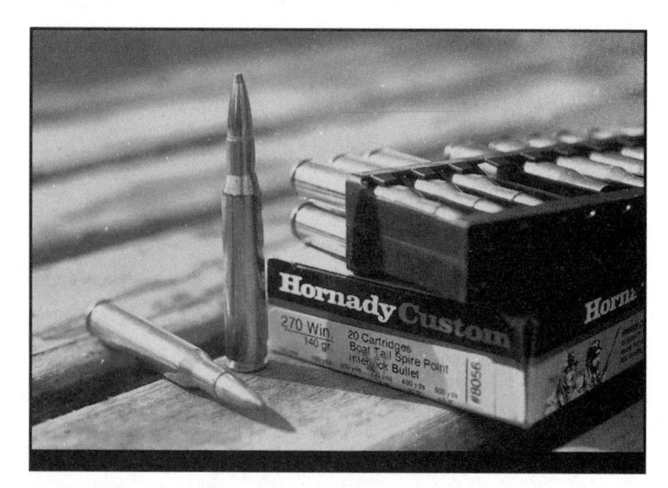

For most big game cartridges like the .270, ignore modest angles at ranges under 200 yards. Steep angles or long yardage may require compensation.

You'll almost never have a steep shot in caribou country!

How much practical difference will canting make? Not much. A cant that escapes your notice won't cause a noticeable shift in bullet impact at normal hunting ranges. As the targets get smaller and the range longer, and as you impose stricter accuracy standards, cant starts to matter.

When I was on the Michigan State University rifle team, I marveled at a colleague who shot very well but used a cant that would have spilled coffee. Standing, he looked straight ahead through sights that fell into his natural line of vision when he rotated the rifle on his shoulder. The adjustable butt let him do that without changing the contact angle of the butt hook. So his rifle tilted in toward him. Not only the line of sight but the rifle's center of gravity fell near the centerline of his torso. From a mechanical perspective, this made perfect sense. I tried shooting that way and found it darned near impossible.

In traditional bullseye rifle competition, targets are very small and the rifles supremely accurate, so shooters must correct for cant. (Some globe sights for target rifles have tiny bubble levels that show you the slightest cant at a glance, and similar devices are now available for scopes.) On the other hand, my teammate on the smallbore squad didn't have to worry about horizontal angles because his shooting was done up close at one precisely measured distance. It is easy to accommodate cant at a single distance. You simply move the sights to put the bullets where you look. Forget about what the sightline and trajectory do beyond the target. It doesn't matter.

Hunters don't shoot at just one distance, however, or with adjustable stocks. So although small degrees of cant seldom affect field accuracy, it's a good idea to shoot with the sights squarely on top of the rifle. Cant is just one more thing to worry about, one more distraction, a small but thorny threat to the self-confidence that can help you shoot well. It's easy to see if you have a cant. Simply loosen your scope rings and line up the vertical wire with the butt of your rifle. Now tighten the rings. Throw the rifle to your shoulder with your eyes closed. Open them. The wire should appear vertical. If it does not, you're tipping the rifle. Practice holding it with the sights at 12 o'clock, where they belong.

Now, sometimes shooters slip a cant into their shooting routine without knowing it. They mount the scope carelessly and subconsciously adjust their hold on the rifle to correct for a reticle that's tilted. A

This hunter found his elk in a draw and shot down at a steep angle. But the range was modest and the target big. He held his .270 squarely on the chest.

culprit here is the ubiquitous Weaver Tip-Off scope ring. This inexpensive ring has been around a long time, and for good reason. It's strong and lightweight. But because the top half hooks the base on one side and its two screws take up all the slack on the opposite side, tightening a Tip-Off ring can rotate the scope tube down toward the screws. If they're installed on the right-hand side, you put a clockwise tilt into your reticle as you cinch them up. You may have aligned the reticle perfectly with the butt before installing the top part of the rings, but now the crosswire is tipped! Solution: back off on the screws and twist the scope counterclockwise about as far as you think it moved. Tighten the screws again, and check the reticle.

Canting isn't the only way you can complicate your aim at long range. When I was growing up, side-mounted scopes were common. With them, you could affix a scope to a top-ejecting Model 94 or 71 Winchester, so you could use your iron sights *and* a scope without removing the scope. If it holds the scope off the vertical axis of the bolt, a side mount introduces the same sighting error you get with a cant. Side mounts that put the scope over the receiver are best. A Remington 870 shotgun in my rack wears a side mount, but the scope lies almost directly over the center of the bore. Side mounts installed before rifles were routinely drilled for scopes made hash of collector-quality rifles. Many Winchester Model 54s and 70s, and untold numbers of lever-action carbines, were assaulted in the decades of innocence following World War II. After it dawned on shooters that a pre-war hunting rifle might someday be worth more than $100, and that holes in the receiver were like rust pockets on Duesenbergs, drilling stopped.

Top mounts centered above the bore can also give you problems with sight angles if the bases are extra high. It's easy to see this in exaggeration. Picture a scope with ring bases three feet high. If you adjusted that scope to put its bullets on point of aim at 35 yards, you couldn't expect it to give you a 200-yard zero. The reason: Directing the line of sight to intersect bullet trajectory at 35 steps puts the two paths at steep angles to each other. The entering and exiting angles are naturally the same. Your sightline now diverges quickly from the bullet's trajectory, diving under it in a straight but steeply descending path. The bullet will come down to meet it, courtesy of gravity. But that won't happen for some distance.

I once watched an elk hunter miss a huge bull from prone, with a rest. The distance was perhaps 300 yards. He asked me where to hold, and I suggested high behind the shoulder, assuming his rifle was zeroed at 200. I checked it later, shooting to ranges of 400 yards. It was zeroed at 330. The shooter had fired it previously only at 100 yards, and someone had told him that bullets striking 3 inches high at 100 would be dead on at 200. They neglected to point out that the super-high rings used to get his astronomy-size scope clear of the barrel put the line of sight at a steep angle to the trajectory. Result: more distance between the crossings.

Perhaps the most common questions about angles have to do with uphill and downhill shooting. Aiming at some angle other than horizontal affects bullet flight, and the effect is essentially the same whether you're shooting up or down. As you deviate from the horizontal, you diminish the effect of gravity on the bullet over any given shot distance. That is because gravity always acts perpendicular to the earth; it's always pulling things toward the earth's center. Just as a wind at right angles to a bullet has a greater effect on the bullet's flight than

A steep downhill angle didn't save this bull, because the range was short. Rifle: a Remington .30–06.

Shooting uphill or downhill, you may hold the rifle differently, and that can make a big difference in your results. Practice from uncomfortable positions can be valuable.

wind coming from 7 o'clock, so gravity applied at right angles has its strongest influence. Imagine a bullet fired straight up, or one fired straight down through a hole deep enough to reach China. In both cases, gravity's pull would act parallel to the bullet's flight. Both bullets would travel in straight lines until gravity and drag stopped them cold. You could argue that the bullet fired into the sky fights gravity immediately, while the bullet sent to China gets an initial boost from gravity. But truly, the effect of gravity on the nose and heel of bullets shot parallel with gravity's pull is minor. In fact, it's negligible. The drag exerted by atmosphere is hundreds of times more powerful on a fast-moving bullet.

When you shoot at any angle to horizontal, gravity's effect is the same as you might expect if you considered only the horizontal component of the bullet's flight. Say that you're shooting at a steep 45-degree angle at a deer 280 yards away. From geometry, you remember that a right triangle with one 45-degree angle has another also, and that the hypotenuse is roughly 1.4 times as long as either leg. In this case, the hypotenuse represents your bullet's flight path in a triangle whose horizontal leg is the distance your bullet would cover if the target moved along gravity's path until it gave you a flat shot. That leg is the horizontal component of your bullet's flight, no matter whether you're shooting uphill or down. If actual distance to the target is 280 yards, the horizontal leg is 200 yards. Hold dead on, if you're zeroed at 200.

You can determine effective range from actual if you know the shot angle. The actual range is the hypotenuse of a right triangle whose horizontal leg is your effective range. If your math is a bit rusty, divide the actual range by these numbers:

degrees angle	*divisor*
10	1.02
15	1.04
20	1.06
25	1.10
30	1.15
35	1.22
40	1.31
45	1.41

A lot of hunters miss game because they forget to compensate at long range, when a bullet's steep trajectory makes range estimation difficult and where it's easy to build in too much holdover. But a lot of hunters also miss at closer ranges because they overestimate the effect of shot angle. A rule of thumb may be in order: Don't adjust your aim for shot angle if the animal is less than 200 yards away, unless you have to aim closer to vertical than to horizontal. When the shot is 300 yards, and the angle around 45 degrees, hold 6 inches lower than you normally would for a 300-yard shot. At 400 yards, hold a foot lower than you normally would for a 400-yard shot. Steeper angles require a little more adjustment, of course, but it's not often that you'll have to shoot far at a 45-degree angle. For gentler angles, keep your reticle on the animal and shade slightly as you estimate the reduction in bullet drop.

Flat-shooting cartridges offer a bonus if you're not shooting horizontally:

Adjustments in aim for targets at 45 degrees to horizontal

	range, yards	*0*	*100*	*200*	*300*	*400*	*500*
.300 Sav., 180-grain	velocity (fps)	2350					1410
	hold (inches under)		1	4	10	20	33
.308 Win., 180-grain	velocity (fps)	2620					1600
	hold (inches under)		1	3	8	15	26
.30–06, 180-grain	velocity (fps)	2700					1660
	hold (inches under)		1	3	8	15	25
.300 Win., 180-grain	velocity (fps)	2960					1980
	hold (inches under)		1	3	6	12	19
.300 Wby., 180-grain	velocity (fps)	3250					1990
	hold (inches under)		1	2	5	10	17

Truly, errors caused by cant and steep sight angles pale beside those caused by rifles that don't stay still. Holding a rifle steady and executing a shot smoothly are still your most important tasks when the time comes to shoot at game.

Aiming Where They Aren't

The moose stands there, looking vulnerable as before. I let the red dot in my Aimpoint sight sink below the hump, then settle into the black hollow behind the shoulder. When everything seems right, the rifle fires itself.

But then the moose runs, and I must shoot again. Flicking the Blaser's bolt, I'm back on target right away, the rifle moving now. The dot sweeps forward of the shoulder, even with the ear. The .30–06 bucks, and the moose is gone. Another flip of the bolt, and I listen.

"Five, four, five, four."

It is my best score on a Swedish moose target. The two standing shots were probably the fives, though there's no way to tell. After a pair of runs, right and left, the men in the pits see only four holes.

There's no reason you shouldn't tally a perfect score on this target. It is only 100 yards away, and though the bullseye is invisible from the line, it is in the logical place and big enough to hit offhand. When the moose moves, it is at a predictable speed, and smoothly. There's no change of speed or direction, no brush in the way. Still, few shooters keep their bullets inside that five-spot. As a matter of fact, I lost one shot outside all the scoring circles and put several bullets in the three ring. Hardly a stellar performance for someone who's been shooting rifles pretty regularly for 35 years.

Then again, hitting a moving target with a rifle is not so easy as some shooters would have you believe. One fellow told me he routinely killed running pronghorns farther than I like to fire at motionless game. But the next day he crippled a fine buck that was standing still as a post at 200 yards. Another man boasted of killing an elk on a dead run 300 yards off. A little probing revealed that his bullet clipped the skull, nearly 3 feet from where he'd intended to hit. A couple of years ago, I watched a bear guide miss a grizzly three times at less than 80 yards as it motored around an open hillside after my

Elk in timber must sometimes be taken on the move or not at all. To dawdle is to miss your chance.

friend John Chisnall had delivered a near-lethal blow with his .300 Remington Ultra Mag. The bear skidded on its nose when John fired again.

The Swedes know that most riflemen think too highly of their own marksmanship, and that chance hits simply encourage irresponsible shooting. That's why a passable score on the moving moose target is a prerequisite for big game hunters in Sweden. It may not teach you how to hit running animals, but it will give you a healthy dose of humility.

I'm poorly qualified to write about shooting game on the run, because I avoid doing that whenever possible. (All right, it's *always* possible. Sometimes my reactions are faster than my brain.) Actually, I started strong, with a one-shot kill on a whitetail streaking through aspens. It was a grouse shot at 90 yards. Flushed with my achievement, I figured my keen eye and iron-sighted .303 were a match for anything on hooves. Then I missed six consecutive shots at a forkhorn waltzing across an alfalfa field. And missed a wounded buck scrambling for cover. I shot behind a pronghorn and well ahead of a blacktail deer close enough to hit with a medicine ball. I began to have second thoughts about moving targets.

Hitting with a rifle is harder than hitting with a shotgun because not only does your bullet lack the pancake-skillet breadth of a shot charge; it also lacks depth. A cloud of shot may be several feet long by the time it gets near the target. A lot of birds are killed by trailing pellets, so you can afford to be generous with lead. Often you're not as generous as you think, though, and the leading edge of the pattern connects. A bullet gives you no latitude. You're spot-on, or you miss.

With this mandate for precision, it's easy to be stiff and tardy. Shooting moving game with a rifle calls for the same instant response and fluid body movement as using a shotgun. If you dawdle or interrupt your swing trying to refine that sight picture, you'll miss, or lose your chance to shoot at all.

One deer hunter who taught me a lot about shooting in the woods said that it is like boxing: "The first thing to think about is your feet." He pointed out that many hunters fire at running deer before they get their feet planted right. "That's like throwing a punch when you're off balance. It probably won't connect." Moving through cover, he was careful to keep his weight over his feet and never get so "twisted up" that he couldn't shoot to any point on the compass by moving one foot quickly. When he paused, it was always where he had good

The author killed this Alberta bear as it began to run. A foot of lead at 100 yards proved sufficient.

opportunities to shoot, but more importantly, where his feet could rest and pivot easily for a quick offhand shot. "Deer most often break cover when you stop moving," he said. "If you stop where your feet can't help you shoot . . . well, that's just dumb."

My mentor swung as gracefully with his little 6.5x55 carbine as with his favorite 20-bore double. His knees were bent slightly, most of his weight on the balls of his feet as he leaned into the rifle. "Take the shot right away, even if it isn't perfect—unless you've already decided to wait for the animal to stop or cross a better shot alley. The time to decline a shot at moving game is *before* you mount the rifle. Once you've committed, follow through. Raising the rifle and getting your body in motion and focusing on the vitals and taking up trigger slack—that's all part of a shot. The bang is just the finale."

When you spot a moving target, determine right away whether you have a shot or not. Sometimes you don't. It's irresponsible to fling bullets at every running deer in range. But to swing tentatively is to miss. Forget about light brush. If you're in timber, you should have a shot alley picked out before the butt hits your shoulder. As soon as the sight sweeps in front of the target, the rifle ought to fire itself. Your only reasons to abort: if another animal runs across your line of sight or if the target buck switches direction suddenly, forcing you to start over.

Try to apply the same discipline as when shooting at stationary game. I don't shoot unless I think there's a 90 percent chance my bullet will hit vitals. Making that decision instantly is part of shooting well at running game. There's no disgrace in passing up a shot. But if you miss often, you're probably shooting too often. When trees catch my bullets, I'm more generous with my scorecard than I am in open country, because in timber even dead-center shots will occasionally be short-stopped. If it were against the law to hit a tree, you'd never shoot at running animals in the woods.

How much do you lead? That depends on the target's apparent speed (actual speed mitigated by target angle) and bullet flight time (distance x average velocity). A buck peeling out within slingshot range requires insignificant lead, if your swing is fast enough to keep the sight in the vitals, because bullet flight time is so short. Bullet lag will keep the bullet from centering the vitals, but no more than sighting error and rifle wobble. At ranges to, say, 50 yards, your bullet should hit close enough to point of aim to kill quartering big game. Figuring lead takes time; better to shoot fast squarely at the target. A deer running 25 mph straight away requires no lead, no matter the distance, because its path is in line with the bullet's arc. A deer trotting across your front at half that speed must be led (unless it is very close) because its angle of travel is 90 degrees to the bullet's path.

As soon as that deer bursts into view, determine the lead *based on what you know.* Temper your instinctive urge to fire blindly. Calculate. Running deer up close seem jet-propelled, and your tendency is to snap-shoot. You generally have time for an aimed shot, and if you don't, you don't have a shot at all. The same animal farther out requires more lead than you think. Like a goose in the distance, it *appears* to be cruising slowly. Steep shot angles at long range can require *less* lead than seems right, because you're smitten by the animal's actual speed and may not recognize its displacement relative to the bullet's path.

Whitetail hunting in close cover with his Browning Buck Special, the author shows proper form in taking quick aim. Don't try to avoid the twigs. If there's too much brush in the way, don't shoot.

It isn't a good idea to shoot with a stationary rifle. Your reaction time and the rifle's lock time add to the bullet's flight time to extend the lead. Also, the interval between your brain's signal to fire and the striker's release will vary. With a moving rifle, you have only the calculable flight time to fret about. If you prefer to "swing through" with a rifle moving fast enough to pass the target, you'll have extra lead built in, so you won't need as much air between sight and target as with a sustained lead.

I was once asked to settle an argument as to how much lead was required for a hunter shooting a 180-grain .30–06 bullet to kill an elk running 20 mph at 200 yards under calm conditions. Here's my reply:

Assuming the elk is crossing at 90 degrees to the line of shot, and the rifleman is using a sustained lead (both important assumptions), you calculate first the speed of the animal in feet per second. As I recall from my Driver Education days back in the Stone Age, a car traveling 60 mph moves 88 fps. So an elk at 20 mph covers about a third that distance—say, 30 feet—in one second. Now, the fellow didn't give me his bullet's velocity or ballistic coefficient, both of which matter. But we can get close by using a factory ballistics chart, which claims a starting speed of 2700 fps (higher than *my* chronograph registers with most factory loads). Now, because the bullet slows down as soon as it leaves the rifle, we must find its average speed over 200 yards to calculate its flight time. With a 200-yard speed of a little over 2300 fps, we figure an average of 2500 fps. But that's a little skewed, because the bullet spends more time at low speed than at high speed (the deceleration rate decreases). Let's ball-park the average at 2450. Two hundred yards is 600 feet, so the bullet takes only about a quarter-second to reach the bull. During that time, the elk travels a quarter of 30 feet, or about 7½ feet. You'd hold about 6½ feet forward of the point of the shoulder.

If the animal were running at a 45-degree angle, you'd halve the lead.

Here's another example: Say you're shooting a rifle chambered for the hot .338 Remington Ultra Mag, and on the first day of elk season a six-point bull trots across your front at 300 yards. It's an open shot, and you're sitting, with the rifle over a stump and steady enough for a smooth lead. But the elk is moving into a strong, 20-mph headwind. Now, your handloaded 210-grain Nosler exits the muzzle at 3100 fps. Remember, though, that fast bullets lose velocity at a higher rate than slow ones, all else equal, because they meet heavier air resistance. And that loafing elk is covering ground at about 10 mph, or 15 fps. The math: Your bullet, averaging 2700 fps, covers 300 yards (900 feet) in one-third second, during which time the elk moves 5 feet. Your hold, then, should be four feet in front of the shoulder—if you were zeroed for 300 yard and there were no wind. But the more practical zero for an elk rifle is 200 yards, so you'll have to hold about 6 inches above where you want the bullet to hit. And since the 20-mph wind will drift your bullet back toward the animal, you must add the drift into your lead. That's another foot. Swinging five feet ahead then, with the horizontal wire, just below the hump, you squeeze off.

Or you don't shoot at all. Long shots, especially those that require wind or elevation correction, are risky when the target isn't still. Land sakes, a 300-yard *standing* shot is a challenge under most field conditions. If you aren't sure, hold your fire.

This mule deer buck got up in front of another hunter and bounded past this one. A single shot from the man's Weatherby ended the hunt. Being ready is a first step toward better shooting.

A running shot made is a delightful thing, an accomplishment not soon forgotten. The quick, fluid shots up close—shots begun and finished in a heartbeat—are best of all. I've not made many. One I recall happened long ago, when a whitetail buck exploded from cover a few feet away. I fired as if at a partridge, decking the deer instantly. The whole episode lasted less than three seconds, then all was quiet.

Another time, on the last day of a demanding quest for a big eland in Zimbabwe, my tracker and I happened upon a fresh track in dense thorn. We followed. As luck would have it, we'd moved a scant hundred yards when an eland cow sped away, vanishing at spitwad range before I could shoulder my rifle. But such was my destiny that on her heels came a huge eland bull. I couldn't see horns well enough to judge them, but Phillip shouted something in Endebele that I took to mean *"Shoot!"* A patch of shoulder came clear for a millisecond as the scope field swept up from behind, filled with a blur of eland and thorn. I fired with the memory of that shoulder in my mind's eye, and a ton of hard-won trophy collapsed. The Core-Lokt bullet from my .300 Holland had broken the bull's neck.

Such is the stuff a hunter's dreams are made of. Sometimes to make them come true, you have to shoot where the animals aren't.

Recoil

If you've been whacked by a scope during recoil, you've had a shooting lesson. A few lessons like that, and you'll pay attention to recoil. You may change your grip or head placement or body position. Maybe you'll move the scope forward or install a muzzle brake. Perhaps you'll switch to a gentler rifle. If you don't eliminate punishing recoil, you'll start flinching when you expect it, just as you blink when a tree branch swings in front of your face. Flinching moves your rifle. The rifle then sends the bullets where you don't want them.

Recoil hits you hardest when you're planted firmly behind the rifle, as you are on a bench. If your body can rock and sway with the blow, it won't hurt so much.

Sir Isaac Newton described recoil when he figured out that for every action there must be an equal and opposite reaction. You can calculate recoil's kinetic energy easily enough with this formula: KE = MV2 / GC, where M is the rifle's mass and V its velocity. GC is a gravitational constant for earth: 64.32.

Now, mass and weight aren't the same. Mass is really the measure of an object's inertia. The theory of relativity tells us that two objects have equal mass if the same force gives them the same acceleration. Using gravity as the force, we can equate mass with weight. That is, weight is a measure of the force by which gravity draws an object to earth. Because rifles respond pretty much the same to gravity, rifles of the same weight have essentially the same mass.

To get rifle velocity we have to crunch some numbers. We already know most of them. The formula:

V= bullet weight (grs.) / 7000 ×
bullet velocity (fps) + powder weight (grs.) /
7000 × powder gas velocity (fps).

Powder and its gas figure in because like the bullet they are "ejecta" and cause recoil. You can get powder weight from factory rounds by pulling bullets and weighing charges. Gas velocity varies, but Art Alphin, in his A-Square loading manual, says 5200 fps is a useful average. The "7000" denominators simply convert grains to pounds so units make sense in the end.

For a 180-grain bullet fired at 3000 fps from my 8½-pound .30–338 or .308 Norma rifles I'd calculate recoil this way: 180 / 7000 × 3000 + 70 / 7000 × 5200 = 8.5 × V. That simplifies to (77.143 + 52)/ 8.5 = V = 15.19 fps. Then I can calculate recoil using the first formula: KE = MV2 / GC. The result looks like this: 8.5 (15.19)2 / 64.32 = 30.49 foot-pounds of recoil.

Kinetic energy is not "kick." Felt recoil can vary significantly among rifles delivering the same amount of recoil in foot-pounds. There are a couple of reasons for this. One is that while bullet speed figures into the energy calculation, its contribution to rifle "slap," or the blow of quick recoil, does not. Plainly put, a bullet that exits fast dumps all its energy fast too. Pile enough foot-pounds behind that slap, and it becomes a punch.

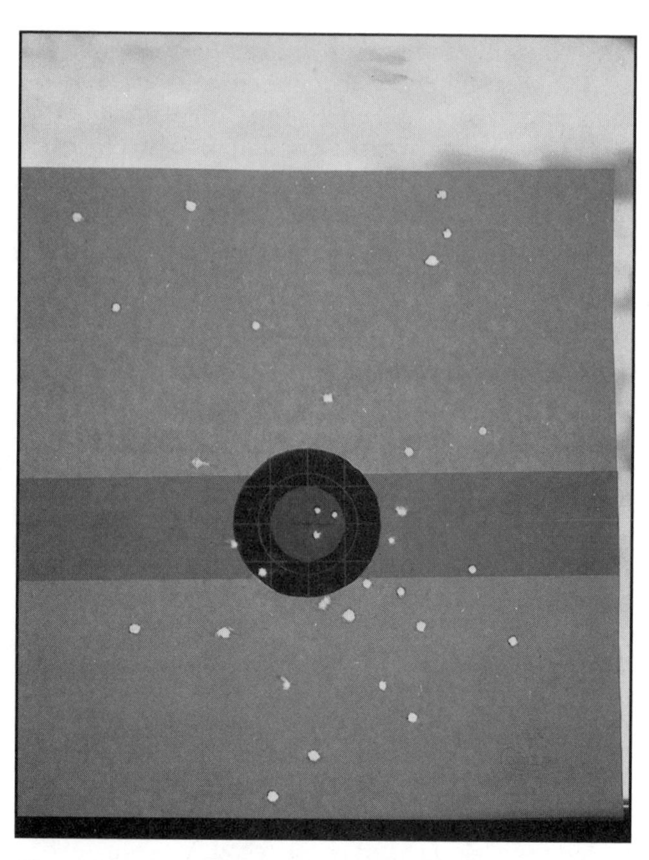

These bullet holes represent about half of the shots fired at this target offhand at 100 yards by shooters who had just zeroed their rifles. One shot per hunter. Such poor shooting is largely due to flinching.

Example: A long-barreled Ruger Number One spitting a 405-grain .45–70 bullet at 1800 fps (don't try this load in your Springfield!) recoils at about 17 fps. So does a .338 Winchester Magnum thrusting a 225-grain spitzer along at 2800 fps. That means you absorb about the same-size bundle of energy from these two rifles. But the .338 is apt to feel more punishing because the bullet leaves faster. All the reactive motion must be completed right away. The .45–70 delivers recoil over a longer period.

Some shooters say rifles chambered for big-bore British rounds merely push you, while sharp-shouldered magnums belt you, as if initiation by cordite somehow made cartridges more genteel. I'm not convinced. There's more than a push to a .600 Nitro in recoil! Even .470s light enough for safaris without gunbearers can get vicious. The short breech sections of doubles put the muzzles closer to your face, and their off-center bullet launch pivots the rifle. True, they have twice the barrel steel up front, and velocities and breech pressures are low. But bullets that weigh more than an ounce, in front of powder columns as long as a half-smoked cigar, ensure that you'll get quite a jab. Inexperienced shooters or faulty locks that fire both barrels in quick succession make recoil more memorable. Also, there's less fore-hand control with a double. While you can point surely and quickly, the double's slender forend and slick barrels won't match the checkered fore-stock of a bolt rifle for grip in absorbing recoil.

It does seem that sharp-shouldered magnum cartridges sometimes kick harder than they should. I think that's largely due to high bullet velocities—the sudden dumping of relatively modest loads of recoil. Combining heavy bullets with high pressures and big charges of powder behind abrupt shoulders makes rifles bounce violently. The .378 Weatherby and cartridges like it can bring on a flinch fast if the rifles don't wear brakes.

A muzzle brake reduces recoil by reducing jet effect at the muzzle. It bleeds gas pressure off through vents instead of letting it erupt suddenly with explosive force. In siphoning the gas, a brake also provides opposing surfaces fore and aft of each vent for the gas to push against. So not all the thrust of escaping gas is rearward. Brakes boost muzzle blast. Noisy rifles are easy to shoot at the bench, where you have ear protection. But you won't wear those muffs when you hunt. Shooting through a brake without them can ruin your hearing. Hunting

This Alaskan moose hunter is fond of his .375. Though it kicks hard, the chance of encounters with bears makes it a sensible choice.

guides, who often find themselves beside a hunter at the shot, by and large hate brakes. Angling vents forward helps reduce noise, though not by much.

A brake can also affect your shooting from low positions. Lying prone in the dust, you'll trigger a tornado that leaves dust on your scope lens and in your barrel, maybe in your eyes. At best, it will hang aloft long enough to obscure your view of the animal's reaction. Without wind it can hang there after the animal and your chance for a second shot are gone.

Brakes needn't be unsightly. Many are barrel-diameter, with unobtrusive vents. Winchester and Browning had to add bulk with their BOSS (Ballistic

The author shot this nilgai in Texas. These elusive animals are also very tough. The .300 Winchester Magnum with 180-grain Fail Safe bullet was a smarter choice than a milder load.

Optimizing Shooting System) device. Apparently a lot of hunters are willing to take recoil over noise, because now the BOSS comes without vents. (There's no change in profile.) Anyone ordering a brake is wise to ask for a cap to protect the barrel threads when the brake is off. A Mark X Mauser in my rack, barreled by Intermountain Arms, has a brake that's installed only finger-tight. At the bench, this .338 is docile but noisy. I replace the brake with a knurled cap for hunting. Not all barrels will shoot to the same point of impact with and without a brake. This one does.

A long barrel reduces felt recoil, partly because it delays the jet effect of powder gas at the muzzle, partly because it thrusts the blast farther from you. It also adds weight to the rifle's front end, counteracting muzzle jump—which means there's less lift to slam the rifle into your chops.

Another thing that keeps recoil from hurting you is a well-designed stock. That stock will have just enough drop at the comb to allow your cheek to settle your eye behind the sight. The comb will be essentially straight from front to rear (no more drop at the rearmost area of cheek contact). It will have a well-rounded top, just thin enough to permit perfect lateral alignment of your eye with the sight. Cast-off (the comb offset to the right) is typically an option only on custom rifles.

Early cartridge rifles from central Europe have "Tyrolean" cheekpieces with pronounced dishing.

They're comfortable if you're shooting deliberately but awkward on the trail and ill-suited for quick shots from unorthodox positions. Combs that slope up at the rear (such as on the Weatherby Mark Vs) seem to reduce damage to my cheek. Proper stock pitch (angle of butt to bore) is important too. Not only does it help you aim quickly; it determines the muzzle's arc after bullet exit. A soft buttpad of generous length and width slows and spreads recoil. It also protects the buttstock better than does a steel plate.

Shooting position and form affect felt recoil. When I was young I wondered why anyone would build a rifle with a crescent buttplate. The butts of old lever rifles dug into my shoulder like barbecue forks when I aimed at imaginary bears above the wood stove in the hardware store. I'd fired a handful of carbines so equipped, and they all hurt me. What I learned later was that people who shot them in the days when crescent butts were in vogue placed the butts on their upper arms, not tight against their tender necks. They bent their heads over the stocks and let the wood caress their cheeks instead of relying on the comb for cheek support. They would have found modern rifles awkward.

It's good form to keep your head erect and your eyes straight ahead when aiming so as to get the clearest sight picture. But for shooters with lots of space between chin and shoulder, there's a problem. We can't get full shoulder contact with the butt while holding our heads up and keeping our cheek in contact with the stock. The result: We move the butt up so not much more than the toe bears on our shoulder. All the recoil, then, lands on that small area, and it can make us wince. It is still better than bending over the stock and squinting up to find a spot of target with a section of reticle.

Recoil seems most severe in the prone position partly because you are normally shooting at an upward angle but also because your body has lots of ground contact and can't move easily. That means you don't act as a shock absorber to help the rifle decelerate; you stop it suddenly. Shooting uphill can remind you of recoil because it puts the scope closer to your noggin. Shooting sharply uphill from the prone position, you have an even chance of drawing more blood from your skull than from the animal.

You'll absorb recoil most comfortably from the offhand position. Your body can flex and rock, damping the jolt. But as with most good news, there's a catch: Flinching causes the worst shooting offhand because there's no brace for your torso to counteract the tug of suddenly-tensed muscles. That crosswire is gyrating wildly, and you're running short of breath trying to settle it. Your trigger finger is threatening to yank the shot just to get it over with. The sear is almost ready to drop. All at once your subconscious resurrects the jab you got after you reached this stage last time. Oh, my, did that hurt! Land sakes, we're not doing this again! Yikes! It's gonna git me! Now! NOW!

A straight stock and light-recoiling round like the .25–06 makes for greater precision downrange. And that can make up for a deficit in foot-pounds.

You may have called the shot wild. Be nakedly honest. If you start lying to yourself about shots, you'd best take up billiards. If you knew the crosswire was off-target, why did you turn the bullet loose? Well, sometimes, it just happens. Near the breaking point of the sear, you'll have trouble holding finger pressure as the reticle wobbles off target and increasing pressure as it comes back. The additional pressure needed to release the sear becomes less and less. You're chopping this onion finer and finer, and the rifle does not always fire just when you want it too. On the other hand, the bullet may have left because you hurried it out the door as your mind told you that this was going to hurt and you'd better get it over with.

On a recoil tolerance scale of gunshy to Godzilla, I'm a notch above wimp. My muscles are scattered in places that don't arrest buttpads. I'm painfully aware when I pull triggers that save for a slice of rubber, all that's between my clavicle and an aspirin-bottle load of IMR 4350 are brass, steel, and hardwood. People make hammers and mallets out of that stuff. Pounding such objects into my body with charges of gunpowder makes as much sense to me as sitting on stumps over dynamite. On the other hand, I like to put holes in things far away, and shooting is the best way to do that. So I work at controlling those muscles that bunch when they sense I'm taking up the last ounce. It isn't always easy.

Some rifles have a reputation for wicked kick. The Winchester 95 in .405 was one. Lots of power coupled with a low, sharp comb made this a molar-masher. Perhaps the most fearsome rifle in my rack is a Mauser in .458. Its stock is proportioned just right for the iron sights, and it has that lively feel of an upland bird gun. It doesn't weigh much more than a bird gun either, and that's a liability during recoil. You don't have to put up with it long; that rifle kicks as fast as lightning on skids. Three shots and my head throbs. My jaw comes out of numb after the fifth, and I start feeling it for cracks. When the sights again settle on the paper, I'd almost trade places with it. If this four-five-eight with the hooves of a mule didn't point like a Rizzini and stick 500-grain solids into chestnut-size groups, I might part with it. Someday maybe it will drop a buffalo and earn its place in the rack with a good story.

Sometimes you'll need quick repeat shots. Heavy recoil slows you down.

The worst thing about recoil is that it shuts down practice. Shooters who don't have fun don't shoot much. Leaning over a rifle on a table, you put your shoulder hard into the butt. The weight of your torso keeps you hard against it, so you absorb all the violence of a hard-kicking rifle. One way to make bench shooting more pleasant and keep flinching at bay is to use extra padding on your shoulder. In T-shirt weather folded bath towels work well because you can just drape them over your shoulder and shed them between strings on a hot day. You'll need a soft pad for your right elbow, too, or the bench top will skin it. An old sweatshirt works well here—better than small pads because you won't tweak your position or put tension in your arm to keep it on the pad.

If you don't think about recoil, it may bruise you, but it won't make you miss. If you think about it, you'll flinch. Repeated flinching is flinching well-practiced—a routine you don't want your body to learn.

PART III

Ballistics Afield

Extending the Rifle's Reach

Most big game is killed within 200 yards of the rifle, and most of it is shot from the side. So if you have a deer rifle chambered for a .308 or .30–06 or a derivative thereof, you have reach enough for the shots that kill game. It's fruitless to speculate on the piles of antlers you might collect if your rifle shot like a lazer to several hundred yards. You can't hit game that you can't see clearly, and you can't hit game if you can't steady the sight. Cartridges and bullets designed for long shooting offer marginal help if you're not an exceptional marksman.

Flat trajectory appeals to any shooter who has tried to hit at long range, because it makes accurate range estimation less critical. But a racehorse cartridge or a super-velocity load may not give you enough less drop to matter. For example, a standard 140-grain .270 bullet fired from a rifle with a 200-yard zero drops 7 inches at 300 and 20 inches at 400. A Federal High Energy or Hornady Light Magnum load will bring point of impact up just half an inch at 300 and 1½ inches at 400. So really, you'll have to hold the same as with the standard load at 300; and, unless you have a very steady rest, a super-accurate rifle, and a high-powered scope, the difference at 400 will also be indiscernible. Fire that same bullet from a .270 Weatherby Magnum (at 3300 fps instead of the standard 2940) and you'll still see 6 inches of drop at 300 yards, more than 16 at 400.

Long, blunt military bullets in cases of modest capacity heralded smokeless powder back in the 1890s. The 7x57 launched a 173-grain bullet, the .30–40 Krag a 220-grain, the .303 British a 215-grain, the 8x57 (then the 7.9x57) a 226-grain. All these bullets traveled at roughly 2200 fps from long barrels. When streets were still mostly dirt and our Civil War still sharp in collective memory, 2200 fps was pretty fast. No, it was *very* fast.

As with all things relative, fast didn't stay fast. When Germany announced a new 154-grain bullet for its 8x57 cartridge in 1905, it raised the standard. This missile had a pointed nose and left the muzzle at a scorching 2880 fps. The U.S., not to be outdone, scrapped the 220-grain .30–03 bullet (adopted from the Krag) and replaced it with a 150-grain spitzer. Velocity: 2700 fps. Both countries, incidentally, made other changes in their service cartridges at that time. Germany increased bullet diameter from .318 to .323, the present diameter of 8mm bullets. A "J" was used to designate the original round (J meaning I for infantry), and an "S" (for Spitzgeschoss; also JS or JRS) given to the new one. American ordnance people shortened the .30–03's case .07, then renamed it the "Ball Cartridge, Caliber .30, Model 1906."

Other nations quickly saw the advantages of pointed bullets at high speed. Flat trajectory made precise range estimation less critical and hitting easier. Soldiers discovered a bonus in reduced recoil. By the time the Archduke Ferdinand was felled by an assassin, most countries had rifles that far outstripped their sights. The .30–06 was claimed to be lethal at 4700 yards, significantly farther than most doughboys could keep a Springfield's bullets on the

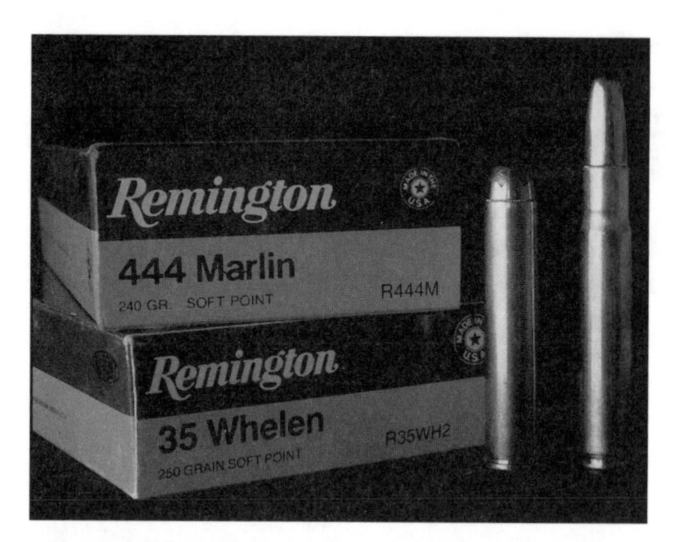

Tube magazines mandate the flatnose bullet of the .444. Its great weight and large diameter make the most of its limited range. Despite a rounded nose, the long bullet of the .35 Whelen flies flat enough for a 200-yard zero.

Most of the hunting bullets we use look like the bullets that Germany designed in 1905. They're pointed and sleek. Choosing between a blunt and a pointed bullet only gets hard when you want to boost bullet weight. At some point, you can't make a pointed bullet longer. Either the magazine won't let you (unless you bite deeply into your fuel space by seating deep) or the rifling twist won't stabilize the bullet (the longer and heavier a bullet, the sharper the spin needed to keep it flying point on). To add heft without adding length, then, you must round the bullet nose, trading aerodynamic shape for higher sectional density.

And what does that swap give you? For a given type of bullet, increasing sectional density gives you

side of a commercial granary. Still, like arrows loosed en masse at Agincourt, a volley of bullets raining on distant enemy positions was bound to be felt. *Effective* range and the range at which a soldier could hit an enemy soldier of his choosing were quite different!

Given advances in artillery, reach was important to generals trying to conserve their troops. So it was with consternation that in battle they found the 150-grain .30 Springfield bullet largely ineffective beyond 3400 yards. The upshot: U.S. ordnance changed the service bullet again, this time to a 173-grain spitzer with 9-degree boat-tail and long, 7-caliber ogive (radial curvature of the nose between shank and tip). Muzzle velocity dropped—but not much. The new "M1" bullet, first issued in 1925, left the barrel at 2647 fps. Its higher ballistic coefficient carried it an astounding 5500 yards.

Then, in 1939, the Army went back to a lighter bullet, a 152-grain spitzer designated the M2. The reason for the switch was the brutal recoil delivered by the M1 load. Even at 2805 fps, the 152-grain bullet was easier to shoot. Like the M1, it had a gilding metal (zinc and copper) jacket to prevent metal fouling. With the M2 we fought the second world war. International conversion to fully automatic infantry rifles later changed what soldiers wanted from bullets, so here is a good place to end this review.

The 7x57 cartridge dates to 1893, but modern loads in modern rifles make it a good choice for deer at all practical ranges.

more penetration and greater retained weight (not necessarily a higher *percentage* of original weight). The long blunt bullet was a logical tool early on, a holdover from a time when black powder and bullets with no jackets kept a lid on velocities. Nineteenth-century hunters who wanted to ratchet killing power up a notch went to a heavier bullet with enough powder to push it as fast as lighter bullets. Nose shape didn't matter because soldiers and hunters alike did their shooting at iron-sight distances. A blunt bullet carried well to 200 yards, and if you made it heavy enough, would penetrate the biggest game end to end.

The advent of smokeless powder did not change hunting bullet preferences overnight. While lightweight full-jacketed bullets could drop enemy soldiers, big game animals would run off if a smallbore spitzer sailed through without expanding. The unpredictable nature of early softpoints promised no better results. Insufficient expansion let some animals escape. Premature upset and fragmentation kept other bullets from reaching vitals. Long, heavy roundnose bullets thus remained popular, especially among hunters who plied the timber. Those who used lever-action rifles with tube magazines were compelled to stick with blunt bullet noses, to prevent one cartridge from setting off another ahead of it during recoil. But even in open country, spitzer bullets took some time to catch on. In the U.S., deer and elk hunters used surplus Krags and 220-grain bullets. Canadians after a winter's meat supply relied on 215-grain softpoints from Short Magazine Lee Enfields. Long, blunt bullets from the .303 and the 8x57, as well as from smaller rounds like the 6.5x54 and 7x57, killed game as big as elephants in Africa.

Between the world wars, effective shooting range was limited largely by sighting equipment. But in the last half-century, advances in optical sights have enabled hunters to aim precisely at long range. A blunt bullet suddenly became a handicap. So bullet companies worked hard to make pointed bullets more reliable in game. By 1947, When John Nosler developed his Partition bullet to get deeper penetration in moose, Winchester and Remington had already come up with softpoints that worked well in deer-size game. Now most popular hunting bullets have sharp noses. And the heaviest factory-loaded bullets aren't as heavy as they used to be. The .30–40 Krag, 8x57 Mauser, and .303 British are all commercially loaded with lighter bullets than they featured at the turn of the century. You can still buy 220-grain .30–06 ammo, but it is an anachronism among myriad loads featuring 150-, 165-, and 180-grain bullets with sleek noses. The heaviest roundnose bullets are still made by Barnes. There you can find a 195-grain 7mm, a 250-grain .308, a 300-grain .338. These are part of the extensive Barnes "Original" line, which gets less press than the Barnes X-Bullets but is still a deadly (and costly) bullet.

According to Hornady ballistician Dave Emary, roundnose bullets not only open reliably; they may track straighter during upset than spitzers. He notes that roundnose bullets are easier to design for expansion and produce more tissue damage in the first inches of penetration. "On balance, they're more accurate too," claims Dave. He adds quickly that any difference in accuracy between blunt and pointed hunting bullets would be very hard to detect in hunting-quality rifles.

If the blunt bullet has an edge in the accuracy department, it has to do with a relatively low length/weight ratio and a short radius on the leading edge of the shank. Square entry of the bullet into the rifling contributes to accuracy as well. "Pointed bullets can be temperamental," Dave says. "If you run into trouble getting spitzers to shoot, load up some roundnose bullets of the same weight. In my 8x57 Mausers, Hornady 170-grain roundnose bullets commonly group tighter than the 150-grain Spire Point." Like many hunters, he prefers flat-base bullets to boat-tails for big game, noting that the tapered heel trims bullet drop only at extreme range, when the bullet has slowed considerably. For example, if you start two 140-grain 7mm bullets—one flat-base and one boat-tail—at 2700 fps, both reach the 100-yard mark clocking about 2500 fps. At 200 yards, the boat-tail bullet is clipping along at 2320, just 15 fps faster than its flat-base counterpart. At 350 yards, the boat-tail bullet is leading by 35 fps. The difference in drop at that range amounts to only about half an inch. Expect the same difference in wind drift.

Big cartridge cases fueled by slow powders enable you to use heavier bullets than do smaller cases of the same neck diameter. For example, if you fire a 180-grain bullet from a 22-inch barrel in .300 Savage, you'll do well to get 2350 fps. With the larger .308 Winchester case, it's no trick to clock 2650 fps. A .30–06 will give you 2750 with the same bullet, a .300 Winchester 3050, a .300 Remington Ultra Mag 3250. The Ultra Mag will drive a 200-grain Nosler Partition at over 3000 fps, so is really better served with the heavier bullet. On the low end of the scale, you're

The author shot this heavy-antlered caribou with a Lex Webernick rifle in .260 Remington.

probably smart to load 150-grain bullets in the .300 Savage and 165s in the .308. Not only will the lighter bullets offer less resistance to the modest powder charge; their shorter shanks will enable you to seat them farther forward, boosting the usable powder capacity of the case. Result: a faster launch, flatter flight over normal hunting ranges. As for delivered energy, a gain in speed can more than offset a loss in mass. Within a limited range of bullet weights, faster, lighter bullets deliver more energy and flatter trajectory within the first 200 yards or so. A slower, heavier bullet of the same profile and loaded to the same pressure has the advantage at long range.

Everything worth anything comes at a price. Throat wear accelerates as you boost the speed of any bullet—a function of increases in friction and pressure—but to a greater degree as bore size shrinks relative to the powder charge. A 120-grain bullet is a long one for a 25-caliber cartridge. It's a short bullet in a 7x57. The advantages of bigger bore diameter include a bigger bullet base for the powder gas to push against. When the bullet starts to move, it releases pressure much more quickly than would a bullet of smaller diameter. That's why I can drive a 300-grain bullet at 2450 fps from my .411 Hawk, but can't match that speed with a 300-grain bullet from a .338 Hawk. To the shoulder, the cases are identical, with the same powder capacity. Giving up sectional density as you boost bullet diameter, you get a faster start.

Target shooters who compete at 1000 yards have long favored big cases with itty bitty mouths. The .300 Weatherby necked to 6.5mm had a following

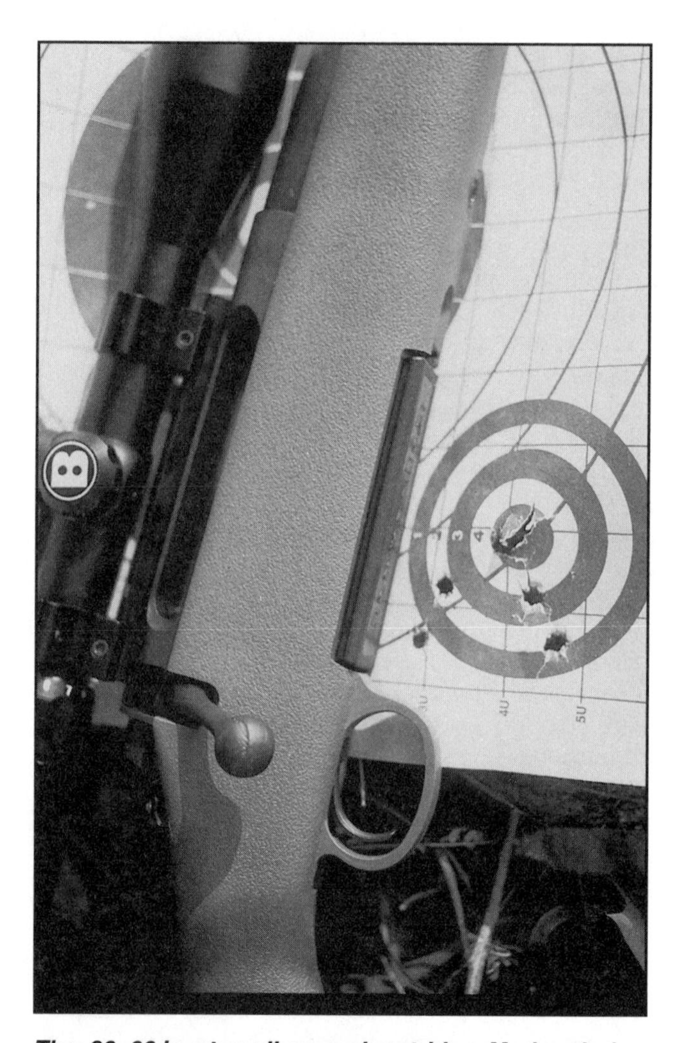

The .30–06 is a top all-around cartridge. Modern bullets, and loads by Federal and Hornady that boost velocity, extend its reach and effectiveness. It's a standard offering in rifles like this Remington 710.

for awhile. Recently, I came close to buying a Model 70 bull gun so chambered, only because the big, grossly "overbore" case has a compelling history. I demurred because rifle builder and fellow shooter Rick Freudenberg has me convinced that a match-grade 6.5/284 can put holes in the middle at long range without making my gums ache. Despite its small bore, it is an efficient round, relatively easy on throats.

But few hunters shoot at game 1000 yards off. Few, indeed can guarantee lethal hits beyond 300 under normal hunting conditions. So the flight of hunting bullets beyond 400 yards or so is really of academic interest. And that should influence the way we think about both bullets and cartridges.

Choosing a Big Game Cartridge

"Sit," I whispered. He did, resting the .300 across his knees.

The elk were less than 100 yards away, across a ravine and shielded by young aspens. We didn't have the wind. It was fishtailing, quartering from our rear. I hoped the bull would come.

He did. The next bugle blasted from the dark gut of the draw. Antler tips winked at us, and he strode into view at 35 yards. "Now!"

The magnum bellowed; the bull went to his knees. But in an instant he was up again. The hunter froze, transfixed by the great antlers—or perhaps astonished that one heavy Nosler in the neck had not been enough. "Hit 'im again!" I screeched. "Shoot, shoot, shoot!"

He did, but the bull had already turned. There were misses and bad hits as we raced across the ravine and up into the meadow on its far side. The Minnesotan was out of breath when the elk gave us one last chance. Again, the flat boom of his Winchester. This time, though, the bull dropped with finality.

When you've seen several dozen elk shot, you should come to some conclusions about elk cartridges. It seems to me that no matter what you shoot elk with, the elk dies or escapes depending on bullet placement. Bullet upset matters too, but not nearly as much.

In fact, hitting the proper spot is the key to clean kills whatever the game you're hunting. The size and speed of the bullet counts for little. I've seen only a few elephants shot, but enough to surmise that the principle holds true even on the biggest animals. Because the elephant has a thick hide and more massive bones than an elk, and because you typically need a long wound channel

A century ago, choosing a cartridge was easy. The .30–30 offered high-velocity (1960 fps!) with smokeless powder. Marlin's 336, its distinctive lever hinge shown here, is a popular .30–30 rifle.

and sometimes must make a frontal shot to the brain, you're better off with a solid bullet than with a softnose. Otherwise, the mandate is the same: Hit the vitals.

When I was young, a lot of deer hunters favored the .30–30. Progressive riflemen hunted with the .30–06 and .270. Now the .30–30 is seldom seen in open country. The .270 and .30–06 have become old hat, and bigger cartridges are popular. If you're hunting deer, though, you're still well equipped with a .270 or .30–06. Which is best? Let's assume that you'll limit bullet weight to 150 grains and that you want the best long-range performer.

The difference in diameter between 150-grain .277 and .308 bullets (actual measure) gives them different C values and flight characteristics, as shown in a PMC ammunition chart listing 150-grain boat-tail softpoint bullets for both rounds. Drop is calculated for a 200-yard zero.

	velocity, fps			*energy, ft-lb*			*drop, inches*	
	muz.	*200*	*400*	*muz.*	*200*	*400*	*300*	*400*
.270	2850	2477	2134	2705	2034	1516	−7.4	−21.4
.30–06	2900	2427	2000	2801	1961	1332	−7.7	−22.5

The .30–06 bullet starts 50 fps faster, but because it is shorter (lower ballistic coefficient), it loses that edge quickly. At 100 yards the bullets are neck-and-neck, traveling 2660 and 2657 fps. At 200 the .270 has a 50-fps lead, reversing the advantage held by the '06 at the muzzle. Double the distance, and the .270 more than doubles its lead. Energy figures reflect that too. But how much *real* difference is there? Not enough to notice. Both bullets have more than enough speed and energy to kill deer farther than you can aim accurately. The differences in drop—.3 at 300 yards and 1.1 at 400—are insignificant.

Cartridges at the top of the performance charts may also show negligible differences if you look closely. For example: The .257 and .270 Weatherby Magnums are both loaded by Norma with 100-grain pointed bullets. Here's how they stack up (a 300-yard zero this time):

	velocity, fps			*energy, ft-lb*			*drop, inches*
	muz.	*200*	*400*	*muz.*	*200*	*400*	*400*
.257	3602	3016	2500	2881	2019	1388	−7.7
.270	3760	3061	2462	3139	2081	1346	−7.6

The results are pretty much the same as in the previous comparison: The longer smallbore bullet overtakes its shorter, faster counterpart. In this case, though, the point of convergence comes not at 100 yards but farther downrange, at 300, where velocities are 2750 and 2751 fps. The .257 Magnum begins to pull away, but at 400 yards has only a 38-fps advantage; 400-yard drop is essentially the same. For deer hunting at long yardage, either round is deadly. For all-around big game hunting, however, there's little question the .270 is superior. First, it delivers higher velocity and more energy over normal hunting ranges. Second, like the .30–06 in our first example, it offers the advantage of handling bullets not available or practical in the smaller bore. The .257 Weatherby Magnum is most competitive with 87-grain bullets, by most standards not heavy enough for reliable killing of elk, moose, and big bears.

Some riflemen prefer lighter bullets for their mild recoil. Recoil energy depends on the weight of the rifle and the weight and velocity of the ejecta (the bullet and any powder that leaves the case in solid form). In pairs of cartridges differing only in bullet diameter, recoil is identical or nearly so, given similar rifles. If you've shot a .300 Weatherby with a 200-grain bullet, you know about what it's like to shoot a .340 Weatherby with a 200-grain bullet. Here's how the two stack up ballistically (300-yard zero):

	velocity, fps			*energy, ft-lb*			*drop, inches*
	muz.	*200*	*400*	*muz.*	*200*	*400*	*400*
.300	3060	2668	2308	4158	3161	2366	−9.8
.340	3221	2688	2213	4607	3208	2174	−9.9

In this case, the smallbore bullet overcomes a significant deficit quickly, to catch the bigger slug just past the 200-yard mark. At 400 yards it has built a lead of nearly 100 fps and an energy advantage of almost 200 ft-lbs. Drop at 400 yards is the same. If you want to use a 200-grain bullet for long shots at big game, you'd do well with either cartridge. The .300 gives you the option of switching to *lighter* bullets.

Choosing among cartridges by the numbers, you have to watch for pitfalls. Ballistic performance in catalogs isn't always what you'll get from your rifle. Even when a manufacturer is scrupulously honest, his ballistics charts can be suspect. Bullet speed varies with barrel length and bore dimensions, bullet type, test temperatures, elevation, even the rifle's support (you'll lose a few feet per second taking a rifle from a machine rest and firing it from the shoulder). The most potent of big game cartridges differ so little in their performance, one from another, that you may find as much variation in the cartridge you choose when you switch bullets or the mercury drops to zero on opening day!

Not all factory loads give the cartridge a workout, or show its potential. Some rounds, like the 7x57 Mauser, were developed when rifles were not as strong as they are now. For decades the 7x57 was factory-loaded to the modest pressures dictated by these early rifles. Only recently have munitions companies offered modern-rifle loads for this 1893 veteran and more ancient cartridges like the .45–70. Among late-vintage sleepers is the .264 Winchester Magnum. If you don't think this round will kick 140-grain bullets out the muzzle faster than the listed 3030 fps, you're saying that the .264 case is about the same size as the .270's. Not so! The .264 Magnum has much more capacity, and you can drive 140-grain bullets from 26-inch barrels at well over 3200 fps without blowing any corks. I'm afraid I can't say why the factory loads are so anemic, or why catalog charts are so hard on this greyhound. When the .264 was introduced in 1958, 140-grain bullets were listed at 3200 fps, 100s at 3700.

Some cartridges that look similar show big differences in ballistics charts. The 7mm Remington and 7mm Weatherby Magnums are essentially the

Jim Nyce is pleased with this caribou, killed with a .30–06 Ultra Light rifle at 200 yards. He hunted in the Northwest Territories with Barry Taylor's Arctic Safaris.

Jim Morey took this mule deer with a Dakota rifle in central Wyoming. Don Allen's Dakota line of cartridges is based on the rimless .404 Jeffery. Most fit .30–06-length actions.

Light-recoiling cartridges like the .243, .260, 7x57 and 7mm-08 have plenty of reach for long shots at whitetails and work just as well in the woods.

same dimensions, but the Weatherby produces higher numbers. Reason: It is loaded by Norma, a Swedish company that need not hew to pressure standards published by SAAMI and adhered to by most U.S. ammo firms. Also, Weatherby Mark V rifles have longer throats that most of the rifles chambered in 7mm Remington Magnum. Roy Weatherby gave his rifles long throats to keep pressures in check when he loaded hot to boost bullet speed. In long-throated rifles, the 7mm Remington can be handloaded to perform with the 7mm Weatherby Magnum (Hornady Light Magnum fodder comes close). Norma's Weatherby ammo is loaded to take advantage of long throating. Weatherby cartridges loaded by Remington and Federal are not so potent, but neither do they get sticky in the chamber when you leave the ammunition in a Jeep or Land Rover under the hot sun, then load it in a short-throated rifle. Federal High Energy cartridges, charged with powder in two stages, give you the most horsepower. While I've not run pressure tests with this ammo, Federal people tell me that High Energy rounds all fall within SAAMI specs for the cartridge. The point is to be aware that throat length influences breech pressure, and cartridges loaded for rifles with long throats may be too rambunctious when the bullet starts close to the lands.

Ballistic performance isn't the only thing about a cartridge that matters. Recently I fired a high-performance cartridge in a new rifle. Chronograph readings stayed right on target. However, the bullets didn't. My first 100-yard group measured nearly 5 inches! If you wanted a flat-shooting, hard-hitting big game cartridge, you'd think after looking at the charts that this cartridge was a great choice. Sadly, the rifle and factory-loaded ammo I tested were not compatible. More serious problems may have been affecting the accuracy as well. Speed and power are of no account if you can't direct the bullet.

Finally, there's the issue of "shootability." Ballistic muscle is mainly a function of the cartridge. Bench accuracy depends on the cartridge and rifle. Hunting accuracy depends on your equipment and how you employ it. Making the rifle do what you want it to is important. A "shootable" rifle is proportioned so it is easy to hold, with a trigger that allows you to fire the shot without disturbing the sight picture. A shootable cartridge is one that doesn't beat your gums blue on recoil. If the rifle is poorly designed or the cartridge a brute that makes you flinch, nothing else matters. To kill game, you must hit it without the help of a bench.

Big elk can be taken with lightweight bullets through the ribs, but breaking the shoulder on quartering shots requires strong, heavy slugs. Hence the popularity of the .300 and .338 Winchester magnums.

The Deadliest Big Game Cartridges

A lot of hunters pick their cartridges and loads by comparing numbers in ballistics charts. Here are the current champions, by bore diameter, with popular runners-up for comparison:

cartridge, intro. date	*bullet weight and type (ammo mfr.)*	*velocity (fps)*		*energy (ft-lbs)*	
		muzzle	*400 yards*	*muzzle*	*400 yards*
.243 Win., 1955	100 Power Point (Win)	2960	1993	1945	882
.243 Win., 1955	100 Power Point + (Win)	3090	2092	2121	972
.243 Win., 1955	100 BTSP Lt. Mag. (Hor)	3100	2138	2133	1014
6mm Rem. (.244), 1955	100 PSP C-L (Rem)	3100	2183	2134	1058
6mm Rem. (.244), 1955	100 BTSP Lt. Mag. (Hor)	3250	2311	2345	1186
.240 Wby. Mag., 1968	100 Nos. Partition (Wby)	3406	2415	2576	1294
.257 Roberts, 1934	120 Nos. Partition (Fed)	2780	1970	2060	1030
.257 Roberts, 1934	117 BTSP Lt. Mag. (Hor)	2940	2031	2245	1071
.25–06 Rem., 1969	115 Ballis. Silvertip (Win)	3060	2188	2391	1223
.25–06 Rem., 1969	117 BTSP Lt. Mag. (Hor)	3110	2168	2512	1220
.257 Wby. Mag., 1945	115 Barnes X (Wby)	3400	2504	2952	1601
6.53 Laz. Scramjet, 1995	120 Nos. Partition (Laz)	3550	2694	3219	1854
6.5x55 Swedish, 1894	140 Hi-Shok (Fed)	2600	1860	2100	1080
6.5x55 Swedish, 1894	129 SP Lt. Mag. (Hor)	2750	1995	2166	1139
.260 Rem., 1997	120 Ballis. Tip (Rem)	2890	2131	2226	1210
.264 Win. Mag., 1959	140 Power Point (Win)	3030	2114	2854	1389
.270 Win., 1925	130 Power Point (Win)	3060	2110	2702	1285
.270 Win., 1925	130 Power Point + (Win)	3150	2161	2865	1348
.270 Win., 1925	130 SST Lt. Mag. (Hor)	3215	2384	2983	1640
.270 Wby. Mag., 1945	130 Nos. Partition (Wby)	3375	2458	3288	1744
7mm-08 Rem., 1980	140 PSP C-L (Rem)	2860	2094	2542	1363
7mm-08 Rem., 1980	140 TB High E. (Fed)	2950	1900	2705	1120
.280 Rem., 1957	140 PSP C-L (Rem)	3000	2102	2797	1373
.280 Rem., 1957	140 TB High E. (Fed)	3150	2050	3085	1310

7mm Rem. Mag., 1963	140 PSP C-L (Rem)	3175	2243	3133	1564
7mm Rem. Mag., 1963	139 SST Lt. Mag. (Hor)	3250	2475	3259	1890
7mm Dakota, 1993	140 Barnes X (handload)	3300	2417	3385	1816
7STW, 1989	150 Power Point (Win)	3280	2314	3583	1783
7mm Wby. Mag., 1945	140 Nos. Partition (Wby)	3303	2434	3391	1841
7.21 Laz Tomahawk, 2000	140 Nos Partition (Laz)	3379	2598	3550	2100
7mm Ultra Mag, 2000	140 Nos. Partition (Rem)	3425	2534	3646	1995
7.21 Laz. Firebird, 2000	140 Nos. Partition (Laz)	3750	2905	4372	2625
.30–06 Springfield, 1906	180 SP C-L (Rem)	2700	1466	2913	859
.30–06 Springfield, 1906	180 Bronze Point (Rem)	2700	1899	2913	1441
.30–06 Springfield, 1906	180 Power Point + (Win)	2770	1997	3068	1594
.30–06 Springfield, 1906	180 TB High E. (Fed)	2880	1940	3315	1505
.300 H&H Mag., 1925	180 Fail Safe (Win)	2880	1952	3315	1523
.300 Win. Mag., 1963	180 Fail Safe (Win)	2960	2110	3503	1780
.300 Win. Mag., 1963	180 Power Point + (Win)	3070	2236	3768	1999
.300 Win. Mag., 1963	180 SP Hvy. Mag. (Hor)	3100	2275	3840	2068
.300 Win. Short Mag, 2000 180	Fail Safe (Win)	2970	2120	3527	1804
.300 Dakota, 1993	180 Barnes X (handload)	3150	2409	3967	2319
7.82 Laz. Patriot, 1999	180 Nos. Partition (Laz.)	3184	2493	4052	2485
.300 Wby. Mag., 1945	180 PSP C-L (Rem)	3120	2181	3890	1902
.300 Wby. Mag., 1945	180 Nos. Partition (Fed)	3190	2400	4055	2305
.300 Wby. Mag., 1945	180 Nos. Partition (Wby)	3240	2449	4195	2396
.300 Wby. Mag., 1945	180 TB High E. (Fed)	3330	2410	4430	2320
.300 Ultra Mag., 1998	180 Swift Scirocco (Rem)	3250	2495	4221	2487
.30-.378 Wby., 1997	180 Barnes X (Wby)	3450	2678	4757	2865
7.82 Laz.Warbird, 1995	180 Nos. Partition (Laz)	3550	2810	5038	3157
.338 Win. Mag., 1958	250 Partition Gold (Win)	2650	1960	3899	2134
.338 Win. Mag., 1958	250 N. Part. High E (Fed)	2800	2080	4350	2395
8.59 Laz. Galaxy, 2000	250 Nos. Partition (Laz)	2761	2128	4232	2515
.330 Dakota, 1993	250 Barnes X (handload)	2775	2110	4250	2462
.340 Wby. Mag., 1962	250 Nos. Partition (Wby)	2941	2197	4801	2678
.338-.378 Wby Mag., 1998	250 Nos. Partition (Wby)	3060	2297	5197	2927
8.59 Laz. Titan, 1995	250 Nos. Partition (Laz)	3150	2494	5510	3453

You'll notice from this sampling (it is not a comprehensive list of loads) that performance hinges on more than the cartridge name. For example, a 6mm Remington standard load generates 150 fps less speed than a Hornady Light Magnum load for the 6mm Remington. A similar discrepancy shows up with the .243 Winchester. Power Point Plus loads for the .243 and other cartridges in the list give you more muscle than the standard load, but not as much as the Light Magnum or Federal High Energy recipes. Muzzle data for any cartridge firing a certain bullet weight used to be pretty much the same, no matter the make. Not so now. Besides variations in bullet shape, there are significant differences in fuel charges. On the other hand, the high-octane loads don't always yield a noticeable advantage downrange. For instance, Hornady's Light Magnum .25–06 cartridge pushes a 117-grain bullet only 50 fps faster than the standard Winchester load

The Swedish 6.5x55 was among the first smokeless cartridges. Though small by today's standards, it still accounts for more Scandinavian moose than any other cartridge.

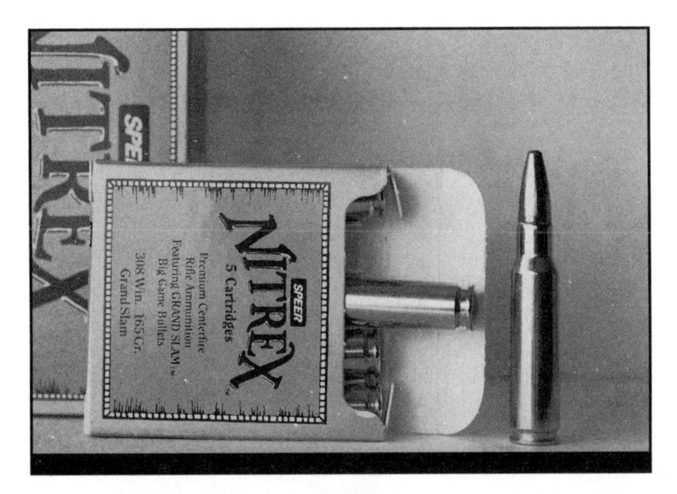

The .308 Winchester, developed as the 7.62x51 NATO in the 1950s, shoots flat enough and hits hard enough for any North American big game. It has modest recoil and, with the right loads, fine accuracy.

launches a 115. At 400 yards, the lead has been reversed, Winchester's bullet edging the Hornady by 20 fps. Energy at that distance is essentially the same. Bullet profile seems the most telling difference here, overcoming the slightly greater sectional density of the heavier bullet. Also at work is the greater rate of deceleration of the faster bullet, as it meets greater air resistance.

Another thing to note is that lightweight bullets driven faster than heavier bullets can produce and retain higher levels of energy, provided they have a streamlined shape. The 6.5x55 Swede, with its 129-grain and 140-grain bullets, shows this very well. The lighter bullet from the Hornady Light Magnum load gives you not only a flatter arc, but roughly 60 foot-pounds more smash, both at the muzzle and at 400 yards. Then again, the Light Magnum Load for the .257 Roberts squanders much of its 160-fps advantage. By the time that 117-grain bullet reaches 400 yards, it is moving only about 60 fps faster than the 120 Nosler that came out of the gate at a lope.

The difference in energy is an insignificant 40 foot-pounds.

You can harness the energy of a magnum's additional powder to kick a bullet faster, or to drive a heavier bullet. For example, the .300 Winchester Magnum will push a 180-grain bullet about 3000 fps, as factory loaded. The .30–06 gets less than 2700 fps with the same bullet. The Magnum delivers 3600 foot-pounds at the muzzle, compared to 2900 for the .30–06. But the .300's real advantage isn't a harder hit; it's the flatter flight that makes distant targets easier to hit, and the high terminal speed that opens the bullet more violently. The king-size dose of energy *released in bullet expansion and fast travel through tissue* results in more destruction around the bullet channel, presumably with no appreciable loss of penetration.

If you'd rather use the Magnum's additional fuel to launch a heavier bullet, you lose the advantage of flatter flight. Assuming the light and heavy bullets are the same shape, the heavy one will hold up better at extended range because it has a higher sectional density, which boosts ballistic coefficient. At normal hunting ranges, the trajectory of a heavy bullet will be close to that of a light bullet started at the same speed. For instance, factory-loaded 165-grain .30–06 bullets exit the rifle at 2800 fps, about the same as 200-grain bullets from a .300 Winchester Magnum. Energy of the 165-grain bullet: 2870 foot-pounds. The 200-grain bullet delivers 3480, about 600 more. Zeroed at 200 yards, these rifles will put the two bullets within one-tenth vertical inch of each other at 300 and 400 yards! The advantages of the heavier bullet: deep, reliable penetration through big muscles and bones.

Some hunters talk about weight as if it were the most important variable among bullets. It does not always matter that much. To show how small a difference bullet weight can make in killing game, here's a look at two Nosler Partition bullets loaded by Remington in its .300 Ultra Mag (250-yard zero).

	velocity, fps			*energy, ft-lb*			*drop, inches*	
	muz.	*200*	*400*	*muz.*	*200*	*400*	*300*	*400*
180-grain	3250	2834	2454	4221	3201	2407	−3.0	−12.7
200-grain	3025	2636	2279	4063	3086	2308	−3.4	−14.6

Which would you choose? Either would work fine on the big game for which this cartridge was designed. The relatively even rate of velocity loss shows these bullets are close in ballistic coefficient (.474 to .481). There's no noticeable difference in bullet drop to 300 yards, and the 2-inch disparity at 400 is of no account when you're shooting at a moose. From the Ultra Mag case, these two bullets perform about the same, the 200-grain bullet offering a little extra weight to counter the higher impact speed of the 180. You'll work hard to show any real difference in the effectiveness of these bullets.

Another comparison worth making is between bullets of the same weight but of different nose shape. For example, Winchester has two 150-grain .308 loads that clock 2900 fps at the muzzle. Disparities in nose profile show up downrange (200-yard zero):

	velocity, fps			*energy, ft-lb*			*drop, inches*	
	muz.	*200*	*400*	*muz.*	*200*	*400*	*300*	*400*
Power-Pt. Plus	2900	2241	1678	2802	1672	938	−8.9	−27.0
Partition Gold	2900	2405	1962	2802	1972	1282	−7.8	−22.9

You might choose a Partition Gold bullet over a Power Point on the basis of this chart—and forget that most big game is shot closer than 200 yards, where either bullet would produce a quick kill. Accuracy also matters, and maybe the Power Point Plus load shoots better in your rifle. I once chose a Power Point bullet over a Nosler Partition for an elk hunt simply because the Power Point gave me exceptionally tight groups. I killed a bull with one shot. If you expect fast, close shots at tough game in timber you'll want the Partition—for its penetrating qualities, not because of its superior flight characteristics.

How tall a cartridge stands no longer gives you even a ballpark idea of its performance, because there's been a move lately to the .404 Jeffery case, which is larger in diameter than the .30–06 and .308 cases and the belted hulls based on the old .375 H&H Magnum. The .300 Dakota, a useful but undersung cartridge for North American big game hunting, generates about the same muzzle speed as Federal's load for the longer .300 Weatherby Magnum. The abbrieviated 7.82 Lazzeroni Patriot and .300 Winchester Short Magnum match the ballistic muscle of the longer .300 Winchester Magnum—just as that "short" magnum set new performance standards when it was fashioned from the full-length .300 Holland hull in 1963. A short powder column is considered by benchrest shooters to be more desirable than a long one, because there's more immediate and more uniform ignition in the short case. Accuracy is generally better and pressures lower for the same charge and case volume because less powder starts through the neck as ejecta. Benchrest shooters have proven that a stubby case works better, and hunters are getting more and more opportunities to try this design.

Note that Remington's .300 Ultra Mag performs about the same as Weatherby's load for its .300 Magnum, despite the fact that the Ultra Mag has 13 percent more case capacity. Again, loading has to do not just with how much the hull will accept, but how much pressure is acceptable. As capacity increases, efficiency drops. The .30–378 Weatherby Magnum and 7.82 Lazzeroni Warbird, for instance, offer little more punch than the .300 Weatherby and .300 Ultra Mag, at considerable cost. You'll use over 100 grains of powder to fuel either of these cartridges (as much as 120 grains of the slowest propellants). In a .300 Weatherby or a .300 Dakota, you'll get about the same velocity with 20 percent less powder.

The Knockdown Myth

In Sweden, where thousands of moose are taken each year, sportsmen favor the old 6.5x55 cartridge over more potent rounds. From what I could tell, the .30–06 is next in popularity, then the .308 Winchester. No need to endure the pummeling of a .30 or .33 magnum, or to rattle your brain with the blast from a brake. Nor are huge scopes necessary where most game lives. On a short hunt in Sweden once, I used a .30–06 with 2X Aimpoint sight to shoot two moose. Both animals dropped to one bullet. So did an Alaskan moose I shot a few years ago. As a guide, I've watched hunters tip over elk with lightweight bullets from .270-class rifles. I've also seen elk run after being hammered repeatedly with bullets that dumped frightful doses of energy. The difference: bullet placement. Only after you've mastered shooting technique can you expect better results with a turbo-charged round that shoots flatter than the curve of the earth and delivers a ton of energy in the next time zone.

The notion that your bullet must carry a certain level of kinetic energy to kill big game is flawed. I read once about a couple of fellows who, on a dare, killed an elephant with a .22 rimfire. That is, they downed this huge beast by swatting it with less than 140 foot-pounds of energy. A stunt? No doubt. But it points out the absurdity of measuring lethality in foot-pounds.

Sometimes, you say, you can't insert a bullet like a needle in a vein and you need extra energy to knock an animal down. Well, the .458 Winchester Magnum, a popular elephant round, turns up about 5000 foot-pounds at the muzzle. The bullet energy-to-animal weight ratio of a .458 to an elephant is about the same as that of a .22 Long Rifle to a big whitetail deer. If a .458 can be counted on to stop an elephant, why doesn't the .22 deck big deer?

Few sportsmen these days hunt elephants; those who do select solid bullets for deep penetration. This tusker dropped to a 500-grain solid from a D'Arcy Echols rifle in .458 Lott.

I'll recount here a clever experiment one fellow devised to test knockdown power. He fashioned

A 140-grain Nosler Ballistic Tip from a .270 collected this deer at 125 yards. The deer fell because its vital organs stopped working. The .270 has been a top seller since its introduction in 1925.

handles for a thick steel plate so he could support it in front of him. It was a big plate, and to help hold the weight, he rested its edge on the edge of a table. Then, from close range, a friend fired into the plate with a 500-grain softnose from a .458. Despite previous firings to ensure that the bullet would not penetrate, this exercise held some risk; it is not a safe practice to shoot toward anyone! But the results are insightful. The bullet spent its 2½ tons of kinetic energy on the plate without knocking the man down!

Why? Because when it comes to bullets, there's no such thing as knock-down power. Animals fall down when they are fatally stricken—either weakened by blood loss or denied the function of a vital organ. A bullet can appear to slap the animal to earth when it hits a supporting bone or destroys nerves that control the legs. A dump truck can knock down a deer, but a bullet doesn't kill in the same way.

The reaction of game to a hit, like the lethality of any hit, depends on placement. It also depends on what the bullet does. Boosting power (energy) increases the likelihood of a quick kill only if the bullet can translate that energy into the destruction of vital organs. More power can mean deeper penetration (a longer wound channel) or more violent expansion (a wider wound channel) or both. You may get more value from one than from the other.

The notion that very fast bullets kill quicker has some merit. High velocity means more violent bullet expansion, all else equal. Quick energy release can be instantly lethal, provided you get adequate penetration. The results of explosive bullet action on small animals are often used to argue the case for high-speed bullets—in other words, for a bomb-load of energy dumped all at once. But you wouldn't want to fire a rifle that promised big game kills as

The light-recoiling, flat-shooting .270 Winchester excels as a mule deer cartridge but is also popular with elk hunters. Quick-opening bullets kill deer like lightning; Fail Safes were designed to penetrate.

spectacular as those delivered to pint-size varmints by a .22–250. Besides, on big game we save the meat.

One whitetail I shot popped into view 80 yards away in a woodlot. I fired as soon as I saw rib, and called a good hit. But the buck ran off as if not hit at all. I eased over to where the deer had stood and found nothing but a faint scuff-mark. No blood. No hair. Carefully, I circled the scuff-mark. Eventually I found a bit of lung tissue the size of a BB. Tracks put me on an exit trail. On the third try I caught the buck's right-angle turn and found blood. Shortly, I came upon my prize, dead. The 150-grain .300 Savage bullet had penetrated both lungs.

Was this a quick kill? Yes. The buck had died in mid-stride only seconds after the hit. Would it have been a quicker kill with a faster bullet? Maybe. In my experience, explosive bullet action in the vitals can drop an animal that might stagger off or run when hit in the same place by a slower or less frangible bullet. But the outcome is the same: an animal soon dead.

I borrowed a .30–30 rifle once to hunt blacktail deer. As luck would have it, I spied a dandy four-point as it bounced from a blackberry thicket only 30 yards off. He gave me four shots loping across my front. I led him too much at first, then corrected and hit his shoulder. My last bullet caught him in the rear ribs as he cleared a deadfall going away. He lay dead perhaps 40 yards beyond. A more powerful cartridge, especially one firing a fast-opening bullet, might have put the skids under that buck with the first hit. The second might have sent it into a somersault. But those pokey .30–30 bullets still did the job.

I think the allure of powerful cartridges has to do, at least in part, with the perverse notion that recoil tests manliness. That is, if you shoot a big cartridge, you're more of a male than someone who shoots a small cartridge. As Jack O'Connor might have said, this is the purest of applesauce. You can set a very powerful rifle against a sack of rolled oats and shoot it comfortably all day long. What does this say about rolled oats? Shooting is not about absorbing recoil; it's about hitting a target. There are, of course, legitimate reasons for more powerful rounds. But you might do some math before selling that '06.

Despite a strong market for high-performance big game rounds, there's a counter-culture building that may give us more options like the Winchester .300 Short Magnum and John Lazzeroni's short line: the 7mm Tomahawk, .308 Patriot, and .338 Galaxy. There's nothing at all wrong with the proven .270 and 7mm Weatherby Magnums, the 7mm Remington Magnum, or the newer 7mm Dakota. Most shooters can learn to tolerate kick from the .300 Dakota and .300 Winchester Magnum. You may agree with me that the .300 Weatherby, .338 Winchester, and .330 Dakota with heavy bullets give you all the pounding you thought you didn't deserve. And if you've shot very much, you'll concur that recoil affects your own ability to shoot accurately.

Without accuracy, bullet speed and energy accomplish as much as a locomotive jumping the rails.

Elk hunters have come to demand flat flight and lots of energy from their bullets. Big cartridges like the .30–378 Weatherby can reach far, but they can't knock an elk down.

The two Nosler Partition bullets at left lost front lead during penetration, but those nose fragments caused lethal damage inside, and the heel drove deep. The bullet at right lost its core: not good.

What About the .30–06?

My friend found the bed early on opening day, in fresh snow. The dung was steaming. He took the track, wisely keeping his eyes on the timber ahead, letting his peripheral vision follow the trail. In short order he was rewarded with a glimpse of elk hide. Largely hidden by brush, the animal had no help from the wind. Vern couldn't see antlers through the scope, but inspection with the binocular revealed a tine. He eased the rifle back up, found a patch of shoulder with the crosswire and triggered a shot. The bull sped away, denying him another chance. Calmly he chambered a second cartridge, one he knew he would not need. His .30–06 had already killed this elk.

Some hunters say the .30–06 lacks power for sure kills on elk, moose, and big bears. My friend disagrees. Me too. The .30–06 is not only a versatile round, it goes elk hunting more than any other cartridge. Professional hunters in Africa have said they'd rather see a .30–06 in a hunter's duffle than a .300 magnum, because it kills big animals quickly without savaging the hunter. Hunters who shoot .30–06 rifles are less apt to flinch, so they put their bullets in the vitals. On a safari in 1911, author Stewart Edward White claimed killing kongoni at 566 and 638 paces to test his Springfield with military spitzer bullets. His bag on that trip: 185 animals, almost all taken with the '06.

The .30–06 was conceived in 1900, when U.S. Ordnance engineers at Springfield Armory also

The .30–06 remains a top choice for deer and elk among knowledgeable shooters.

began work on a battle rifle to shoot it. The .30–40 Krag was then our official military round and had served ably in the Spanish-American War. But Paul Mauser's new bolt rifles were upstaging the Krag-Jorgensen, which was also a costly rifle to build. And rimless cartridges were proving superior to the rimmed .30–40.

Springfield Armory fielded a prototype rifle in 1901, and just two years later the Model 1903 Springfield was in production. Its 30-caliber rimless cartridge headspaced on the shoulder, like the 8x57 Mauser. The .30–03 was longer than both this German round and the .30–40 Krag. Powder capacity and operating pressure exceeded the Krag's. The .30–03's 220-grain bullet at 2300 fps was the ballistic equivalent of the 8x57's 236-grain bullet at 2125.

Keen to extend the reach of the 8mm, Germany came up with a new load a year after the .30–03 appeared. A 154-grain spitzer at 2800 fps shot flatter than other infantry bullets of the day. The Americans were obliged to catch up, and U.S. Army Ordnance was quick to introduce the "Ball Cartridge, Caliber .30, Model 1906." It drove a 150-grain bullet at 2700 fps, boosting effective range. The case could have been left unchanged; however, someone decided to shorten it .07, to .494. Consequently, all .30–03 chambers were a tad long for the new round. Soon all .30–03 rifles were recalled and rechambered to .30–06.

The first .30–06 bullets were jacketed with an alloy of 85 percent copper, 15 percent nickel. Satisfactory in the Krag, these jackets did not hold up at .30–06 speeds, and severe bore fouling resulted. Tin plating didn't work either; but a bullet jacket comprising zinc and copper in 5–95 or 10–90 proportions reduced fouling. These alloys became known as gilding metal, which would also be used in sporting ammunition.

The high velocity of the 150-grain bullet was supposed to give it a maximum range of 4700 yards. But troops in World War I found the real limit was nearer 3400. To increase reach, the Army again changed the load, incorporating a 173-grain spitzer with a 7-caliber ogive and 9-degree boat-tail. Muzzle velocity was trimmed to 2647 fps, a minor concession given the substantial increases in sectional density and ballistic coefficient. Introduced in 1925, the new "M1" round extended maximum range to an impressive 5500 yards. Even armchair ballisticians can spot this as an arbitrary figure. In battle, most debilitating hits come at much shorter ranges. Still, development of a 30-caliber load with 3-mile reach had to affect morale on both sides of the trenches.

Oddly enough, the Army saw fit to change bullets again. Apparently the 173-grain bullet gave soldiers a bit too much recoil, so in 1939 it yielded to a 152-grain replacement at 2805 fps. With the "M2" .30–06 cartridge the U.S. fought World War II.

After the war, surplus military '06 ammo sold cheap. But using it entailed some risk. Corrosive primers deposited salts in the bore, causing rust. Though Remington developed non-corrosive "Kleanbore" priming in 1927 (and commercial rounds featured non-corrosive priming exclusively from about 1930), military cartridges had the corrosive FA 70 primers as late as 1952. Since the Korean War, the only domestic .30–06 ammunition with harmful priming was a run of Western Match cartridges with Western "8 one-half G" caps. These were both corrosive and mercuric.

Factory hunting loads for the .30–06 include bullets weighing from 125 to 220 grains. Hornady offers "Light Magnum" loads with 150- and 180-grain bullets that fly about 180 fps faster than bullets from standard loads, without exceeding allowable pressures. Federal's High Energy ammunition gives you the same punch. Among myriad '06 loads available are several with bullets designed for tough game—bullets like the Nosler Partition and Partition Gold, Winchester Fail Safe, Barnes X, Swift A-Frame, and Trophy Bonded. You can buy .30–06 ammo in more

The author killed this Alaskan moose with one shot at 160 yards from an iron-sighted Springfield '03 rifle in .30–06. The bullet was a Trophy Bonded Bear Claw.

out-of-the-way places than perhaps any other kind of sporting rifle ammunition.

By virtue of its service record and a wash of surplus rifles following the Great War, the .30–06 overtook the .30–30 as America's big game cartridge. The '06 became the archetype bolt-rifle round for big game, earning more plaudits as it became available in the Winchester Model 54 and Remington 30S rifles, and later in the Model 70 and 721. Now it is as widely chambered as any sporting cartridge, worldwide—though lately hunters have become enamored of bigger rounds. Here is how the '06 compares with a couple of popular magnum cartridges, in terms of factory-listed velocity and energy. I used factory data for Winchester's 160- and 165-grain Silvertip Boattails, and 180-grain Fail-Safes.

cartridge, bullet weight (grains)			*muzzle*	*100*	*200*	*300*	*400*
.30–06	165	velocity	2800	2597	2402	2216	2038
		energy	2873	2421	2114	1719	1522
7mm Rem Mag	160	velocity	2950	2745	2550	2363	2184
		energy	3093	2697	2311	1984	1694
.30–06	180	velocity	2700	2486	2283	2089	1904
		energy	2914	2472	2083	1744	1450
.300 Win Mag	180	velocity	2960	2732	2514	2307	2110
		energy	3503	2983	2528	2129	1780

In terms of kinetic energy, the 7mm Remington Magnum has only a slight edge on the .30–06. Winchester's .300 delivers at 400 yards the energy an '06 carries only to about 300. Given a 200-yard zero, the 7mm Magnum bullet sags 7.2 and 20.6 inches at 300 and 400 steps, compared to 8.2 and 23.4 inches for the 165-grain .30–06 bullet. At 300 yards the 180-grain bullet from the .300 Magnum dips 7.1 inches, while the same bullet from a .30–06 falls 8.7 inches. At 400 paces the .300 dominates: 20.7 to 25.5 inches of drop. To at least 300 yards, you can hold the same with a .30–06 as with these two popular magnums and expect a kill. At 400 you'll have to shade a trifle higher.

It's useful to note that the .300 Winchester pushes 180-grain bullets about as fast as the 7mm Remington drives 160s—and that with 175-grain bullets the 7mm Remington Magnum actually generates *less* speed and energy than the Hornady Light Magnum or Federal High Energy 180-grain .30–06 load. While the big belted case lets handloaders give the magnums an edge, I'm convinced there's no hunting task that the 7mm Remington Magnum can do better than a .30–06. At 500 yards, a 180-grain spitzer from this old cartridge still has as much energy as a .30–30 bullet at 150.

Most hunters who give the .30–06 a chance stay with it. Not long ago, I guided a fellow who had served as a sniper in Viet Nam and was used to shooting at extreme range. He might have chosen a super-magnum for long shots at elk. His preference for a .30–06, zeroed at 100 yards, might have surprised some hunters. It didn't surprise me. This marksman has learned just how potent the old .30 Government can be.

Far and away the most popular bullet weight among elk hunters who favor this .30-bore is the 180-grain. It's a logical choice, with enough speed for flat mid-range flight, enough sectional density to penetrate on oblique shots. Ballistic coefficients in the middle .400s give you high retained energy and minimal bullet drop at long range. The only other practical choices would be 165-grain and 200-grain

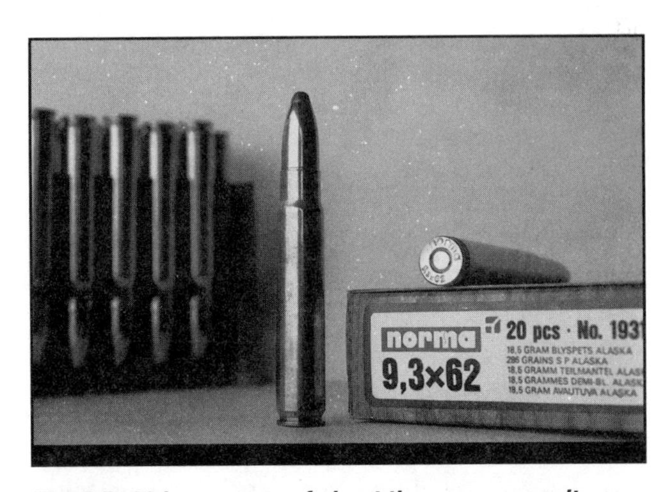

The 9.3x62 has a case of about the same capacity as a .30–06 and shoots a .366-diameter bullet. It has a long and distinguished record in Africa.

bullets. Bullets with bonded cores (the Swift A-Frame, Trophy Bonded) or captive heels (Nosler Partition, Winchester Fail Safe) don't need a lot of starting weight to ensure deep penetration, so a 165-grain bullet is no longer a "deer-weight" bullet. Excepting perhaps bullets like the Barnes X made without lead cores, the 150-grain weight is overrated. Here's why.

Consider a .30–06 bolt rifle with a low-mounted scope zeroed at 200 yards. A factory-loaded 180-grain Remington Core-Lokt will strike one inch above sightline at 50 yards. It will hit 2.4 inches high at 100 steps and 2 inches high at 150. At 250 yards it will drop 3.7 inches, at 300 yards 9.3 inches and at 400 yards 27 inches. Compare those numbers with these, from a factory-loaded 150-grain Core-Lokt: +.8 at 50, +2.1 at 100, +1.8 at 150, −3.3 at 250, −8.5 at 300 and −25 at 400. There's no tangible difference in trajectory, though the 150-grain bullet leaves the muzzle at 2910 fps, rather than 2700. At 300 the 180-grain hits an inch lower, at 400 just 2 inches. At all yardages, however, the 180-grain bullet packs a *20 percent weight advantage.* A minute-of-angle rifle spreads its bullets across 3 inches of target at 300 yards, 4 inches at 400. Thus, differences in point of impact between 150- and 180-grain .30–06 bullets at the longest practical hunting ranges and with very accurate rifles lie within half the extreme spread of a machine-rest group! Add wind, shooter wobbles, and aiming error, and you get groups big enough to obliterate any difference in point of impact due to weight.

By the same logic, you can argue persuasively for 200-grain bullets. A 200-grain Nosler Partition can be driven at 2650 fps from the .30–06. Given that velocity and a ballistic coefficient of .481, it crosses the 200-yard line at nearly 2300 fps and reaches 300 yards still clocking over 2100. With 2330 ft-lbs of energy at 200 steps and an even ton at 300, it is a first-round pick for heavy game. Standard rifling twist for the .30–06 is one turn in 10 inches, sharp enough to stabilize bullets up to 220 grains.

If you handload heavy bullets, you'll find the .30–06 case too small for some slow powders that seem suited to the job. Just 62.5 grains of H4831 fills a Winchester case to the mouth. You can seat bullets on a full case, but I prefer to keep powder-crushing to a minimum. Slightly faster propellants like IMR or Hodgdon 4350, Winchester 760, Hodgdon 414, even Hercules RL-19, work fine with 180- and 200-

The .30–06 has been chambered in most of the world's great rifles, like this lovely 1961 Mannlicher-Schoenauer.

grain bullets. For 150- to 165-grain bullets, IMR's 4320, 4064, and 4895 give good results, as does Winchester 748 and Hercules RL-15. Because of the .30–06's modest case capacity, you're wise to boost charges gradually when developing loads. Performance and pressure can change quickly.

One foreign .30–06-size round that deserves mention is the 9.3x62, developed in 1905 by Otto Brock of Berlin. It is not loaded in the U.S., and only Barnes and Speer supply its .366 bullets; but in Europe and Africa the 9.3x62 is a star. Launching 285-grain bullets at 2360 fps, it is in the same league as the .35 Whelen. A fellow I know in Zimbabwe used his 9.3x62 for years on cape buffalo and thought it adequate. It killed them quickly from the side; it even stopped them with frontal shots. The notion that we need bigger cartridges to shoot mule deer and elk seems odd.

Naturally, every case has a practical limit as regards bore size and payload. For the .30–06, it seems to me a .358 bore and a 250-grain bullet mark the ceiling. Include, if you will, the .366 bore of the 9.3x62, with 270-grain Speers. Both these combinations allow a muzzle velocity of 2600 fps. If you want to launch bullets heavier than 270 grains, you'll have to go to a .375 bore and accept much lower speeds. Given ordinary ballistic coefficients, you need 2500 fps to justify a 200-yard zero, and 2600 is much better. Otherwise, the bullet arc is so steep that you get a high bullet strike at the top of the bullet's travel, and your bullet quickly drops out of the vitals beyond 200 yards. The .25–06, .270, .280, and wildcat .338–06 make good use of the .30–06 hull. So too the .35 Whelen.

Hunters who use magnums because they shoot flatter than cartridges with the capacity of the .30–06 aren't gaining much. The .30–06-class cartridges shoot almost as flat as their magnum counterparts. Consider these differences in inches of bullet drop at 300 and 400 yards, given a 200-yard zero:

	300-yard drop	*400-yard drop*
.270 Win., 130-grain bullet (3100 fps)	6.3	18.5
.270 Wby. Magnum, 130-grain bullet (3300 fps)	5.5	16.0
.280 Rem., 150-grain bullet (3000 fps)	6.6	19.3
7mm Rem. Magnum, 150-grain bullet (3200 fps)	5.8	16.7
.30–06, 180-grain bullet (2700 fps)	8.5	24.4
.300 Win. Magnum, 180-grain bullet (3000 fps)	6.5	19.1
.338–06, 225-grain bullet (2600 fps)	9.3	26.7
.338 Win. Magnum, 225-grain bullet (2900 fps)	7.2	21.0
.35 Whelen, 250-grain bullet (2500 fps)	10.1	29.4
.358 Norma Magnum, 250-grain bullet (2700 fps)	8.6	24.8

At 300 yards, differences are 2 vertical inches or less. Unless you're an exceptional marksman, you can't hold within 2 inches at 300 yards from hunting positions, so you can ignore these disparities. At 400 steps the .270 and 7mm magnums show insignificant advantage. Bigger bores increase the spread, but 5 inches is still a mighty small difference. Remember, too, that 400 honest yards is a very long shot. You won't shoot that far often, if at all.

Long ago I lay down on a cold rock high on a mountain and tried to steady the crosswire of my scope on a deer standing in canyon shadows far below. It was a long shot. I held left for the soft north breeze, but not enough. The second bullet also missed. I hit that buck in the heart with my third shot, by holding 2 feet high and almost that far into the wind. Because this buck was too far off for a first-round kill, I wouldn't take the shot now. On the other hand, my .30–06 had plenty of reach and punch. The failing was in my aim.

The biggest elk I've killed fell to an '06, as did a fine high-country mule deer with six points on a side. My rack holds five .30–06 rifles now, and I've sold several I'd like to buy back. Depending on what you hunt most, the .30–06 may or may not be among your favorites. But you won't find a more versatile cartridge. And there's nothing you're likely to hunt that's too big for a one-shot kill with a .30–06.

The .300 Savage came about as a short-action alternative to the .30–06. Here the author's wife loads a Model 99 Savage, a brilliantly-designed lever action that survived for nearly a century.

Cartridges Outside the Mainstream

Usually, people follow other people. It's risky to try something different. Pioneers are seldom given the credit they deserve for their initiative and perseverance and willingness to absorb risk and ridicule. Among my favorite pioneers was Charles Newton, who kept developing new rifles and cartridges despite barrages of bad luck. Many of his cartridges were far ahead of their time. Brilliance won't keep you from going broke. Newton died during the Depression.

We haven't seen Newton's like for quite awhile. After the 1930s, wildcatters fashioned new cartridges by changing shoulder and body tapers, necking cases up and down, and trimming long hulls to work in short actions. Perhaps the best-known of many wildcatters was P.O. Ackley, an able gunsmith who also gave his name to a host of Improved cartridges shortly after World War II. A typical Ackley Improved case has a 40-degree shoulder and just enough body taper to facilitate feeding and extraction. The object for the changes: boost capacity to get more powder in the hull and achieve higher velocities. You could fire-form an improved cartridge simply by shooting the parent cartridge in the improved chamber. Headspacing was not altered, so the case would simply iron itself out to new dimensions. Ackley also developed true wildcats that could not be shaped simply by fire-forming. However, not all chambers cut for a particular Ackley round can be assumed identical.

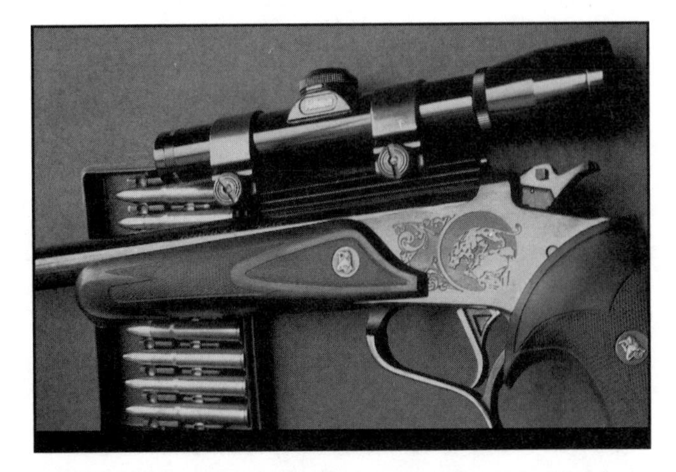

The .30–30 is as hum-drum as cartridges get—in rifles. But in pistols it's a fire-breather. Stout single-shot handguns like this Thompson/Center Contenter handle loads with 200-yard reach.

Body taper, shoulder angle and even headspace measure can differ. And not all Improved cartridges came from Ackley's shop. Fred Huntington gave us the RCBS Improved cases, commonly with 28- and 30-degree shoulders because he found the steeper 40-degree slant an impediment to smooth feeding.

That is probably why sharp-shouldered cases are hard to find among commercial cartridges. The .284 Winchester, with its 35-degree shoulder, looks like an Improved case. So does the defunct 7x61 Sharpe

The author favors cases of modest size. One of his wildcat rifles was built by Charlie Sisk in .338–08.

& Hart, with its *44-degree* slope. But most factory rounds have shoulders of 17 to 25 degrees, and enough body taper to guarantee easy feeding and extraction. A sharp shoulder that prevents a round from sliding smoothly home can cost a trophy on a hunt, a life on a battlefield. A dirty or dented case with little taper can yield the same result. Sharp shoulders and straight cases came into vogue because rifle shooters sought higher ballistic performance from existing cartridges. The .219 Donaldson Wasp is actually an Improved .219 Zipper, the K-Hornet an Improved .22 Hornet. You might say that the .300 Weatherby is an Improved .300 H&H. Case capacity for Weatherby's magnum is about 92 grains of powder, for the Holland, 80 grains. That 15-percent gain in powder capacity nets you less than 15 percent more speed from a 180-grain bullet if you load both rounds to equal breech pressures with the most suitable powders.

Despite the efficient shape of modern big game rounds, some hunters favor Improved cartridges. Jay Postman of RCBS says the .257 Roberts Improved heads the list of Improved rounds used by shooters ordering RCBS dies. "In fact," he adds, "it's so popular we include it with factory-loaded cartridges in our D series dies." Most Improved cartridges require a G series die set from RCBS, at a retail cost nearly double that of the D die set. Should you own a rifle chambered for an Improved round the folks at RCBS haven't yet seen, dies will cost more. Jay points out that new handloading tools are

The .338–08 is a very efficient round, made by necking up the .308 Winchester to 33 caliber. These moly-coated 210-grain Nosler Partition bullets can be driven as fast as 180s in a .308.

just one consideration if you're about to become an Improved shooter. Before you commit to rechambering all your rifles for Improved cartridges, here are some other things to remember:

1. Rechambering often reduces the resale value of rifles.

2. A barrel stamped for an Improved cartridge may be chambered for one of several that differ in dimensions. Improved chambers should be checked for headspace before you load commercial ammo to fire-form. If in doubt about which dies to order, send fired cases to RCBS.

3. You can expect lower velocities from factory ammunition in the Improved chamber because

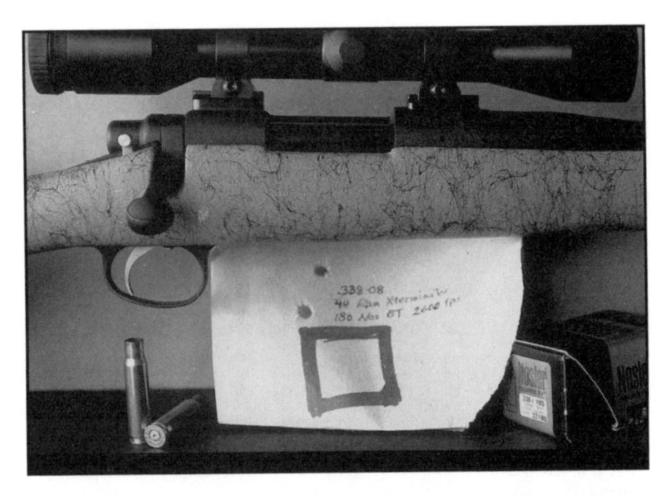

Wildcats aren't necessarily more accurate than standard cartridges. But 1½ inches at 100 yards is precision enough for big game.

The Winchester Model 88 has a front-locking bolt and is strong enough for any short-action cartridge. Wildcatters can use it for rounds based on the .308 and .284 cases.

some of the powder's energy will be used to expand the case.

4. While factory-loaded ammunition may shoot accurately enough for big game hunting, fire-formed cases should be the rule afield. Otherwise, why not stick to a commercial chamber?

5. Specify throat dimensions, or at least be aware of them before you rechamber. They affect bullet seating depth, which affects case capacity, accuracy, pressure, and magazine fit. There are no industry standards ensuring uniformity among reamers for Improved cartridges.

Besides a little extra velocity, are there any advantages to Improved cartridges? One Jay mentions is a more positive stop for retaining proper headspace. "I prefer a .35 Whelen Improved over the standard .35 Whelen when shooting cast bullets because there's not enough pressure with reduced loads to iron a gentle shoulder tight against the chamber wall," he explains. "That means a few firing pin strikes could hammer the case far enough forward to cause excessive headspace."

A contemporary of P.O. Ackley, Roy Weatherby improved the .300 H&H case, then successfully sold high velocity in a commercial line of cartridges. Beginning in the early '40s, he fashioned the .270, 7mm, and .257 short magnums (trimming the Holland case), then came up with his .300 Weatherby Magnum and later the .340. Roy had other wildcats that didn't survive to present, including the .220 Rocket. His .375 died out but has returned as factory chambering in the Weatherby Mark V African. The .240 Weatherby magnum is made from a .30–06-size case; there is no other commercial round on this belted hull. By 1953 Roy had designed the .378 Weatherby magnum on a new belted case that would later be used for the .460 and .416, the .30–378 and .338–378.

Rocky Gibbs moved to Idaho from California during the March blizzard of 1955 and began testing his new wildcat cartridges on a 500-yard range he set up on 35 acres just out of Viola. Gibbs developed a stable of potent wildcats on the .30–06 case. Like

Ackley's Improveds, Gibbs' cartridges showed minimum taper and sharp shoulders. But case forming involved more than a pull of the trigger. The Gibbs shoulder was located farther forward than on the '06 case, changing the headspace. Case forming meant over-expanding the neck, then reducing it in a Rocky Gibbs sizing die to form a false shoulder. Or you could use the hydraulic case-forming tool marketed by the enterprising Gibbs.

Some wildcat cartridges have been adopted by major arms and ammunition companies. The .257 Roberts (on the 7x57 case) and the .22–250 (on the .250 Savage) were snapped up by Remington, which much later brought the .35 Whelen (on the .30–06 hull) into its fold. It also picked up the .25–06. Oddly enough, the popular .338–06 remained an orphan until 1999, when Art Alphin of A-Square succeeded in getting standard dimensions approved by SAAMI (the Sporting Arms and Ammunition Manufacturers Institute). Two years later Weatherby announced that it would chamber the .338–06 cartridge in its Ultra Lightweight Mark V.

Since the introduction of the trim six-lug Mark V action in 1997, I had urged Weatherby to add the .338–06 to the list of chamberings in its svelte 6¾-pound Lightweight rifle. The Ultra Lightweight's debut in 1998 gave big game hunters a 5¾-pound option. Weatherby chose the Ultra Lightweight as the first commercial home ever for the .338–06. The rifle had sold well for Weatherby in other chamberings. The folks at Weatherby used their noggins in its design, sticking with a 24-inch barrel to keep muzzle velocity up and give the rifle fine balance. I've shot the Lightweights and was so impressed that I bought one. The Ultra Lightweight is essentially the same, with a top-quality synthetic stock, alloy bottom metal and some extra fluting to trim ounces. Affordable dies can be had from RCBS and Redding. You get your first 20 rounds of .338–06 brass free: they're packaged with the Weatherby rifle.

The .338 bore is just now gaining popularity among U.S. hunters. Overseas, it's had a long history. The .318 Westley Richards has logged nearly

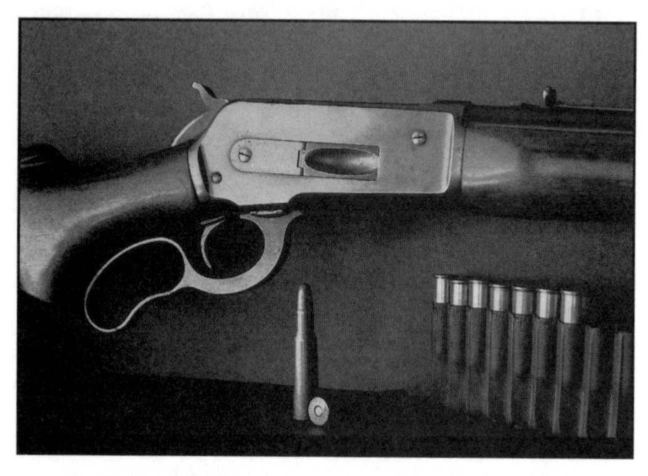

The .348 Winchester, designed for the Winchester 71 rifle (and no other!), has been blown out to form even more potent rounds by Payne, Busha and others.

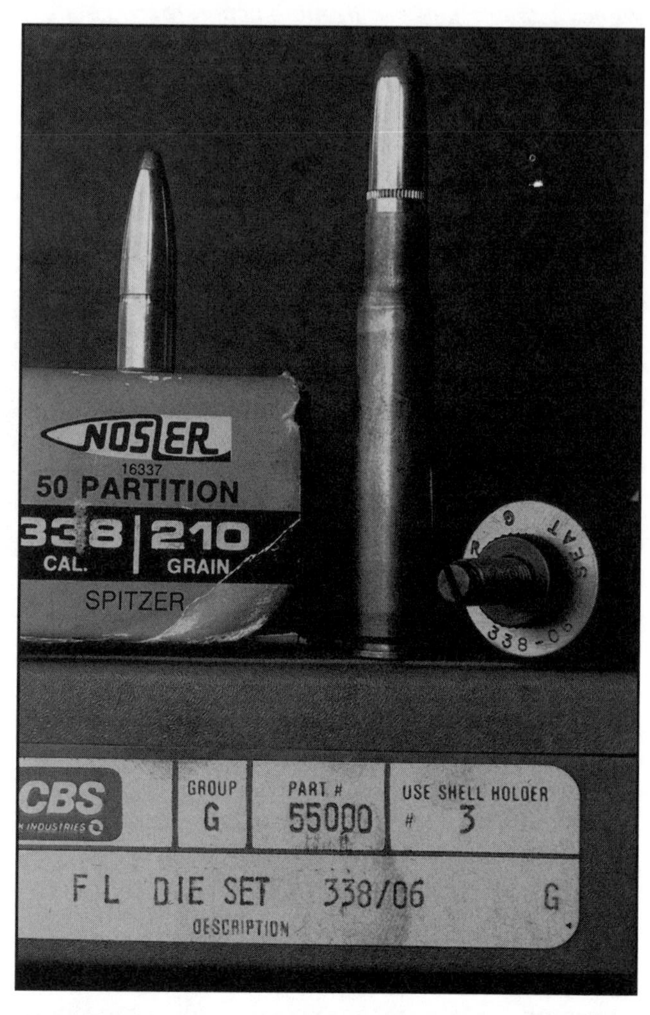

The .338–06 dates back decades, but no commercial chambering was available for this wildcat until Weatherby offered it in the Ultra Lightweight rifle in 2001.

a century as a workhorse round in Africa (bullet diameter for the .318 is .330, not .338 as is the .338–06, .338 Winchester, and .340 Weatherby.) The .333 Jeffery came along in 1911, a year after the .318 appeared. Its 250-grain .333 bullet moved out around 2500 fps, or 100 fps faster than an equivalent bullet from the .318. It, too, became a hit in Africa, and inspired American wildcatters Charlie O'Neil, Elmer Keith, and Don Hopkins to design the .333 OKH just after the second world war. They fashioned it after the British .333 Rimless Nitro Express, or .333 Jeffery (the .333 Flanged NE was the rimmed version). The OKH group simply applied the Jeffery's geometry to come up with a cartridge that performed at about the same level. The .333 OKH was a .30–06 necked to accept .333 bullets, then available from Kynoch. One was a 300-grain steel-jacketed roundnose with a sectional density of .376. This missile couldn't be moved very fast from the original OKH case. A logical solution was to load it in a necked-up .300 Holland hull, which held more powder. In a classic mismatch, the long 300-grain bullets did not stay together when driven into heavy game at high speed. They were designed for the .333 Jeffery and muzzle velocities of around 2100 fps. That is, they were best suited to the '06 case. Later OKH rounds were necked to .338 so they could use American bullets. They became the .338–06 and .338 Winchester Magnum.

By 1960, two years after its introduction, the .338 Winchester Magnum had pretty much erased interest in the .338–06. Even by today's standards, the .338 Winchester packs a punch. It kicks 200-grain bullets along at 2900 fps, 225s at 2800 and 250s at 2700. Federal offers it in High Energy loads: a 225-grain Trophy Bonded at 2940 fps and a 250-grain Nosler Partition at 2800. Both shoot as flat as the 180-grain bullet from a standard .300 H&H load and carry more energy. At the muzzle, both the 225- and 250-grain spitzers pack over 4300 foot-pounds. Hornady's .338 "Heavy Magnum" load is similar. It pushes a 225-grain Spire Point at 2950 fps. The .338–06 won't do that well, because the .30–06 case is smaller than the .338 Winchester's. It comes very close, though, and doesn't kick as viciously. I've chronographed these loads in my .338–06, a Mauser with a 25-inch Douglas barrel.

.338–06

powder charge, type	*bullet weight, type*	*velocity (fps)*
54 IMR 4064	200 Nosler Ballistic Tip	2830
54 RL-15	200 Jensen	2810
52 AA 2520	200 Nosler Ballistic Tip	2779
62 H414	210 Nosler Partition	2780
60 H414	225 Nosler Partition	2698
52 RL-15	225 Nosler Partition	2608
53 Scot 4065	230 Hawk	2627
63 RL-19	230 Hawk	2695
51 IMR 4064	250 Elkhorn	2500
61 RL-19	250 Swift A-Frame	2463
61 H450	250 Nosler Partition	2432
62 H4831	250 Winchester Silvertip	2459
59 H4350	265 Herter	2387
60 AA 3100	275 Speer	2380
56 Scot 4351	275 Speer	2422

Weatherby's six-lug Lightweight rifle, introduced in the late 1990s, was immediately available in popular rimless chamberings. An Ultra Lightweight followed. It was later bored for the .338–06.

With bullets of modest weight, the .338–06 can be loaded to within 100 fps of the .338 Winchester Magnum factory loads. You can get 3300 foot-pounds at the muzzle with 210- and 225-grain bullets.

Among the most appealing .33s I've used lately is the wildcat .338–308, a short hunting cartridge suitable for deer and elk. It can be loaded to push a 200-grain bullet as fast as a .308 launches a 180, with a trajectory to match and significantly more energy. A friend owns two rifles built for this short .33. One, a Model 70 Winchester, has a 24-inch barrel. The other, a Model 88 lever-action, has a 22-inch tube. The shorter barrel leaks about 70 fps. I've shot these rifles, getting minute-of-angle groups with the bolt gun and almost that level of accuracy with the 88. I generated this data with a rifle built by Charlie Sisk of Crosby, Texas. Barrel length: 23 inches.

.338–08

powder charge, type	*bullet weight, type*	*velocity (fps)*
42 RL-7	180 Nosler Ballistic Tip	2745
47 IMR3031	180 Nosler Ballistic Tip	2767
41 IMR4198	180 Nosler Ballistic Tip	2720
46 A2460	180 Nosler Ballistic Tip	2740
49 4320	180 Nosler Ballistic Tip	2700
46 H322	180 Nosler Ballistic Tip	2775
45 VV133	180 Nosler Ballistic Tip	2760
46 2015	185 Barnes X	2730
47 A2520	200 Hornady	2645
46 IMR4895	200 Nosler Ballistic Tip	2615
44 H322	200 Nosler Ballistic Tip	2610
49 RamTAC	200 Hornady	2620
49 H380	225 Armfield	2350
49 Win760	225 Hornady	2340
50 A2700	225 Speer	2350

Small Bores, Long Reach

The new cartridge had a huge case for its 7mm bullet, and its blistering speed limited bullet drop at long range to only a third of what shooters were used to. About that time a 30-caliber cartridge with an even bigger case gave the same zip to heavier bullets.

I'm talking, of course, about the 7x57 Mauser and .30–40 Krag.

In 1892, the 7x57 was brand new, a small-bore cartridge that used smokeless powder to launch a 175-grain roundnose bullet at a sensational 2300 fps. Given a 100-yard zero, the bullet fell about 2 feet from line of sight at 300 yards—flat-shooting indeed compared to the 6-foot drop of a 405-grain .45–70 bullet started at 1300 fps.

The .30–40 Krag, also one of the first smokeless military cartridges, fired a 220-grain bullet at just under 2000 fps. Drop at 300 yards was nearly a foot greater than that of the 7x57 bullet, but the full-jacket .30–40 bullet had more weight. It could also drive through 58 seven-eighths-inch pine boards at 15 feet. Compared to the .45–70, it was a brickload of lightning.

Our standard big game cartridge is no longer the .45–70. Nor is it the .30–40. For a long time it was the Krag's successor, the .30–06. But now American big game hunters expect a lot more out of their cartridges than they did at the turn of the century, or when I grew up marveling at the reach of the .270.

As always, where there's demand, manufacturers soon get busy building supply. Winchester perceived a market for a long-range big game cartridge

This antelope hunter can use the reach of his Winchester M70 in 7 STW. The scope is a Zeiss 5–15x.

in 1925, when it introduced the .270. Not for another 30 years would the company take such a bold step. But in 1959, on the heels of its first short belted magnum (the .458), Winchester put another fast, flat-shooting cartridge on gunshop shelves. It was the .264 Magnum, listed at 3200 fps with 140-grain bullets and a blistering 3700 with 100-grain bullets. Hunters didn't buy the long-barreled Model 70s chambered for it (or the Featherweights with 22-inch barrels). Advertised as a plains rifle, the .264 could indeed shoot flat to kill pronghorns and mule deer at long range. The 140-grain Power Point was a

Right to left: 7mm Weatherby and 7mm Remington Magnums, 7STW. The 7mm Weatherby came along in the early 1940s, the Remington in the early '60s, the STW in the late '80s.

reliable bullet. But there was no talk of bigger game. Winchester apparently assumed that hunters would shoot coyotes with the frangible 100-grain bullets. Most hunters, however, decided the .264 had nothing over the .270 except noise and recoil. It ate barrel throats fast too. If they were going to buy a magnum rifle, they wanted it to kill bigger game than deer.

Cannily, Remington stepped up three years later with its 7mm Magnum on the same case. There's only .020 difference in bullet diameter between the 140-grain .264 bullets and 140-grain 7mm bullets. Theoretically, you can push the 7mms slightly faster because they have a bigger base, but the .264 bullets have higher sectional density. The huge and almost instant popularity of the 7mm Remington Magnum was due largely to a sympathetic press and company advertising that billed the new offering as a deer-elk cartridge, not a deer-coyote cartridge. It arrived, too, with the introduction of Remington's 700 rifle, a more attractive choice than its predecessor, the 721.

Factory .264 ammo (now featuring only the 140-grain bullet) now has a catalog velocity of 3030 fps—slower than the 130-grain .270 load! But this data falls far short of the round's potential. Here are some of my .264 loads:

.264 Winchester Magnum

powder weight, type	*bullet weight, type*	*velocity, fps*
68 IMR 7828	120 Speer	3314
66 H4831	125 Nosler Partition	3420
66 IMR 7828	129 Hornady	3288
67 IMR 7828	129 Hornady	3263
65 H4831	129 Hornady	3324
64 H4831	140 Sierra	3184
64 H4831	140 Hornady	3266
71 H570	140 Hornady	3231
73 H570	140 Hornady	3346
63 H450	150 Acme	3051
68 H570	160 Acme	2960

The 7mm Remington Magnum only got more popular as the .264 faded. In surveys I've taken among elk hunters, only the .30–06 is more popular than the 7mm Magnum. The big 7's instant success took a lot of shooters by surprise, because Roy Weatherby had fashioned a nearly identical cartridge in the early 1940s. In fact, the 7mm Weatherby Magnum, as loaded by Norma, outperforms the Remington cartridge (both can be taken to about the same ceiling by handloading). But Weatherby's cartridges were proprietary. No commercial arms factory offered Weatherby chamberings, and you had to buy expensive Weatherby ammunition for custom-built rifles. The success of Remington's 7mm was due in part to the brilliant advertising campaign at its introduction, the concurrent unveiling of a new Remington rifle to chamber it (the Model 700) and the hunting public's readiness for an all-purpose magnum round that didn't hurt in recoil as much as a .338.

The 7 STW arrived to satisfy hunters who wanted a flatter-shooting 7mm. It began life in 1983, a wildcat derivative of the then-new 8mm Remington Magnum. Gun writer Layne Simpson, who is credited with developing the cartridge, told me that there was no interest from Remington at the time. "Marketing

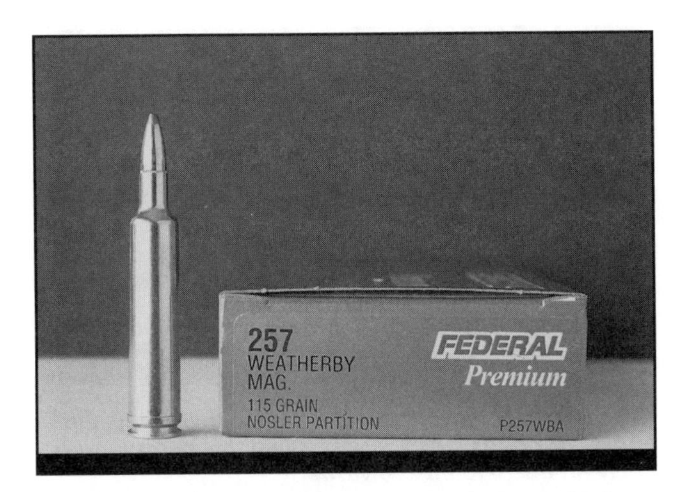

Roy Weatherby's .257 Magnum was among the first of his short magnums and still sells well. But it lacks the versatility of the .270 and 7mm versions, with their heavier bullets.

people may well have thought it would drain sales from the 7mm Remington Magnum," he said. U.S. Repeating Arms, (Winchester's title for its firearms enterprise) actually chambered rifles for the 7 STW before Remington manufactured it. Now both firms offer the 7mm STW as a chambering, besides selling loaded ammunition. It's a full-length magnum, with a case length of 2.850 like its grandparent, the .300 H&H. That long hull gives it significantly more capacity than the 2.500-inch 7mm Remington Magnum.

When Remington adopted this new round, the name in circulation at Ilion was 7mm Remington Maximum. But Layne had already dubbed it the STW—Shooting Times Westerner—after the magazine that helped him initiate and publicize his work.

The STW was not really new. Warren Page (longtime shooting editor at *Field & Stream*) favored 7mm wildcats. The long 7mm Mashburn was based on the full-length .300 H&H case blown out. So is the STW's 8mm parent. The wildcat 7mm/.300 Weatherby used by 1000-yard benchrest shooters since the 1940s has roughly the same case capacity, just a different shoulder.

According to Layne, Remington decided to adopt the 7 STW partly because the company was selling lots of 8mm Magnum brass and few 8mm Magnum rifles. (In fact, that chambering is so unpopular it has been dropped from all but custom-shop rifles at Ilion.) The firm eventually concluded that 8mm cases were turning into STWs inside RCBS dies! Shooters were taking advantage of super-slow powders like H1000, RL-22, and IMR 7828. Warren Page didn't have those propellants in the 1960s. Advances in bullet design also contributed to making this high-capacity 7mm more effective on game, and gave it greater accuracy. Without fine accuracy, a speedy bullet is like an Indy race car without tie rods.

"The first 7 STW ammo came from A-Square," recalled Layne. "Remington followed months later. Within a year after the cartridge was available from Remington, one gunsmith told me he did 600 rechamberings and that Remington sold 600,000 rounds." Current loads from Ilion include 140-grain pointed Core-Lokts and 140-grain Swift A-Frames, both clocking 3325 fps at the muzzle. Shooting factory-loaded Core-Lokts in a Winchester Model 70 Sporting Sharpshooter, I recorded higher average velocities. Several rounds sped over the screens in the mid 3400s. Accuracy was outstanding: Half-minute three-shot groups. A friend reported getting an occasional 3-inch group at 500 yards with his Hart-barreled Remington in 7 STW.

Randy Brooks at Barnes Bullets shared with me 7 STW data from that company's American Fork, Utah test tunnel. Though some of the velocities suggest room for improvement, Randy cautions handloaders to approach them as if they were maximum loads, seating X-Bullets .1 off the lands to keep pressures within reason and achieve the best accuracy.

7mm STW

powder weight, type	*bullet weight, type*	*velocity, fps*
80.5 N204	100 Barnes X	3827
80 H4350	100 Barnes X	3843
88.5 Vit N165	100 Barnes X	3862
86 H1000	120 Barnes X	3444
76 IMR 4350	120 Barnes X	3457
81.5 Win WMR	120 Barnes X	3494
76 Norma MRP	140 Barnes X	3218
74 H450	140 Barnes X	3229
73 RL-19	140 Barnes X	3242
80 H5010	160 Barnes X	2972
72 Win WMR	160 Barnes X	2984
87.5 AA 8700	160 Barnes X	3053

There's no magic in the 7mm STW. Kicked from this big case, a bullet simply flies flatter and hits harder than the same bullet from a 7mm-08, 7x57, .280, or any short 7 magnum. At 300 yards it delivers what a .280 carries to 200. Sometimes its extra reach proves invaluable. I watched a companion hit a fine pronghorn buck at about 375 yards with his STW. He didn't need all those foot-pounds for the light-boned creature; but the bullet's fast, flat flight made accurate range and wind calculations less critical. There was little time for the shot, and he made it quickly. Another fellow shot an elk at nearly 500 yards with this cartridge. He used a Harris bipod and a Shepherd range-compensating scope. He hit the bull twice, a little too far back. In this case, high remaining energy helped destroy enough vital tissue to bring the elk down. The recovered Core-Lokts were picture-perfect mushrooms.

You might wonder, as I have, why Remington doesn't offer heavier bullets in factory-loaded 7 STW ammo. The answer probably has more to do with marketing than with meeting needs in the field. If you boost bullet weight, you must reduce velocity, and velocity is what sold this big 7 in the first place. A heavier bullet—say, 160 grains—would make the round more useful in some types of hunting, but it could not be driven nearly as fast as a 140. Take the 7mm Remington Magnum as an example. It launches a 140-grain softpoint at 3175 fps. Substitute a 160-grain bullet, and velocity drops to 2950 fps. Now, given a streamlined shape, the heavier bullet begins to catch up as soon as it leave the muzzle, and downrange its higher ballistic coefficient will eventually bring it even with its lightweight counterpart. At 500 yards, the 140-grain and 160-grain 7mm Magnum bullets are both traveling about 2040 fps, and the 160 has an edge in energy: 180 ft-lbs. But at shorter yardage the lighter bullet gets to the target sooner and with more foot-pounds of energy. It generates less recoil in the bargain.

Bigger bores accommodate any weight change more easily. That is, you can add 20 grains to the weight of a 30-caliber bullet without sacrificing so much speed. The additional weight has less effect on the bullet's length and bearing surface than does a 20-grain increase in a smaller bore. So pressures aren't as sharply affected. Nonetheless, a heavy-game bullet in the 7 STW would make sense. Remington already loads the 160-grain Swift and the more aerodynamic 160 Nosler Partition in its 7mm Magnum ammunition. The bigger STW case would seem even better suited to these bullets and would boost the versatility of the cartridge.

If the STW could use a heavier bullet, surely the 7mm Ultra Mag deserves one. Introduced late in the year 2000, the 7mm Ultra Mag is the logical followup to the .300 Ultra Mag, announced two years previously. Remington brought the .338 UM ahead of the 7mm, partly, no doubt because the 7 STW was available, and the company had no big .33. The 7 UM and .375 UM appeared at the same time. Neither is much more useful than the belted magnums the bigger, rimless Ultra Mag case was designed to replace. The .375 UM kicks harder than a .375 H&H but for thin-skinned game probably won't be any more lethal than the Holland round. The 7mm is simply overbore. You can get 100 fps more than from an STW, perhaps exceed 3500 fps with a hot 140-grain handload. But few if any hunters are handicapped driving 140s at 3200 from 7mm Remington Magnums, or 3350 from their STWs. If bigger cases have an advantage, it's in their ability to keep velocities high with heavy bullets. In the 7mm UM, Remington has the ideal engine for a 160-grain payload. I mentioned this to Remington management and got reluctant agreement. "But we'll probably use 175-grain bullets, not 160s."

The .280 Remington is among the most versatile big game cartridges. Terry Moore killed this caribou with his, with one shot at 200 yards.

Now, this didn't make sense to me. Boosting bullet weight to 175 grains from 140 constituted a huge jump. Adding that much weight meant increasing shank length significantly too, which would eat into the space in the case, reducing powder capacity. A longer shank would mean more friction in the bore, stacking more resistance on a bullet already hard to move for its greater mass. I suggested that a .284 bore is best served by bullets between 140 and 160 grains in weight. "If I need a 175-grain bullet, I'll go to a 30-caliber cartridge that can drive a 180-grain bullet more efficiently."

"That's a good argument," was the reply. "But we have a huge stock of 175-grain 7mm bullets."

The Biggest .30s

There's something about the .30 magnums that makes perfect sense. They can drive 180-grain bullets about as fast as 7mm cartridges with the same powder capacity can hurl 160s. A 180-grain bullet at 2650 fps has proven adequate for all North American big game, as thousands of .30–06 shooters can attest. A .30 magnum that drives it 3000 fps doubles the range at which you can deliver a ton of energy to a tough animal. The '06 carries a ton to 200 yards, the .300 Winchester Magnum to 400.

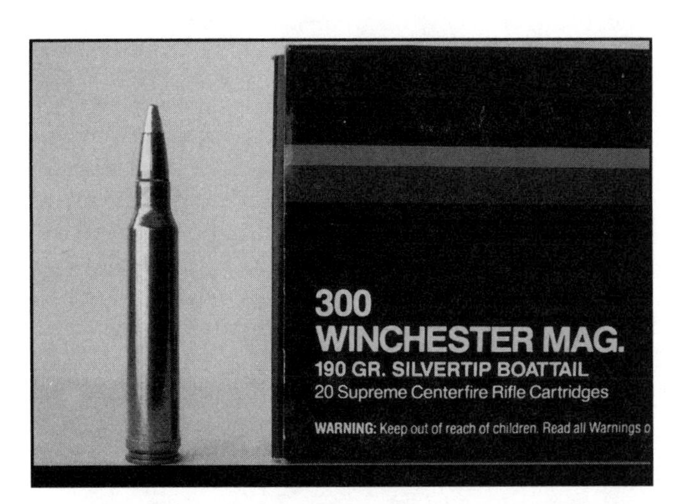

The .300 Winchester Magnum, introduced in 1963, has a 2.62-inch case, slightly longer than the cases of its siblings, the .264 and .338. Its neck is shorter. This round has become a hit with hunters.

But the .300 Winchester was surprisingly late in coming. In 1920 the British firm of Holland & Holland came out with the .300 H&H Magnum, or Super .30. It had the long case of its parent, the .375 H&H, circa 1912. The .300's belted hull had been designed for the long sticks of cordite powder popular in English cartridges of that time. Smoothly tapered, with an 8½-degree shoulder, it fed through magazines like grease itself. Western Cartridge Company introduced this round to American shooters in 1925. For a decade thereafter, no commercial rifles were chambered for it. Then, in 1935, Ben Comfort won the 1000-yard Wimbledon Match at Camp Perry with a 1917 Enfield barreled to .300 H&H. Two years later, Winchester listed the cartridge as a charter offering in its Model 70 rifle. The 70, and Remington's 721 (introduced in 1948) had magazines that would handle the .30's 3.60-inch overall length. Other actions had to be opened up.

There aren't many factory loads for the .300 H&H now. Remington lists none. You can get a Winchester Supreme load with a 180-grain Fail Safe at 2880 fps. It kicks 3315 foot-pounds out the muzzle, and delivers about 1525 to 400 yards. Given a 200-yard zero, this bullet hits 8 inches low at 300 steps, 23 inches low at 400. Federal's Premium Safari load launches a 180-grain Nosler Partition to deliver about the same performance. At one time, you could buy .300 H&H ammo with 150-grain bullets at nearly 3200 fps, and 220-grain bullets at 2620. The

The .300 H&H has a long, gently sloping shoulder to accommodate Cordite powder that was used in the early 20th century. Introduced in the U.S. in 1925, the .300 H&H has fathered many magnums.

The .300 Winchester Magnum is the most widely chambered belted .30. This is a Ruger 77.

Among the first affordable rifles to chamber full-length magnum cartridges in the U.S., the Remington 721 appeared in 1948, 11 years after Winchester's Model 70.

180-grain spitzers shoot 15 percent flatter than those from standard .30–06 loads, and deliver a 10 percent advantage in wind—but with 35 percent more recoil. Expect about 26 foot-pounds of kick from a Super .30, compared to 17 for a same-weight .30–06. For some reason, I don't feel that much difference. Perhaps it's because this magnum does not jab; rather, it seems to push.

When Roy Weatherby trotted out his line of long-range cartridges during the 1940s, the .300 was the fourth in line, after the .257, .270, and 7mm. The

.300, however, was the only one with a full-length case. Roy had simply reduced the body taper of the .300 Holland & Holland, then given it a new shoulder with a double radius. The .300 Weatherby magnum held considerably more powder and could drive a 180-grain bullet 3250 fps, about 200 fps faster than a .300 H&H.

Neither Winchester nor Remington pursued another .300 magnum for decades. When Winchester did unveil one in 1963, it wasn't what shooters expected. They assumed the company would neck down the .338 hull, as many wildcatters had been doing since the .338's introduction five years earlier. It then would have been similar in form to the .308 Norma Magnum, still a new cartridge. But Winchester instead trimmed the case to 2.620 inches, not the 2.500 inches of its other short magnums. The .300 Winchester had a shorter neck than the Norma or wildcat .30–338—a liability, if you were a student of cartridge design but in practical terms of no consequence. The .300 Winchester Magnum held more powder, but in actions sized to accommodate the .30–06 or other short magnums you couldn't use the extra capacity because you had to seat bullets deep to clear the magazine.

From left, the .308 Norma, .300 Winchester, .300 H&H and .300 Weatherby Magnums. The H&H came first, followed by the Weatherby, then the "short" magnums for .30–06-length actions.

Remington decided it needed a big .30 only after Weatherby had raised the performance bar with its .30–378. Introduced at the 1996 SHOT Show, this chambering wasn't cataloged until 1997. By then dealers were piling up orders, and Weatherby's plant in Maine scrambled to fill them. Shooters (and collectors, and people who just wanted to own a .300 magnum with a case the size of a bottle of bore cleaner) flocked to buy Weatherby Accumarks in .30–378. In those first months, Weatherby had 10 times as many orders as it had expected. Norma, a relatively small ammunition firm despite its long-standing prominence in the U.S., couldn't make cartridges fast enough. To compound the problem, Norma took longer than predicted in developing loads. Many new .30–378 rifles remained idle for want of ammunition. Production caught up with demand eventually, and now both rifle and ammo are readily available.

Remington's search for a high-performance .30 became known among industry pundits long before the .300 Ultra Mag was announced in 1999. Based on the .404 Jeffery, a rimless British round dating to 1910, the .300 Ultra Mag is not just a necked-down version. Ilion engineers fashioned the new cartridge specifically for the long-action Model 700 receiver. The case is about as big in diameter as reliable feeding allows (with some changes in the action dimensions). The rim is slightly rebated to fit magnum bolt faces, eliminating the need for more tooling or another shellholder on the handloader's bench. Though the case is long, it is just short enough to permit ordinary seating depth without jamming bullet noses against the front of the magazine.

The Ultra Mag has 13 percent more case capacity than the .300 Weatherby. But you won't get 13 percent more velocity. As case volume grows, cartridges become less efficient. You need more space and more powder per increment of performance than you did for the last increment. In fact, there's negligible difference between the .300 Ultra Mag and the .300 Weatherby Magnum as factory loaded—provided

The .300 Remington Ultra Mag appeared in 1999. It's a rimless case with 13 percent more capacity than the .300 Weatherby but can fit in the same actions and is loaded to deliver similar performance.

you compare Remington's Ultra Mag load with the Norma ammo that has for decades defined .300 Weatherby performance. The Ultra Mag launches a 180-grain Nosler Partition or Swift Scirocco at 3250 fps, while the Weatherby Magnum gives a 180-grain Partition or pointed softpoint bullet 3240 fps. Remington also loads a 180 Swift Scirocco and a 200-grain Nosler Partition in the .300 Ultra Mag. The Nosler leaves the rifle at 3025 fps, compared to 3060 for the same bullet in a Norma-loaded .300 Weatherby. No practical difference.

The picture changes when you supplant the Norma .300 Weatherby load with Remington's factory offering. According to its own charts, Remington stokes the .300 Weatherby cartridge to only 3120 fps with 180-grain bullets, or 130 fps off the Ultra Mag's speed. Hornady's catalog shows 3120 for the Weatherby as well. Federal loads both the Ultra Mag and the .300 Weatherby, at 3250 fps and 3190 fps respectively, with 180-grain bullets. Federal's 200-grain Trophy Bonded bullet clocks 2900 fps. There's one .300 Weatherby load that outperforms the Ultra Mag: Federal's 180-grain Trophy Bonded High Energy. It registers 3330 fps at the muzzle.

If you consider handloading potential, the Ultra Mag has the edge on the Weatherby. It's bigger. Purists also like the idea that, as a rimless case, it headspaces on the shoulder. This Remington brainchild gets about all the bullet velocity that can be milked from a rifle action of ordinary dimensions. In factory-built Remington 700 rifles, shooting the Scirocco bullets, I've found it very accurate and surprisingly easy to shoot. In sum, the .300 Ultra Mag amounts to an improved .300 Weatherby Magnum. The Weatherby is still a great cartridge, with a field record that few others can match. And you won't see a bit of difference between it and the Ultra Mag on the mountain. On paper, advocates will forever argue the merits of each.

The .30–378 Weatherby Magnum is the most powerful 30-caliber hunting round. The big .378 case was developed in 1953—essentially a rimmed .416 Rigby.

A muzzle brake softens recoil by diverting exiting gas before the bullet leaves. A brake is noisy but almost necessary for extended shooting with big .300 and .338 Magnums.

Here are some handloads I've fired in a Remington 700 rifle built by Charlie Sisk (26-inch barrel):

.300 Remington Ultra Mag

powder weight, type	*bullet weight, type*	*velocity (fps)*
92 Reloder 19	150 RWS Cone Point	3524
88 Scot 4351	150 Hornady Spire Point	3590—maximum!
89 H450	165 Speer	3411
85 H4350	165 Barnes X	3400
94 Reloder 22	165 Hornady Spire Point	3551—fast, accurate
Remington factory	180 Nosler 180 Partition	3266 load
96 H1000	180 Winchester Fail Safe	3315—maximum!
93 Reloder 25	180 Swift A-Frame	3271
90 IMR 7828	180 Sierra Match	3301—accurate
89 Accurate 3100	185 Berger Match	3264—accurate
85 H4831SC	200 Nosler Partition	3006
101 Accurate 8700	220 Herter	3022
102 H870	220 Nosler Partition	3111—accurate

It's not fair to compare the .300 Ultra Mag to .30–378 Weatherby Magnum, which requires a bigger action and bolt face, limiting it to the likes of the Weatherby Mark V rifle. The parent .378 cartridge was introduced in 1958, and soon spawned the .460. Later this brawny pair was joined by the .416 Weatherby Magnum. Until the debut of the .30–378 in 1996, no other commercial rounds have shared this hull. The .30–378 is factory loaded with 165-grain Nosler Ballistic Tips at an advertised 3500 fps, 180-grain Barnes X-Bullets at 3450 and 200-grain Nosler Partitions at 3160. Given a 300-yard zero, these bullets strike less than 4 inches high at 200 yards. The 165- and 180-grain bullets hit a mere 7½ inches low at 400, while the 200-grain Noslers drop 9 inches. All three bullets carry well over a ton of energy to 500 yards and land as close to the line of sight as bullets from a .270 with 200-yard zero do at 400 yards!

John Lazzeroni's 7.82 Warbird is very similar in performance to the .30–378 because case capacity is nearly identical. Until 2000, it was available only in Lazzeroni rifles based on McMillan actions. Then Sako chambered its Model 75 for the Warbird, giving shooters a more affordable option than the Lazzeroni or Weatherby. Here is how my .30–378 handloads, and John's 7.82 Warbird factory loads, performed over my chronograph (26-inch barrel):

Boost velocity, and you need stronger bullets to expand predictably and penetrate. Woodleigh bullets from Australia are noted for great weight retention. Right to left: close-range to long-range impact.

.30–378 Weatherby

powder weight, type	*bullet weight, type*	*velocity, fps*
107 H4831	140 Barnes X	3733
105 RL-22	150 Nosler Ballistic Tip	3544
102 AA 3100	165 Trophy Bonded	3536
119 H870	180 Speer Mag Tip	3352
107 RL-25	200 Hawk	3241

The .30 magnums are ideal for elk hunting. They have 400-yard reach and plenty of bullet weight. Recoil ranges from tolerable to brutal, depending on the load and the rifle.

7.82 Warbird (Lazzeroni)

factory loads, 180-grain Nosler Partition	3531
(nine rounds fired, lowest and highest	3554
readings deleted)	3541
	3554
	3531
	3575
	3520

The .30–378 and Lazzeroni Warbird can be loaded to essentially the same velocities. My .30–378 loads here reflect my usual caution in starting low and working up slowly. The Lazzeroni loads are near maximum. Note the low variation in chronograph readings for the Warbird. John Lazzeroni and his staff still handload their ammunition. Consistency results.

I should have bushels of field anecdotes featuring the super-size 30s. The hype that accompanied them surely encouraged hunters to test them at long range. Truly, though, the animals that I've seen shot have been shot at ordinary distances. It's not that the cartridges fall short; rather, most game becomes most visible within a quarter mile. Broken topography and open forest can limit your effective visibility to 100 yards or so. Thickets may give you first glimpse of the animal when it's only a few feet away. If you see a fine buck within range of a .30–06, will you scoot off to find a longer shot alley so you can wring out your super-cannon? Not me. Another variable is marksmanship. A rifle that shoots as flat as the curve of the earth and can put all its bullets inside a teacup two townships distant will hardly guarantee kills. You must hold the rifle steady and execute the shot perfectly to hit at long range. Not many hunters can stretch the .308. I'd bet that a .30–06 shoots flat enough and hard enough to reach farther than most hunters can hit a washing machine every time from hunting positions. This isn't to say big .30 magnums are useless. Their speedy bullets make range and wind estimation less critical. They handle heavy bullets better than ordinary rounds, for game like elk and moose. But with their blast and recoil, and additional rifle weight, they give you mixed blessings. Certainly, you'd have a hard job convincing any experienced hunter that a .30–378, or a .300 Remington Ultra Mag, or a 7.82 Warbird, will bring him more animals than his trail-worn .300 Winchester. A 400-yard shot is still a mighty long shot.

Short but Fast

You get high velocity from long cases with lots of powder. At least that used to be the rule. But looks don't always reflect performance. In 1974 benchrest shooter Dr. Lou Palmisano reshaped the .220 Russian round to form what he and cohort Ferris Pindell would call the .22 PPC. A 6mm PPC came later. These cartridges were short and squat. From the base to the 30-degree shoulder measured barely over an inch, though basal diameter approached that of the .30–06. Palmisano figured a shorter powder column would yield better accuracy. In benchrest shooting, where quarter-minute groups are commonplace, the PPC would have a tough test. The .222 Remington and 6x47 (a necked-up .222 Magnum) had been the darlings of serious competitors for long enough that they held most of the records. Their popularity also meant that the "triple deuce" and 6x47 were chambered in the best rifles and used by the most competent riflemen. The PPC not only had to be superior; it needed time in the hands of shooters who didn't like to lose points experimenting, who might replace a barrel to try a new round but would be a lot more reluctant to change the bolt face on their pet bench rifle.

Palmisano and Pindell were convinced that the PPC case had promise. Soon their champion-level shooting and careful handloading produced winning groups. Other shooters followed. In the 1975 NBRSA championships, two of the top 20 shooters in the sporter class used PPCs. By 1980, 15 of the top

Short magnums like the 8.59 (.338) Lazzeroni Galaxy have more capacity than the longer .30–06, and hit as hard as belted magnums.

20 were so equipped. In 1989 *all* of them used the short, fat rounds. Even more impressive: the top 20 entrants in the demanding Unlimited class shot PPCs. And 18 of the 20 best in Light Varmint and Heavy Varmint categories favored them too. Sako eventually chambered rifles for the PPCs, but despite vigorous campaigning by Palmisano, American gun companies demurred. The East Coast physician then took it upon himself to supply the benchrest community with PPC brass, importing more than a million .220 Russian cases! And at this writing, 25 years later, the short, squat profile of the PPC has started to show up in hunting camps.

John Lazzeroni has been a pioneer in bringing the short powder column to big game cartridges. Ironically, he is perhaps best known for his long magnums—huge beltless rounds that are inefficient but rule the ballistics charts. His 7.82 Warbird is essentially the equivalent of the .30–378 Weatherby. His 8.59 Titan matches the .338–378. Lazzeroni's long cartridges range in caliber from 257 to 416. The cases are of his own making. "If I'd done a little more homework," admits John, "I might have used an existing case, like the .404 Jeffery. Head dimensions of my rounds are ridiculously close to the .404's."

Since 1997, the Tucson entrepreneur has sold enough of his handloaded high-performance ammo, and his futuristic long-action rifles (based on the McMillan action), to justify exploring short magnum cartridges. He likes what he sees in the ballistics lab and has come up with half a dozen shorties based on his long hulls. John explains: "They have the same beltless heads, just .750 less case up front." The bases for his .243 and .264 short cartridges mic .532—same as that for an ordinary belted magnum like the 7mm Remington or .300 Winchester, and identical to the head on Lazzeroni's long .257 Scramjet. The short 7mm, .300, .338, and .416 Lazzeronis feature .580 heads, like their longer counterparts. There used to be a .264 and a 7mm on a .546 case, but John discontinued them. He dropped the long .264 altogether and gave the 7mm a bigger case. John lists these velocities for his compact versions:

- 6.17 (.243) Spitfire: 85-grain bullet at 3618 fps
- 6.71 (.264) Phantom: 120-grain bullet at 3312 fps
- 7.21 (.284) Tomahawk: 140-grain bullet at 3379 fps
- 7.82 (.308) Patriot: 180-grain bullet at 3184 fps
- 8.59 (.338) Galaxy: 225-grain bullet at 2968 fps
- 10.57 (.416) Maverick: 400-grain bullet at 2454 fps

John Lazzeroni's 7.82 Patriot was developed for Lazzeroni rifles but later found a home in Savage 111 rifles as well.

All these cartridges are short enough to fit in actions for the .308 Winchester. Of course, the bolt face and magazine must be fitted to the fatter case. Or you can buy John's latest rifle, the L2000SA. Appropriately dubbed the Mountain Rifle, it weighs only 6½ pounds. The McMillan action gets the same treatment as on long-action models: an oversize Sako-style extractor, custom bottom metal, a three-position safety, fluted Schneider cut-rifled barrel and Jewell trigger. John bores out the standard 6–48 scope mount holes, then drills and taps them to 8–40 for dead-center spacing and a more secure hold. Except for the magazine spring, all steel in this Lazzeroni rifle is stainless, jacketed with a satin-silver NP3 electroless nickel finish. It wears a classic-style synthetic stock, with a long, slender grip, straight comb, and sharp checkering. A soft buttpad makes recoil quite tolerable.

This Lazzeroni rifle is comfortable to shoot, partly because of the stock design but also because of the case shape. According to John, a big benefit of the squat hull is its ability to burn fuel quickly. "The powder is concentrated around the primer, so ignition at the front of the column occurs faster. That means less of the powder starts moving with the bullet. Since powder following a bullet out of the case mouth is considered as ejecta, the less powder you send down the bore, the better. If you have lots of slow powder in a long column, you substantially increase the weight of ejecta, and you feel that weight during recoil."

Sharp reductions in case diameter at the neck, combined with long, heavy bullets, mandate the use of slow powders. But evidently short cases give you the option of coming up a notch in burning rate. John, for instance, uses RL-15 behind 225-grain bullets in the Galaxy. I've chronographed them at 2760 fps—essentially the speed I'd expect from factory-loaded .338 Winchester Magnum ammo. RL-15, while an excellent powder, is faster than most fuels recommended for medium- to heavy-weight bullets in the .338. It seems to work just fine in the Galaxy, producing top-end velocities with no signs of high

The .300 Winchester Short Magnum, introduced in 2000, is a trifle longer than the Lazzeroni short magnums, but not as big in diameter.

pressure. By all logic, it should result in less ejecta weight than, an equivalent load of 4350 in a longer case. A factory test target for the Galaxie showed ¾-minute accuracy.

The short Lazzeroni cartridges preceded Winchester's 2001 announcement of a similar round with a .532 base. Slightly longer and than the Lazzeroni Patriot, the .300 Winchester Short Magnum duplicates its performance—which is to say, it shoots as flat and hits as hard as a belted .300 Winchester Magnum. Like the Lazzeroni, it fits in .308-length actions. Overall cartridge length measures 2.76 inches, half an inch less than that of the .300 Winchester. In fact, a .300 WSM round is shorter than the *hull* of a .300 H&H, the granddaddy of all belted magnums. Winchester's new short magnum outperforms the Holland by nearly 100 fps.

Browning actually came to Winchester with the idea for the round early in 1999. Winchester ammunition engineers determined final dimensions. Browning and U.S. Repeating Arms Company (USRAC, manufacturer of Winchester firearms) redesigned their flagship bolt rifles for the .300 WSM. Browning has tooled up to chamber A-Bolts in .300 WSM. You can pick from Hunter, Stainless Stalker, Composite Stalker, and Medallion configurations, all with 23-inch barrels. The average weight is slightly under 7 pounds. Or choose a Winchester M70 Classic Featherweight; at 7¼ pounds, it is still light enough to carry day after day in rugged country. Its 24-inch barrel and short action make for good balance and quick handling without sacrificing much bullet speed. These Browning and Winchester rifles hold three WSM rounds in the magazine.

WSM factory loads include two in Winchester's Supreme line: a 180-grain Fail Safe at 2970 fps and a 150-grain Ballistic Silvertip at 3300 fps. Muzzle energy for these bullets exceeds 3500 and 3600 ft-lbs. The 150's sleek form means a high ballistic coefficient; it drops only 16 inches at 400 yards with a 200-yard zero. The 180 Fail Safe sinks 21. The lighter bullet also delivers 1940 ft-lbs of energy at 400 steps, 140 more than the Fail Safe. But savvy hunters recognize that sometimes a shot requires the penetrating power of a Fail Safe. The other .300 WSM factory load comes under the Super-X label and features a 180-grain Power-Point. This bullet won't drive as deep or retain as much weight as the Fail Safe, or match the Ballistic Silvertip's trajectory; but it is a dependable bullet, delivering a bit more downrange punch than the Fail Safe. Super-X cases aren't nickeled like the Supreme hulls and they cost less. Like John Lazzeroni's stubby cartridges, the .300 Winchester Short Magnum in all its guises gives you 400-yard reach in a compact package. As Lou Palmisano demonstrated decades ago, there's also potential for fine accuracy.

You can't tell everything about a cartridge by its profile.

The .300 Winchester Short Magnum duplicates the ballistic performance of the .300 Winchester Magnum, which accounted for this fine Utah bull.

.33s: Coming Back Strong

In 1902 Winchester announced a new .33 cartridge for the Model 1886 lever rifle. Its 200-grain flatnose bullet sped away at 2200 fps, carrying well over a ton of energy. It struck a mighty blow indeed, compared to the .30–30, whose original loading called for a 160-grain bullet at only 1970 fps. It beat the .303 Savage, with a 195-grain bullet starting at 1950 fps. It generated 40 percent more energy than the .45–70: 2300 ft-lbs. Downrange the .33 held up well against other cartridges hurling flatnose bullets. But it couldn't stay ahead of the .30–40 Krag. At 100 yards the .33, which started 200 fps faster, had only a 70-fps edge. An energy advantage of 200 foot-pounds had become a 40-foot-pound deficit. At 200 yards the Krag's bullet was ahead by about 50 fps and had widened the energy gap to 170 foot-pounds. Beyond 200, the Krag was clearly superior.

It's easy to see why the .33 Winchester faltered at long range. A Krag bullet, hardly streamlined by modern standards, had a ballistic coefficient of about .300. The .33's bullet—lighter, larger in diameter and with a flat nose—mustered a ballistic coefficient of only .200. Still, the new round should have been successful. In 1902, hunters used iron sights and preferred lever-action rifles. They were used to bullets much larger in diameter than the .33 and equated a big nose with a big hole through the vitals. At the time, 2200 fps was a high-water mark on velocity charts. But Winchester kept the .33 in its line only until 1936, when the Model 71 rifle supplanted the '86. The 71 brought with it a new and more powerful cartridge, the .348 Winchester.

When Winchester introduced its .338 Magnum in 1958, it was chambered in a Model 70 called "the Alaskan." Modestly popular for decades, the .338 is now among the most widely used rounds for elk.

There's nothing anemic about the .33 Winchester, even today. The .35 Remington, introduced in 1906, was designed to give a 200-grain bullet the same speed and is still in production (despite consistently clocking well below factory claims). Alas, the .33 did not share the .35's short, rimless (modern) case, and Winchester made no effort to find a home for it. The .33 had been offered only in Models 1886, 1895, and

Ron Holden killed this wide-antlered bull with a well-placed shot at 300 yards from a custom-built .338. The bullet was a Nosler Partition.

single-shot 1885 rifles. Remington chambered the .35 in the autoloading Model 8 and slide-action 14, then in the Model 81 and 141. It hopped from one rifle to the next (appearing briefly in the Winchester Model 70) and survives in Marlin's sturdy 336. You can even order it in a new Model Seven Remington.

But outliving the .33 counted for little as another competitor emerged to set a trend for American big game cartridges. The .30–03, redesigned in 1906, earned a new name as its 150-grain bullet attained the scorching speed of 2700 fps. The popularity of the .30–06 in post-war hunting camps led to the 1925 introduction of an even racier cartridge, the .270. Ever since, hunters have looked for higher velocity, flatter bullet flight, and bigger energy payloads downrange.

Charlie O'Neil, Elmer Keith, and Don Hopkins experimented with front-end ignition and duplex loads in the .338–06, and in the .334 OKH, a more potent round based on the .300 H&H Magnum hull. They installed a tube in the center of the cases to carry primer spark forward. While these experiments failed to improve on standard ignition, and duplex loads showed no more velocity than they could get with a single powder, the men found that both cartridges were excellent performers on elk-size game. The .334, also initially sized for Jeffery bullets, had about 250 fps on the .333. Rumor was that the higher velocity proved too much for the English jackets, designed for impact speeds of 2000 to 2400 fps.

A paucity of suitable bullets may have hindered other efforts to develop new .33s. But in 1958 Winchester introduced its .338 Magnum, two years after announcing its first belted cartridge, the .458. On the heels of the .338 came the .264. Like its siblings, it had the .532 head of a common ancestor: the .300 H&H Magnum. Winchester's cartridges were "short" magnums, the cases 2.50 inches long compared to 2.85 inches for Holland's "Super .30" (the .300). The new rounds fit in a standard .30–06-length action.

Lapua's .338 was developed as a military sniper round. It performs with the .340 Weatherby. The rimless cases are much heavier. Dakota and Sako chamber the .338 Lapua in commercial rifles.

Even by today's standards, the .338 Winchester Magnum is a handful. It drives 200-grain bullets at 2900 fps, 225s at 2800, and 250s at 2700. Federal offers a couple of High Energy loads: a 225-grain Trophy Bonded at 2940 fps and a 250-grain Nosler Partition at 2800. Both shoot as flat as the 180-grain bullet from a standard .300 H&H load, and of course they deliver more energy. At the muzzle, both the 225- and 250-grain spitzers are packing over 4300 foot-pounds. By the time they pass the 200-yard mark, the higher ballistic coefficient of the heavy bullet has whittled away at the 225's initial velocity margin and delivers 250 more foot-pounds than the 225, which dumps an even 3000. Hornady also markets a high-octane .338 round. Its "Heavy Magnum" load pushes a 225-grain Spire Point at 2950 fps.

Dakota's "Longbow" is a tactical rifle available in .338 Lapua. It's shown here with a Leupold tactical scope.

When I first read about the .338 Winchester Magnum, I was young enough to be awed by its great power. Winchester dubbed the Model 70 .338 the "Alaskan," and for some years it proved a steady but uninspiring seller. Shooters used to killing deer with a .270 or a .30–06 had little need for a round with the brutal recoil of the magnum, which they viewed as exceptionally powerful and best suited to animals the size of brown bears. But over time, elk hunters came to see the value of its reach in open country and of its ability to anchor a bull elk with a raking shot in timber. In surveys I ran during the 1980s and early '90s, the .338 stayed on the list of five most popular elk cartridges. It was the only .33 in the top 10.

I've seen a number of elk shot with the .338. One big six-point came to us bugling, and my client hit him in the chest at 80 yards. The animal collapsed on the spot. Another took a bullet obliquely through the paunch at 150 yards and ran—but then paused on a ridge long enough for the hunter to send his second Nosler through the lungs. The bull died right away. An elk I shot in a fir thicket at 40 yards proved too tough for the bullet that disintegrated on his shoulder, but two followup shots put him down. Another time, I watched a hunter miss a standing bull, then jam his rifle. The bull trotted toward the lodgepoles. As I was hunting too, I pinned my crosswire to the point of his shoulder and triggered a shot just before he vanished. We trailed this five-pointer 50 yards and found him dead. A .338 isn't magic, but it *is* potent.

The .340 Weatherby Magnum appeared four years after the .338, and no doubt because the .338 showed market promise. My field experience with Weatherby's .340 has been limited, simply because I seldom need a cartridge with that much power. Depending on the load, this full-length magnum has about 250 fps on the .338 Winchester with identical bullets. It shoots flat enough to permit a 300-yard zero, and carries a ton of energy to 500 yards. It is one of the nicest-looking cartridges around, tall and leggy with gracefully curved shoulders and a long neck. The .340 is also relatively efficient, a super-achiever that gives you a lot for the powder charge. Experience has shown me it's accurate as well. One of many rifles I regret sending off to another home was an early Weatherby Fibermark in .340. It printed lovely triangular groups just a little over an inch to the side. It seemed to recoil less than most .338s too—thanks largely to the high Weatherby stock comb that supports your face and tapers away from it toward the comb nose. Pre-64 Winchester M70 stocks come at you like a Mike Tyson hook.

There aren't many long-range cartridges with more punch than a .340 Weatherby. When it was introduced in 1962, wildcatters had already necked the .378 Weatherby Magnum's bigger case to .33. But very few .378s were in circulation, even nine years after its debut. A .338–378 certainly had more moxy than a .340, but required a bigger action. The .340 has a loyal following; factory ammo is even available from Federal (a 225-grain Trophy Bonded at 2900 fps). The case holds enough fuel to accelerate 225-grain bullets to 3100 fps. I've loaded 250-grain

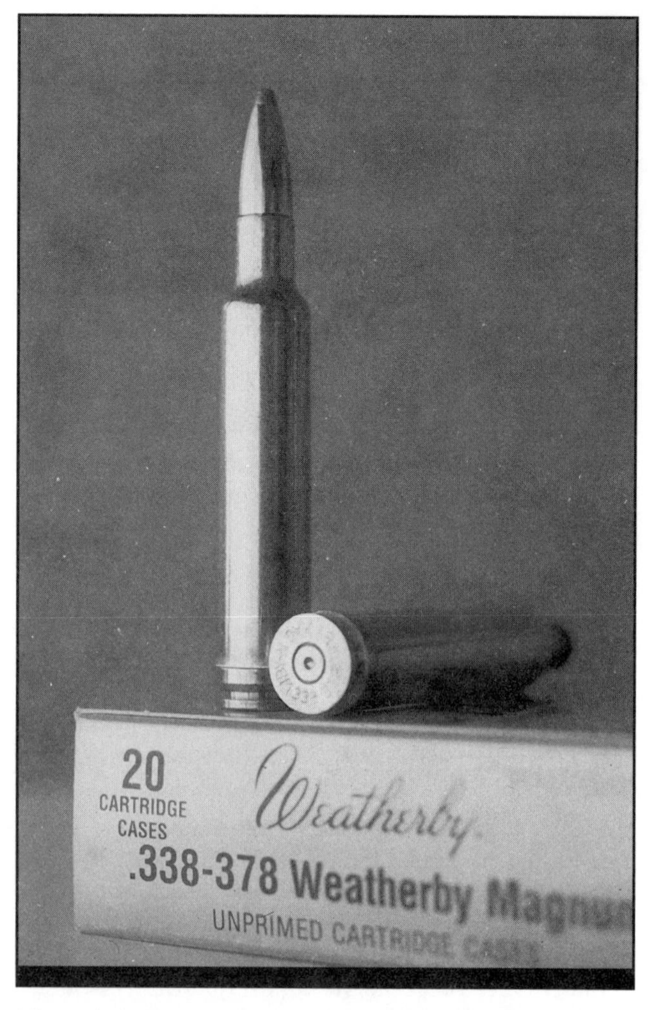

The .338–378 Weatherby is the most powerful of 33-caliber magnums. The Lazzeroni 8.59 Titan delivers about the same speed, energy—and recoil!

spitzers to 3000. They hit like jet-powered wrecking balls.

The .330 Dakota has about the same punch. At 2.540 inches, the .330 hull isn't much longer than a .338 Winchester's, but it holds more powder. Don Allen fashioned the .330 Dakota (and its 7mm, .300 and .375 mates) on a shortened .404 case. It will fit a .30–06-length action. The Dakota 76 action is built specifically for the wide-bodied .404. Dakota's line of cartridges appeared in the early 1990s and has since grown with the addition of big-bore rounds on the full-length .404 case.

Unlike Don Allen, entrepreneur John Lazzeroni had little to do with the shooting industry when he decided the world needed another high-performance .33. In the late 1980s he was experimenting with some potent wildcats, including the .338–378 Weatherby. His 8.59mm Titan was the first of what would become a line of Lazzeroni rounds, all on John's own rimless hulls. The Titan, a .33 with almost as much capacity as a .338–378, went through several renditions before MAST Technologies turned out its first batch of cases. At .576, the Titan's base diameter is essentially that of a .338–378 without the belt. It's about .008 smaller than a .416 Rigby's, a case John might have used if he had known a little more about it.

Within a whisker of matching the .338–378 for case capacity, the 8.59 Titan may have more "awe factor" than utility. John eventually gave shooters a more efficient option. He shortened this big case and came out with 7mm, .30, and .33 rounds that look like giant PPC cartridges. The .33, named the Galaxy, holds about as much powder as the longer .338 Winchester and matches it ballistically.

Anyone in touch with the industry could have predicted commercial production of Weatherby's .338–378. It launches the biggest bullets that can be considered *long-range* hunting bullets, and it sends them off faster than any other round. I saw my first .338–378 many years ago, when it was still a wildcat. Between stages at a smallbore prone match in rural Oregon, one of the riflemen broke out a long-barreled bolt gun not quite heavy enough for an artillery carriage. He bellied onto a sandy rise and proceeded to demolish rocks on a hillside hundreds of yards away. They were big rocks, and they splintered when struck by those heavy bullets. The blast, I remember, rattled the clubhouse windows. Without its muzzle brake, I expect that rifle's recoil would have plowed furrows with the toes of the man's boots!

The .338–378 became available as a commercial load in 1999. Shortly thereafter, I borrowed a Mark V from Weatherby and benched it. The 250-grain Noslers from Weatherby's ammo gave consistent readings over the Oehler chronograph, averaging 3005 fps from the Accumark's 26-inch barrel. That's 55 fps lower than advertised—close enough, on a cold day. I was mightily impressed by the accuracy: about a minute of angle. At 300 yards one three-shot group measured 2½ inches. A friend shot the rifle

Remington's .300 Ultra Mag (left) was based on a .404 Jeffery case (dimensions differ slightly). It spawned the 7mm and .338 Ultra Mags. The .338 UM performs like the .340 Weatherby.

and beat my best 200-yard group with one that crowded 1½ inches. Our bullets seemed to arrive quick as lightning.

This rifle was comfortable to shoot with the factory-installed brake, though under the tin roof even double ear protection didn't tame the blast. Thoughtfully, Weatherby provides a small rod with which to remove the brake, and a knurled cap to replace it. I had to feel this rifle unshackled, so I fired an offhand shot without the brake at a 300-yard target. The kick was not only hard, but quick. I've fired my share of big-bore rifles, including a .700 Nitro Express. The .338–378's recoil is as invigorating as any. It lifted the heavy rifle almost out of my hands—though I confess to using a light grip on the forend—and spun it so the scope banged my skull. Enough of that fun! I was pleased to find the bullet hole just 5 inches to five o'clock of center on the 300-yard target. Here are Weatherby's chart figures for the .338–378:

The drop figures here are predicated on a 300-yard zero. For most cartridges, a 300-yard zero may be too long, because it would put the point of impact uncomfortably high at mid-range. But the .338–378 shoots as flat as a 7mm Weatherby Magnum, an archetype among long-range hunting cartridges. With the 300-yard zero, a .338–378 gives you a mid-range rise of only about 4½ inches, depending on the bullet. So with this powerful cartridge you can hold in the vertical middle of most animals out to 350 yards.

The .338–378 has no close rivals, save the 8.59 Lazzeroni Titan. At 200 yards either will generate a crushing two tons of energy, as much as the .338 Winchester Magnum can claim at the muzzle and about equal to the payload of a .340 Weatherby Magnum at 100 yards. The .338–378 delivers more energy at 500 yards than the .33 Winchester does at the muzzle (and more than the .30–06 at 100 steps). A 250-grain bullet from the .338–378 Weatherby Magnum (starting speed 3060 fps) arrives at 500 yards with 2500 foot-pounds of energy, and it's still moving at 2125 fps! A .338–06 hurling the same bullet (starting speed 2400 fps) brings 1420 foot-pounds to 500 yards; but velocity has dropped to 1600 fps, marginal for bullet upset.

Another hyper-caffeinated .33, the .338 Lapua, was developed as a sniper round, bridging the gap between the .308 Winchester and .50 BMG. Lapua, a 75-year-old Finnish company, is better known in the United States for its rimfire target ammunition. But this rimless .33 strikes me as one of the best-designed of its class. A 250-grain boat-tail bullet in front of 88 grains IMR 4350 and Federal 215 primers clocks 2890 fps. In tests at Quantico, Virginia, a custom-built Keberst rifle gave half-minute accuracy at 1000 meters. The best groups at 500 meters measured just over 1 inch.

Because of its military heritage, the .338 Lapua has probably undergone more long-range testing than any other .33. At 1000 meters this .338 yields

.338–378 factory loads

bullet weight/type	*muzzle velocity/energy*	*500-yard velocity/energy*	*drop, 400*	*drop, 500*
200 Nosler Ballistic Tip	3350 / 4983	2232 / 2213	8.4	22.9
225 Barnes X-Bullet	3180 / 5052	2238 / 2501	8.9	24.0
250 Nosler Partition	3060 / 5197	2125 / 2507	9.8	26.4

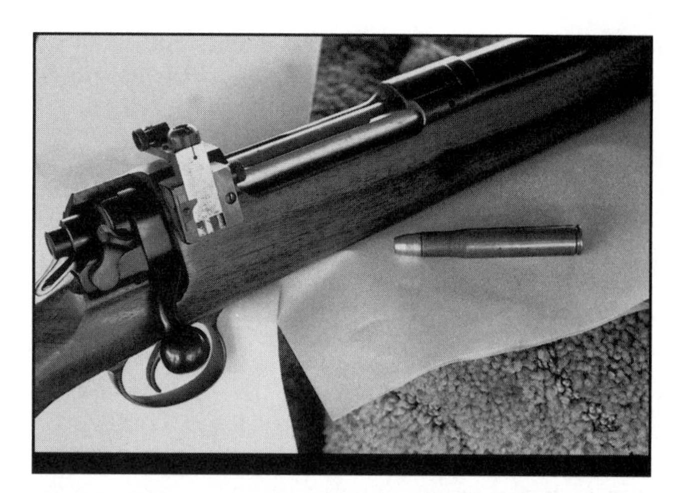

The .404 Jeffery dates to 1910. Its rimless case inspired the .330 Dakota and .338 Remington Ultra Mag rounds. The .404 rifle here was built on a 1917 Enfield action.

1308 foot-pounds of energy, the .308 Winchester just 221. The Lapua bullet drops 13 inches at 300 meters and 370 inches at 1000, while the .308 bullet strikes 16 and 506 inches low. The .338 Lapua shows about half as much wind drift as a .308. The only liability is recoil. With roughly the same case capacity as Weatherby's .340, the Lapua does punish you.

The .338 Lapua is chambered in two commercial rifles stateside: the Sako TRG and the tactical Dakota Longbow. I used a Sako to chronograph handloads. Dakota's Don Allen says his Longbow project began at the request of the Finnish government. It has since sparked interest from the FBI, and several U.S. police departments now have Dakota sniper rifles. They're built on the Dakota 76 action with a Model 70-style trigger set at 2.5 pounds, and can be had with either a blind magazine or a single-shot platform. The stainless Lothar Walther barrel (1-in-10 twist) is 28 inches long and .950 at the muzzle. A McMillan A-2 stock with adjustable cheekpiece is built to use with the bipod supplied. The rifle, 51 inches long with its muzzle brake, weighs 13.7 pounds. Don Allen says the Longbow can deliver half-minute accuracy.

Norma provides Lapua brass to Dakota, which also offers loaded ammunition. Bullet choices include the 230-grain Fail Safe, 250-grain Swift A-Frame, and 300-grain Sierra MatchKing.

The Lapua's head is the size of a .416 Rigby's, though the body is shorter. It requires a Rigby-size bolt face. Mechanisms designed for our short magnums must be opened up to feed the Lapua ammo. On my scale, Lapua hulls weigh half again as much as .340 Weatherby cases. Don Allen cautions against equating web thickness with case weight, however. Often brass is heavier simply because it is denser.

The most recent .33 at this writing is Remington's .338 Ultra Mag, the anticipated sequel to the .300 Ultra Mag. As loaded by Remington (a 250-grain Swift at 2860 fps), it comes within a quick breath of matching the .330 Dakota. The case is longer, though. Handloaders should like it and get more out of it. Remington claims that its Ultra Mag load delivers 12 percent more muzzle energy than Federal's High Energy .338 Winchester Magnum ammo, and 25 percent more than standard loads for that round.

cartridge comparisons: .33s

cartridge, bullet weight and type	*velocity, fps*		*energy, ft-lb*		*drop, in*
	muz.	*300 yds*	*muz.*	*300 yds*	*300 yds*
.33 Winchester, 200 Hornady flatnose	2200	1222	2149	663	21
.338–308, 210 Nosler Part.	2600	1982	3152	1831	10
.338–06, 200 Hornady Spire Point	2700	2005	3237	1784	9
.338 Winchester Mag., 230 Fail Safe	2800	2202	3980	2472	8
.340 Weatherby Mag., 225 Barnes X	3001	2434	4499	2959	7
.338 Rem. Ultra Mag., 250 Swift A-Fr.	2860	2244	4540	2794	8
.338 Lapua, 250 Sierra	2970	2502	5284	3770	6
.338–378 Wby. Mag., 250 Nosler Part	3060	2475	5197	3401	6
8.59 Lazzeroni Titan, 250 Nosler Part.	3100	2509	5344	3494	6

Note: This data was taken from catalog listings and exterior ballistics charts. Commercial ammo may not match catalog claims. Changes in barrel length typically amount to 30 fps per inch, but that figure varies. Most high-velocity cartridges shown here were chronographed in 26-inch barrels, but the .338–308 was clocked in a 22-inch barrel, the .338–06 in a 24-inch. Lapua data came from a 24-inch barrel.

The Biggest Bores You Should Need

Thirty-five-caliber rifles were mildly popular for a time, long ago. Winchester had a .35 cartridge, along with its .33. Remington's short rimless .35 has plodded across the decades much more successfully than rounds like the .35 Newton, a powerful cartridge years ahead of its time. American hunters were hooked early on by the .30-bore's faster bullets, however. Higher velocity meant extra reach and energy, while big bullets just left big holes. Slower progressive-burning powders in big cases, plus strong bullets in 7mm and .308, enabled shooters to get game-stopping punch with flat trajectory and relatively light recoil.

By last count there were seven rifles with .358 bores loafing in the overcrowded corners of my office. One is a particularly lovely rifle, built on early Model 70 metal and stocked by Gary Goudy. The French walnut is nicely detailed, perfectly laid out and features strong but not overbearing figure, black on dun. But what's different—and, to me, appealing—about this rifle is its chambering: .350 Griffin & Howe.

The .350 G&H was one of the first truly powerful .35s to come along. Reportedly, the maiden rifle was commissioned right after the Great War by Leslie Simpson, a widely traveled and well-known big game hunter. Simpson wanted an American round that would match the power and exceed the range of British big-bore cartridges he'd used successfully in Africa. Griffin & Howe necked the full-length .375 H&H case to .35 to give a 275-grain bullet 2440 fps and a 220 bullet 2790. Legend has it that the light bullet jackets of the day fragmented when hunters fired .350 G&Hs into big animals up close.

The black-powder .35 cartridges that preceded the Great War had all the snap of a funeral hymn. The .35–30 Maynard, introduced with the 1882 Maynard single-shot rifle, nudged a 250-grain bullet along at around 1200 fps. A more potent .35–40 in

From left: the .35 Whelen is a necked-up .30–06. Remington has loaded it commercially since 1988. The .358 Winchester, on the .308 case, is more potent than it looks. It followed the .348 Winchester, loaded for the company's Model 71 rifle, which was produced from 1935 until the mid '50s.

the same rifle gave that bullet 1350, for about 1000 foot-pounds of energy. In 1905 Winchester announced its semi-rimmed .35 Self-Loading and the rifle to fire it. That cartridge lasted only until 1920, evidence that even in gentler times few shooters wanted a 180-grain flatnose bullet clocking 1452 fps. Like the later and similarly anemic .30 Carbine, it proved more popular with law enforcement people than with hunters.

But the New Haven firm had already fielded a much better .35. In 1903 it began offering its 1895 lever-action rifle in .35 Winchester, which booted a 250-grain bullet out at nearly 2200 fps. Muzzle energy totaled 2670 foot-pounds, making this rimmed round an excellent choice for any North American big game at woods ranges. The .35 Winchester survived until 1936, when it was replaced by the company's .348.

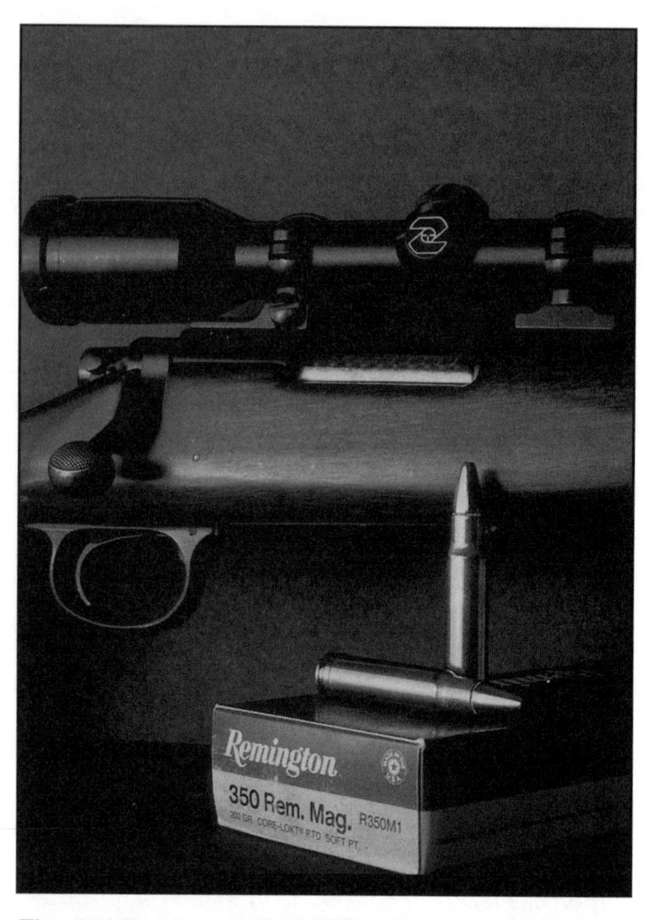

The .350 Remington Magnum was introduced in 1965. Its 2.17-inch case was shared with the 6.5mm Remington Magnum that came a year later. The .350 is essentially a short-action .35 Whelen.

In 1907 Winchester had another go at a .35 for autoloaders. The straight-hulled .351, like the .35 Self-Loading, had limited case capacity. Still, its 180-grain bullet delivered much more energy: 1370 foot-pounds. This cartridge featured a .351 bullet, not a .358 as loaded in other .35s. Winchester's improved Model '07 autoloading rifles became popular with police and were even used by French forces during both world wars. The .351 stayed in production until 1957.

Remington's first smokeless .35 has been a steady success. Though few hunters pay it much attention these days, the .35 Remington has logged nearly 100 years on the production line. Developed for the firm's Model 8 self-loading rifle, it later appeared in the Model 81, and in the slide-action 14, 141, and 760 rifles. I'd like to have back the 760 .35 I once owned. With a 200-grain roundnose bullet at 2210 fps, this cartridge yields over a ton of muzzle energy and in Marlin's immortal M 336 lever rifle has taken countless whitetails and black bears. Remington also loads a 150-grain pointed Core-Lokt, but its sleek form and higher speed give deep-woods hunters less benefit than the higher sectional density of the heavy bullet. In handguns like Thompson-Center's Contender, the .35 Remington is a typhoon, generating about all the recoil most shooters can handle. It is efficient in short barrels.

The most potent .35 during the pre-War years appeared in 1915. The rimless .35 Newton drove a 250-grain bullet at over 2700 fps, matching the ballistic performance of the later .338 Winchester Magnum! Its failure at market was due largely to the ill luck that plagued Newton's efforts to sell his own bolt rifles.

Another useful .35 appeared in 1922. The .35 Whelen may have originated in Colonel Townsend Whelen's shop or in that of his friend James V. Howe of Griffin & Howe. Soon after World War I, Whelen served as commanding officer at Frankfort Arsenal, and Howe worked there as a toolmaker. Surely both men saw the potential in necking the .30–06 to accept .358 bullets. This cartridge and the .400 Whelen that followed gave Springfield owners an easy route to bigger bullets and more up-close punch. Case forming was a snap because headspace stayed the same. A military bore pitted by potassium chlorate primers could become, at little expense, shiny and new and .35.

The .35 Whelen remained a wildcat until 1988, when Remington began loading it with a 200-grain bullet at 2675 fps and a 250-grain at 2400. Initially the heavier bullet (in my view the most practical for big game) had a round nose. Later it was replaced by a pointed Core-Lokt. This bullet's higher ballistic coefficient puts it neck-and-neck with the 200-grain at 300 yards. Striking energy for the heavy bullet is higher at *all* ranges. I've found the 250-grain bullet about as flat-shooting as a factory-loaded 180-grain spitzer from a .30–06 out to 200 yards, making it the hands-down choice for stout animals like elk.

The Model 700 Classic that Remington used to launch the .35 Whelen is still among the top picks if you're looking for an inexpensive but accurate and well-balanced hunting rifle. Its 22-inch barrel is about right for the .35 Whelen. I used such a rifle in British Columbia some years ago, hunting mule deer with my friend Stuart Maitland. One day, after glassing clumps of tag alder high on a steep, snowy hill, I spied an antler fork. Eventually the body of a huge buck took shape. I tried a stalk but immediately ran into trouble on the icy slope. Clutching alders to stay on the mountain, I backed out—just in time to catch the buck sneaking away over a ridge. My offhand shot caught him in front of the ham, the 250-grain Core-Lokt exiting high between his shoulders and chipping the base of a thick antler. He stayed on his feet, however, and plunged downslope. I missed a hurried followup shot, then charged after him. He bedded shortly, and another bullet ended the chase.

While bolt rifles are the choice of most hunters now, tube-fed lever-actions were big sellers up to World War II. Winchester's last new rifle of this type was the 71, introduced in 1935 and chambered in .348 Winchester. This cartridge featured flatnose bullets that weighed 150, 200, and 250 grains. The 200-grain load survived longest, though the 250 was clearly most practical, given the bullet's shape and the 71's mechanism and iron sights. It delivered nearly a ton and a half of energy at the muzzle, making it a favorite of elk hunters. The 71 was a woods gun from butt to muzzle, well-balanced, smooth-shucking and with a handsome, muscular profile that reflected its Model 1886 heritage. Offered as both a rifle with a 24-inch barrel and a carbine with a 20-inch, the Model 71 appeared just two years before Winchester's Model 70, a rifle that would become the firm's flagship and an archetype for post-war big game guns. The Model 71 survived into the 1950s, but a trend to scope sights and long shooting made it an anachronism almost before the Model 70 arrived. Browning resurrected the 71 briefly a few years ago. Winchester loaded the ammo.

The .358 Winchester was chambered in the company's Model 88 lever action in the 1950s. Neither generated brisk sales. The .358 is an efficient killer on game as big as elk at ranges to 200 yards.

About the time .348 rifles became history, Winchester announced the .358, an unpretentious round based on the rimless .308 Winchester (military 7.62 NATO). Chambered in the Featherweight Model 70 and a new hammerless lever-action 88, the .358 caught on slowly. It delivered essentially the same ballistic performance as the .348, but it was considerably smaller. A 250-grain roundnose loading has vanished, while a 200-grain bullet stoked to 2500 fps (2750 foot-pounds) remains. The .358 is a fine close-cover round for animals as large as elk and moose, especially now that there's a wide selection of game-worthy 35-caliber bullets. The .358 is about as much muscle as can fit in a short rifle action, but it has fallen on hard times. Savage chambered it in the Model 99 for a while, and Ruger built some .358 Model 77s. The Browning BLR chambered it. Mannlicher-Schoenauer manufactured .358 carbines during the late 1950s.

Norma of Sweden developed a .358 short magnum and made it available to U.S. hunters in 1959. Market response was predictably flat because no American rifles were chambered for it. A year later the Swedish firm of Schultz & Larsen began to list the .358 Norma Magnum, as did Husqvarna. Sales of these rifles lagged, partly because in 1958 Winchester had announced its .338 Magnum, which didn't quite match the Norma .35 ballistically but still dished out enough energy for any North American game.

The .358 Norma Magnum is a fearsome round, cracking a 250-grain bullet down the bore at 2800 fps for well over two tons of muzzle energy. It is a trifle less potent than the .375 H&H but can be used in an action of standard length. The recent bloom of controlled-expansion 35-caliber bullets gives shooters many options. I prefer the 225- and 250-grain bullets to anything lighter, because the Norma's case is plenty big enough to kick them along chalk-line straight to great distances. My own .358 Norma is built on a Mark X Mauser action and wears a 2.5X Lyman All American scope. Its 25-inch Douglas barrel helps deaden the kick, but after an hour at the bench with this rifle I'm ready to take up croquet.

In 1965 Remington surprised the shooting fraternity with a *true* short magnum, one that fit rifle mechanisms designed for the .308 Winchester. The .350 Remington Magnum featured a .300 H&H case trimmed to 2.170 (the .300 H&H is 2.850, standard short magnums like the 7mm Remington and .338 Winchester 2.500). The .350 made its debut in the Remington 600 carbine, which initially had an 18-inch barrel. In 1968 barrel length was increased to 20 inches. The .350 was the most potent of several Model 600 chamberings. Late-model .350s wore an attractive laminated stock and were known as the Model 660. Due in part to their non-traditional lines, including a dogleg bolt handle that *nobody* found attractive, these potent carbines sold poorly. They were discontinued in 1971. Remington kept the .350 Magnum alive by chambering it in the Model 700, and Ruger listed it for a time in its Model 77. Factory charts show a 200-grain bullet at 2710 fps. Now .350 Model 660 carbines have come back into favor and bring big money.

Winchester has been loathe to abandon the .35 bore. Lukewarm reception of the .358 close after the demise of the Model 71 brought no joy to New Haven, but in 1980 the company announced the new .356 Winchester. Made for a revamped Model 94 with thick action walls and low-angle ejection, the .356 is essentially a rimmed .358. Its heavier case and deep-seated bullet keep it from matching the .358 ballistically, as does the 94's rear lockup. Still, a 200-grain bullet leaves the muzzle at 2460 fps, 10 percent faster than a .30–30 pushes a 170-grain bullet. Both the .356 and its .30-bore sidekick, the .307 Winchester, are limited by tube magazines to flat-nose bullets, a significant drawback for 200-yard shooting.

The latest .35 to get attention from American shooters is the .358 STA (Shooting Times Alaskan). Like the 7 STW, the Alaskan is based on 8mm Remington Magnum brass. Both the STA and STW are full-length magnums. The .358 STW outperforms its forebear, the .350 Griffin & Howe, and even the ferocious .358 Norma Magnum.

One reason 35-caliber cartridges have languished is that until recently very few first-class bullets have been available to handloaders—or even in

Medium-fast powders like these are appropriate for .35s from the .358 Winchester to the .35 Whelen.

commercial loadings. Early .35s were typically roundnose missiles built to open in deer at .35 Remington velocities. Now hunters have a plethora of fine .35 bullets from which to choose. I used a 250-grain Speer Grand Slam in a .35 Whelen Improved to shoot an elk at about 200 yards. The bull fell over right away. The Improved fire-formed easily, and I got good accuracy; but because the Whelen has a relatively straight case to begin with, and a small shoulder, "improving" the hull by reducing taper and steepening the shoulder to 40 degrees gave me little net return.

If you're willing to go to more trouble, you can make a very potent .35 from a .30–06 case. The .35 Brown-Whelen appeared in 1967, when JGS Die & Machine of Coos Bay, Oregon ground a reamer from a hull supplied by Keith Stegall. C. Norman Brown of Anchorage, Alaska apparently determined the dimensions. The Brown-Whelen has more case capacity than the .35 Whelen because its shoulder is farther forward. In case forming you must first expand a .30–06 neck to .375 or .400, then neck it back down to .358 to establish the shoulder, then fire-form. An alternative is to neck the case up to .358, then seat bullets out against the lands to maintain proper headspace during fire-forming.

A modern version of the Brown-Whelen is the .358 Hawk. The Hawk's body is blown out to .454 at the shoulder, adding .013 to the shoulder diameter of an '06. Capacity is about the same as that of the Brown-Whelen, though the shoulder is not as steep. Handloads I've tested have clocked 2850, 2760, and 2700 fps with bullets weighing 200, 225, and 250 grains respectively, in a 26-inch barrel.

Heavy, large-diameter bullets can't be driven as fast as smaller bullets, or as flat. But they make big holes, and bullet weight retention isn't a burning issue. A .348 Winchester bullet that loses a third of its mass during penetration weighs more in its spent condition than an *unfired* .270 bullet. Modest velocities also reduce the chance of bullet failure. Big case mouths make fast powders efficient, so hunters didn't need long barrels. The .35s kill reliably at the ranges most big game is shot, and without as much meat damage as caused by high-speed bullets.

Shotguns and Shotshells

England nursed the shotshell into its present form. Its beginnings might be traced to the ramming of nails and rocks into the mouths of culverins. A spray of projectiles multiplied the chance for a hit at close range. Later, in sophisticated cannon, grapeshot gave more uniform coverage and greater range. The principle applied to hunting guns found its niche in bird coverts. Multiple balls lost their energy quickly and were hopelessly outclassed by rifle bullets for big game; but a cloud of fine shot could down a flying bird flushed close.

A shotgunner's range jumped significantly with the development of choke, or the constriction of a smoothbore barrel. A landmark discovery at a time when firearms design was in a growth spurt, choke boring came along just after the U.S. Civil War. William Pape, an Englishman, investigated the effects of bore constriction on shot patterns in 1866, but little came of the notion until American market hunter Fred Kimble began his experiments four years later. Kimble's first trials left him discouraged: Patterns from a tight choke proved open and erratic. To salvage the barrel, he reamed out the choke. Accidentally or on purpose, he didn't remove it all. His next patterns jumped to nearly 100 percent (pellet count in a 30-inch circle) at 40 yards. Kimble switched to choked barrels for his waterfowl hunting, once killing 57 bluebills with 57 shots. Slaughter like this would not play well now, but it showed the effectiveness of choke.

Traditionally, American chokes have been reamed from a thickened 2- or 3-inch section of barrel near the muzzle. Longer, more gradual chokes have been shown to increase pellet friction and shot deformation; short chokes require less constriction for each desired boost in pellet count on paper. The adjustable Poly-Choke of the 1950s enabled hunters to change the constriction in the barrel by twisting a dial. Effective, but ugly and unsuitable for double-barrels, the Poly-Choke eventually gave way to the interchangeable choke tubes now so popular. These slip inside the barrel so can be used with doubles. They're lightweight so don't affect a shotgun's looks or balance, and they're quick and easy to install.

Common American choke designations—full, modified, and improved cylinder—do not indicate muzzle diameter, only relative constriction. The English measure choke in points. A 12-bore barrel .730 inch in front of the chamber has a constriction of 40 points if the muzzle shows an inside diameter of .690. Such a choke, roughly equivalent to "full" in American nomenclature, is designed to give 70-percent patterns at 40 yards. To further complicate things, American shotgunners sometimes speak of "three-quarter" choke (30 points on the British chart) and "half" choke (20 points). Three-quarter translates to improved-modified. Half is modified, a constriction that should give 60-percent patterns. Quarter-choke, or a 10-point squeeze at the muzzle, ideally produces a 50-percent improved-cylinder

Modern high-performance shotshells have traditional high-brass heads—though the height of the brass has nothing to do with shell strength. Sealed plastic crimps feed smoothly and keep water out.

pattern. Skeet #1 and #2 correspond to cylinder (no choke) and improved-cylinder. They are no longer popular American terms because skeet guns vary so much in their choking.

By 1880, English sportsmen had established the "proper" way to shoot birds. English gunmakers provided their tools—sleek flintlock fowling pieces that were as much art as firearm. In the U.S. shotguns were considered tools, made to be used more than admired. Form, fit, and workmanship mattered less than durability.

In the birthing days of cartridges, waterfowlers on both sides of the Atlantic endured paper-hulled shotshells that swelled when wet and refused to chamber. Damp powder caused misfires; soaked hulls fell apart. Disgruntled shooters eventually turned to metal shotshells, made like the rifle cases for big-bore guns used by African explorers of the time. These drawn-brass cases came to be known as "solids" (not to be confused with the non-expanding or full-patch hunting bullets of the same name). They were frightfully expensive, worth the cost if dangerous game was the quarry but sure to hike the price of a duck dinner beyond reason. They survived for a time as shotshell hulls because the people who used them were wealthy. The cumbersome 10-, 8-, and even 4-gauge shotguns popular then often went to the marsh on the shoulder of a servant, so a few extra shillings for waterproof shells didn't bankrupt most hunters.

Less expensive ammunition arrived in the "Perfect" or thin brass hull, made by attaching a tube of drawn brass to a standard shotshell head. The case was crimped just enough to hold a top wad, which was shellacked to hold it and seal the crimp. Manufacturing cost was substantially less than that of a fully drawn case. The thin brass made the internal diameter of the hull larger than that of standard paper cases, however, necessitating the use of different wads and shot and powder charges. A 12-bore Perfect cartridge might have the same charge weights and deliver the same performance as a 10-gauge paper shell. Though Perfect cartridges filled a need, they eventually fell victim to escalating prices of copper and zinc in Europe during the 1930s.

A Belgian shotshell, developed in 1935, mirrored the Perfect. It featured a brass head and thin-walled zinc body with a full crimp in the manner of blank rifle rounds. There was no top wad. Because the crimp consumed more of the case than did the short crimp of Perfect cartridges, the loaded shell had to be shorter to fit existing chambers. Overall length on a 12-bore zinc shotshell: 2 inches. The thin case wall kept internal capacity about the same as that of a longer paper hull. While zinc's high malleability made it ideal in the chamber, where crimp release and expansion against the chamber were primary concerns, their softness led to denting and deformation in pockets and cartridge bags. The solution did not lie in steel cases, which proved too hard to swell readily and seal powder gas. In 1938 Germany's Rottweil Company fashioned an aluminum shotshell tough enough to withstand normal battering in pocket and bag but malleable enough to hug the chamber upon firing. War interrupted the production of these and other metal shotshells in Europe. In 1947 Imperial Chemical Industries Limited, an English firm, came up with one-piece .410 cases of fully-drawn brass. They had a full crimp. The Italians made a 12-bore shotshell of similar construction but with a roll crimp and celluloid top wad. High brass prices canceled development of both cartridges.

Meanwhile, paper shotshells had improved, though they were still not waterproof. The standard British shotshell began life as layers of paper treated to repel water. These were glued together and rolled into tubes of the proper size. After drying at 100 degrees Fahrenheit, the tubes went to a conditioning room to receive a charge of 9 percent moisture. (Uniformity in hull performance depends on uniform

Jerry Krueger admires a pheasant taken on his South Dakota farm. Late-season birds get up far away, so Jerry recommends heavy charges of big shot.

moisture content.) The tubes were next cut to proper length, polished inside and out, checked for size and fitted into case heads that had been stamped from circular sheets of brass, then cupped, annealed, tempered, drawn, trimmed, and annealed again.

Early paper shotshells included an iron reinforcing cup fitted to the inside of the head. Some cases had a tubular lining of varnished metal coiled around the powder. This extra metal reduced the risk of "cut-offs" or case separations that usually resulted from the wetting and weakening of the paper just in front of the head. Cut-offs raised pressures to dangerous levels because the entire front of the shell might be forced into the bore and through the choke. Intentional cut-offs later became known as "cut shells," a poor-man's deer cartridges in the American Midwest. Scoring the case with a pocketknife prompted it to separate at that point. Instead of opening, the crimp would become the leading edge of a mass of shot, wad and case plunging down the barrel and emerging as a slug. Accuracy was poor, pressures predictably high.

Paper shotshells received a paper base-wad to position the powder and to reinforce the primer cup and cushion the head upon firing. Powder, wads, and shot came next. The first shotshells typically had four wads: an over-powder card, a felt wad, an over-felt card and an over-powder card. The over-powder card protected the powder from grease in the felt, which sealed the gas behind the wad column and cushioned the shot during launch. The best felt wads were fine-textured and light-colored. Grease was commonly applied to felt to help it in its sealing mission. The over-felt card kept the shot from "setting back" into the felt during the initial thrust of firing. An over-shot or top wad, used with roll crimps, capped the pellets in place. Some shooters thought top wads backed into shot patterns, disrupting them; but in 1926 and 1927 British shooting authority Major Gerald Burrard disproved that idea. He set up a screen with a 6-inch hole in its center 6 feet from the muzzle of a shotgun and fired at it 200 times. All top wads fell on the near side of the screen and no shot strayed outside the hole. Top wads became obsolete when the full crimp became standard in shotshell design.

The felt in the middle of traditional wad columns was an expensive component, and for years shooters sought substitutes. Cork initially proved too weak, and it varied in weight and firmness. But in the early 1930s a "pneumatic" cork wad appeared—a disc with a hole in its middle. The compression of firing forced air in the hole to expand the wad and cushion the shot charge. Given uniform pressures and a small hole, the wad worked. High pressure or a large hole would allow powder gas to blow through the over-powder card, fragmenting the cork and muscling into the over-cork wad to disrupt the shot column. The Air-Cushion wad, devised by Imperial Chemical Industries, Limited in 1936, comprised a stout paper tube crimped on both ends. A thin card in the center served as a check in case the rear crimp failed. An extra-thick over-powder card eased strain on the tail crimp. Though the Air-Cushion wad worked, it never became popular stateside.

In the early 1960s, plastic hulls and wads began to replace paper, card, and felt. The first plastic components were thin over-powder cups. Alcan's Plastic Gas Seal expanded to hug the sides of the shell and barrel, reportedly boosting shotshell efficiency 10 percent. By the mid-1960s one-piece plastic wads had appeared, and soon plastic shot collars were inserted to keep the shot from scrubbing itself flat against the bore. The next step was a one-piece wad-collar combination. Slots in the wad let it flex to cushion the thrust of acceleration. Slits in the collar allowed the unit to open like a parachute to air resistance when it cleared the muzzle. The plastic dropped to earth while the shot sped on.

Compression-formed plastic hulls had by that time largely replaced paper. Plastic made the perfect case: hard, thin, strong, lightweight, scuffproof, waterproof, and self-lubricating. It fed smoothly through autoloaders and did not swell when wet. A plastic hull incorporated its own base wad and could be reloaded many times. It sealed the chamber tightly but yielded readily to the extractor. Perhaps best of all, it was cheap—cheaper even than paper. Now only Federal, of the major ammunition firms, continues to manufacture paper shotshells. They're traditional and, to those with long histories afield, a link with days past. Freshly-fired paper shotshells also smell better than anything else in the world. Ironically, they are much more costly and labor-intensive to produce than plastic hulls.

Over the years "high brass" and "low brass" in shotshell heads have come to mean more and less powerful, respectively. But given one-piece plastic bodies, the brass doesn't have much to do with reinforcement. In fact, the brass head was successfully eliminated from plastic hulls by ACTIV, a Puerto Rican firm now in West Virginia. A thin disc molded into the base is the only metal component. High brass on shotshells can be likened to dual tailpipes on automobiles.

Shot pellets are much simpler to make than rifle bullets. The first were chipped from sheets of lead, but by the start of this century most came from shot towers. Molten lead poured through sieves atop these towers become round in free-fall. The sieve perforations determine shot size. Large shot—buckshot—is commonly formed in molds, like lead pistol bullets and balls for muzzleloaders. Steel shot is made like ball bearings: Short chunks of wire are fed between two massive plates that roll them around until they're spherical. Steel shot size is determined by wire diameter, the length of each chunk and plate spacing.

Remington's tower operation is typical for modern birdshot manufacture. About 140 feet high, it supports a cauldron of molten lead at 750 degrees Fahrenheit (plus or minus no more than 15 degrees). The lead contains measured amounts of arsenic and antimony. Pipes feed the lead "soup" to two colander pans, whose holes can be varied to change shot size. The lead droplets plunge 133 feet into 6 feet of water. Bucket belts carry the cool pellets to a polishing machine. Next they're piped to the top of a series of sloped plates and released. Truly round pellets bounce higher than malformed pellets as they pick up speed over the plates. Gutters at the bottom of the plates catch misshapen pellets for recycling. Round ones hop over the gutters into bins bound for the loading room. Remington can produce about 1.2 billion shot pellets daily.

The first birdshot manufactured in the U.S. was essentially pure lead with a touch of arsenic to help make it spherical as it fell. But pure lead proved too soft to withstand the suddenness of acceleration, even with low-pressure loads. "Setback" deformed the pellets. Patterns showed that these deformed pellets quickly veered away from boreline. Full-choke guns were putting half their shot outside a 30-inch circle at 40 yards! Harder shot followed with the addition of antimony. Though some hunters claimed soft shot "killed better" because it upset in birds like an expanding rifle bullet, the truth was that, especially at long range, more hard shot could be delivered to the target. Hard shot also penetrated farther. More and deeper wounds meant more birds in the bag.

Shot with antimony dates back to World War II. It is commonly called "chilled" shot, though there's no chilling in manufacture. Only 1 percent antimony is needed to harden large pellets, 2 to 3 percent for smaller shot. Magnum shot designed for high-pressure loads has more antimony than standard chilled shot: from 2 percent in BB and No. 2 sizes, to 6 percent in No. 7½ and No. 8. Harder pellets deliver measurably denser patterns because scuffing or flattening of a pellet's surface is most responsible for its tendency to fly wide. Plating makes shot even more scuff-resistant. Patented in England in 1878, plated shot didn't break into the U.S. market until the 1950s. Now it is all but standard in large pellet sizes in magnum loads.

Buckshot: Tests Are Mandatory

Buckshot, also commonly plated now, carries deadlier patterns and more energy per pellet than it did before the days of plastic shot collars and granulated plastic filler in the shot column. (The filler or buffer helps cushion the big pellets during launch.) But its range is still limited. Round pellets lose their speed quickly, and they don't penetrate nearly as well as bullets. Besides, there's no way to "place" buckshot; a hunter can only direct the swarm, as he might a cloud of smaller birdshot. The problem with buckshot is that there are very few pellets per shell (8 to 27 in 12-gauge loads, depending on shot size), and the swarm gets thin in a hurry. Beyond 30 yards, the gaps between big pellets can be so great as to preclude multiple hits in deer vitals. Smaller shot yields more strikes but less energy per strike and less penetration.

The advent of choke tubes enabled hunters to "tune" their barrels for buckshot instead of having to buy additional barrels. My shooting trials with various sizes of 12-gauge buckshot from a range of chokes showed that there's no rule about buckshot to which you can't find an exception. My tightest pattern (all pellets in 14 inches at 30 yards) came with a charge of No. 000 buck from an improved-cylinder choke. But No. 00 and No. 0 gave better results in a full choke, and No. 1 and No. 4 worked best from a modified barrel. If you experiment with buckshot before season to find the most effective combination of choke and load, you can carry choke

This Mossberg turkey gun wears a camo finish. Its scope allows you to adjust point of aim to the pattern's center. So equipped, the shotgun would also be ideal for use with buckshot.

tube and buckshot afield for a quick switch from slugs to buck.

When you use a rifled tube you're stuck with slugs. Don't even bother to try buckshot. In my tests with a rifled shotgun, not one pellet bigger than No. 1 hit within a 30-inch circle scribed around point of aim at 30 yards. Moving up to 20 yards, I fired a load of No. 0s; the closest pellet to center was 20 inches out. At 7 yards the pattern from a load of No. 4 buckshot measured just shy of 30 inches wide. I fired the same charge of No. 4s through an improved-cylinder bore and got a 15-inch pattern. Rifling seems to work like a grass-seeder with buckshot, rendering it useless for big game.

But from the right barrel buckshot can be deadly on deer up close. The biggest pellets have greater diameter than an 8mm rifle bullet and almost as much weight as a .243 bullet. Here's a close look (British designations in parentheses):

buckshot dimensions and performance per pellet

shot size	*load*	*gauge*	*diameter (inches)*	*weight (grs.)*	*muz. velocity (fps)*	*muz. energy (ft-lbs)*
000 (LG)	std.	12	.36	71	1325	276
000	mag. 3″	12	.36	71	1225	236
00 (SG)	std.	12	.33	61	1325	237
00	mag.	12	.33	61	1290	225
00	mag. 3″	12	.33	61	1210	198
0	std.	12	.32	50	1275	180
1 (Spec.LG)		12	.30	40	1250	139
1	mag.	12	.30	40	1075	102
1	mag. 3″	12	.30	40	1040	96
1	std.	16	.30	40	1225	133
2	mag. 3″	20	.27	30	1200	96
3 (SSG)	std.	20	.25	25	1200	80
3	mag. 3″	20	.25	25	1150	73
4	std.	12	.24	20	1325	78
4	mag.	12	.24	20	1250	69
4	mag. 3″	12	.24	20	1210	65
4	mag.	10	.24	20	1100	54

Simple arithmetic shows that a standard 12-gauge load of 16 No. 1 pellets packs 2,224 foot-pounds of energy at the muzzle, while a 20-pellet magnum load delivers only 2,040. If the company chronographs are right, you're giving up lots of velocity for just four more pellets in the magnum load. Are they worth the price?

They could be. Pattern density is as important when you're shooting deer as it is when you're shooting birds. Magnum loads offer more pellets for denser patterns. More pellets mean more weight, though, which raises breech pressures unless you reduce velocity. Sometimes the velocity reduction is so small as to leave the ballistic advantage with the magnum load. A standard load of No. 00 buckshot, for example, launches its nine pellets just 35 fps slower than a magnum load with 12 pellets. Obviously, the best choice is the magnum load, provided it patterns well from your gun.

Patterning is useful for bird hunters, but essential if you plan to use buckshot for deer. You may find the choke you like for bird hunting will throw acceptable patterns with every size of buckshot. Odds are, though, that it will perform well with only one or two sizes—maybe none at all.

It's not enough to count pellet holes in a 30-inch circle at 40 yards to come up with comparative percentages. In most cases you'll want the smallest patterns you can get, but if those tight patterns show uneven or inconsistent shot distribution, you must consider other combinations of load and choke.

Some buckshot patterns I've fired have produced two clusters of holes on the paper: one above point of aim, one below. There's nothing in the middle! Just as useless is the pattern with a wandering cen-

The author swings his Model 12 Winchester in typical whitetail cover, where buckshot is still a good choice for fast deer. This gun has a Poly-Choke, which can be tuned to the load.

ter. If you shoot without sights you'll appreciate having the highest pellet count right in front of your bead. Sights or a scope give you the prerogative of adjusting your line of vision to accommodate the shot charge, but the densest part of your pattern should still be near its middle and always in the same spot in relation to your bore.

All modern buckshot loads have ground plastic buffer sifted between the shot to cushion set-back when the pellets accelerate and reduce the contact of pellets with the bore and each other. This keeps the shot round so it flies true. For the same reason, Winchester Double-X Magnum and Remington premier buckshot loads have plated pellets. Winchester's plating is copper alloy, Remington's a nickel compound. Even unplated buckshot has antimony added for hardness, again to curb flattening. Incidentally, buckshot at Remington and Winchester is swaged. At Federal it is formed in a centrifuge—as is that firm's birdshot.

While most hunters prefer big buckshot, denser patterns of smaller pellets ensure better coverage at the target and penetrate almost as well. Recently I ran penetration tests with all sizes at 40 yards, using spaced particle boards behind the target. I found that No. 000 buckshot penetrated more boards than did a .22 long rifle bullet (a high-speed solid!). Predictably, smaller shot showed progressively less penetration. But even the No. 4 pellets went through three three-eighths-inch particle boards. From these tests, I concluded that within 40 yards any buckshot would adequately penetrate the chest cavity of a deer from the side. As the No. 4 shot drove only half as deep as No. 000, there's a significant advantage to the bigger pellets when that whitetail is quartering, and I would use only No. 1 or bigger pellets for deer hunting. Plated pellets recovered from the tests showed less deformation than unplated, but did not penetrate significantly farther.

How far from the gun is buckshot lethal? A 36-caliber ball from a black-powder rifle can kill at 100 yards if it's properly directed, and a No. 000 buckshot pellet is the same size. There's some difference in speed; but the meaningful difference has to do with shot placement. You can't aim buckshot any more than you can birdshot. Multiple hits kill the game. A single pellet in the brain or heart or spinal column will drop a deer, but that's a chance hit. You can't count on it because you didn't engineer it. Next time that errant ball may hit the jaw or lodge in the flank, and you won't find the crippled animal.

Because multiple hits are important, buckshot makes most sense at ranges under 40 yards; and 30 is better. Using the choke best suited to each shot size in my gun, I can keep 60 percent of the pellets (No. 1 and bigger) in a 12-inch circle at 30 yards. That's about the size of a deer's lung. Allowing for some imprecision in aiming and a slight shifting of the densest part of the pattern, one shot to the next, I can comfortably assume half my pellets will reach a deer's vitals at 30 steps. That's enough for quick, sure kills, and I needn't rely on chance hits.

Testing buckshot loads for penetration is as important as testing patterns. The author has used particle boards in series to measure penetration and compare deformation of plated and unplated pellets.

Buckshot remains popular in areas where deer are shot very close to the gun, and often on the run. It is more choke-sensitive than birdshot. While generally giving the densest patterns with tight chokes, the largest sizes sometimes work better with slightly less constriction. Hunters who choose buckshot for deer are smart to try several loads. Interchangeable or adjustable chokes can help tailor a gun to a buckshot load that shows promise—besides delivering the versatility of several bird guns in one package.

Shotguns for Big Game

These days, most shotguns are designed for bird shooting, with specific models and accoutrements offered for use by deer hunters. But at one time, smoothbores were the guns of choice for big game hunters. In fact, the first elephant guns were smoothbores. Pushing west from Zanzibar in the early nineteenth century, Arab traders plied the African bush for slaves, buying from village chiefs the prisoners of tribal wars. When they ran out of prisoners, some of the chiefs sold their own people. Then they helped the Arabs hunt elephants with crude smoothbore muskets.

The tactics were simple: Sneak up close to an elephant and shoot it in the knee with a load of chain links or whatever other metal dross you'd stuffed in your muzzleloader that day. An elephant's bone structure is such that it cannot move on three legs, and a successful shot would immobilize the poor beast. With your primitive missiles, however, you couldn't be assured of a proper hit every time, so immediately after the shot you ran as hard as you could away from the elephant. If you heard no footsteps and felt no trunk around your middle you'd creep back to within a few yards of your quarry and shoot repeatedly into its vitals until it died. It was an inefficient, cruel, and dangerous way to hunt.

British explorers of that time had better ammunition: lead balls the size of the bore. Samuel Baker used a 4-gauge gun firing quarter-pound projectiles. When choke boring made its debut the balls had to be a gauge smaller than the bore (13-gauge balls in a 12-gauge bore for example) to slide easily through the choke. Accuracy was abysmal, but these balls weighed more than rifle bullets and proved deadly at ranges of 30 yards or so.

While rifles soon took over on the ivory trail, shotgun development continued. The self-contained cartridge pioneered by Lefaucheaux in 1836 was first applied to smoothbores. So was Charles Lancaster's innovative breech-loading mechanism of

From left: unfired and expanded hourglass sabot slugs and traditional Foster slugs. The Fosters open a huge wound channel. Sabot slugs give better long-range accuracy from rifled bores.

1851. In the 1860s, just before Fred Kimble showed American shooters the advantages of choke, a Prussian officer named Schultz was working on a new shotgun powder. (In 1864 he patented the process of nitrating wood pulp. DuPont manufactured Schultz's powder 20 years before smokeless. It proved popular with British shotgunners as late as World War II.

By the turn of the century, only a few people were still using shotguns on big animals. European farmers who had no rifles in the root cellar kept wild pigs out of their potatoes with "pumpkin balls" from shotguns. Subsistence hunters without the scratch to buy rifles used them too. These balls flew erratically, however, because they were sized under bore diameter.

English paradox guns of the late nineteenth century delivered better accuracy with pumpkin balls. Their thick-walled barrels had some provision to spin the ball—full-length segmented rifling, an oval bore or a rifled section near the muzzle. Paradox guns (usually doubles) could be choked for use with shot. A paradox was the gun to take if you were heading into the equatorial bush for several months, with no idea of what you'd have to shoot.

Shotgun cartridges with pumpkin balls were made by American manufacturers until World War II. Better big game ammunition had been available in Germany since the mid-1890s in Germany, when the great designer Wilhelm Brenneke fashioned the first successful shotgun slug. Still loaded by RWS, this is a bore-size finned projectile with a sandwich of two card wads and one felt wad screwed to its base. The wads make it nose-heavy so it flies like a shuttlecock. The fins are angled as if to spin the slug. No doubt some spinning occurs, but weight distribution has a great deal to do with the slug's accuracy: it flies nose-first because the nose is heavy. The Brenneke slug, modified in 1931 for paradox guns, became available with a steel tip in 1935 for deeper penetration. Current versions have a one-piece plastic base wad.

About this time, American Karl Foster designed a hollow-base slug with a heavy nose and angled fins. The skirt and fins allow it to squeeze down in a choke, so it can be sized to fit the bore, sealing powder gas. While hollow-base bullets expand to fit the bore, the Foster doesn't bulge much because the wad behind traps powder gas. The cavity does keep weight forward, so this slug behaves in flight much like the Brenneke. The Foster slug was, for decades, *the* slug you meant when you said "shotgun slug." All major ammo firms loaded it, and in all popular gauges.

No matter their shape, heavy Foster slugs are more effective than light ones. When I was a sprout, rumor had it that the one-fifth-ounce .410 slug killed deer like lighting because it went faster than the fat slugs from bigger bores. While the .410 does launch its Foster 200 fps faster than does a 12-, 16-, or 20-gauge shell, the little extra speed in no way makes up for a huge weight disadvantage. At 1800 fps, the .410 slug is moving 400 fps slower than a .30–30 bullet, and its muzzle energy of 650 foot-pounds is less than that of a 12-gauge slug at 100 yards.

In southern Michigan, where I grew up, most deer hunters used slugs. Rifles (all except the .22

Dave Henderson used a Remington shotgun and Remington sabot slug to take this southern whitetail. The red dot sight gave him quick but precise aim.

rimfire) were not allowed, and buckshot had a bad reputation for crippling. But few hunters could afford (or even find) Brenneke slugs. At $5.25 a box of 25, even Winchester and Remington Fosters seemed expensive! Sabot slugs, like automobile air bags, Microsoft, and rap music, were far in the future.

"What's a cut-shell?" The willowy teenager was, by all indications, new to hunting.

The proprietor peered over grimy reading glasses. Thick as an oak, with massive arms and a bull neck, Mike wore a perpetual grimace behind iron-gray stubble shaded by a Tigers cap.

"A cut-shell is a poor man's slug," he explained, slipping off the glasses to better focus his glare. "You take a knife and score the hull around the wad. When you shoot, the whole front half of the shell goes downrange. The hull acts like a shot cup put in backwards, so the shot stays together for a few yards."

"Doesn't sound very accurate."

"Not safe either. You're squeezin' a big plug of shot, wad, and hull through the forcing cone and choke."

"But don't slugs have to squeeze down in a choke too?"

"Well, yes. But not that much. Besides, a slug is pretty much pure soft lead, and it has ribs on the sides that mash down easily in the choke."

"I thought those ribs were to make it spin."

"Nope. The ribs you see on most slugs don't help much with spin. What makes a slug accurate is its heavy nose and hollow base. The weight's all forward. Think of a shuttlecock or a spear. They're not as accurate as rifle bullets, but they fly pretty straight over short distances. A slug's base is supposed to expand to fill the bore—like Minie balls soldiers used during the Civil War. Of course, in a rifle that idea makes more sense than in a shotgun, where you have a wad in front of the powder. There's probably some set-back as the slug gets going, like tires squat when you pour the coal to a hot-rod." Mike looked at the gangly youngster. "Ever do that?"

But the kid wanted to know more about slugs. "What's a Foster slug?"

"What we've been talkin' about." Mike explained that Karl Foster was a Winchester ballistician who came up with his slug about 1933. "The only other kind of slug is the Brenneke. That came out of Europe. We didn't see it here until after the second world war. It's got ribs like the Foster, but the wad is screwed onto it, and there's no hollow base."

It was 1963, and coming up on deer season. The most coveted whitetail guns among farm boys back then were Ithaca's Model 37 Deerlayer pump and Browning's Auto 5 Buck Special. Both came with short, cylinder-bore barrels and iron sights. These were costly smoothbores, and scarce among the people I knew. Most deer hunters simply stuffed their pheasant guns with slugs. Coarse beads on full-choke barrels hardly gave you long-range accuracy. Still, some remarkable shooting was done with the worn pumps and autoloaders pulled from mud-room racks. One farmer I know killed a buck, off-hand, at 147 paces with his 16-gauge Remington

Winchester's sabot slugs and a Browning bolt-action shotgun with rifled barrel and scope give you the reach of many traditional woods rifles. Three-minute accuracy is commonplace.

11–48. "It was the last of 10 shots Bill and I poked at that deer," chuckled Harvey. "It *had* to be good."

In this enlightened time, we might not feel justified launching slugs from a smoothbore at a deer that far away. But in those days, lumbering Fosters from 12- and 16-bore repeaters collided with a lot of bucks many corn rows from the muzzle. Almost nobody used buckshot. We were dimly aware that it was favored in the Deep South, where hounds chased deer at full throttle past waiting hunters. Michigan woods could be as thick as Mississippi swamp; but sometimes you needed a slug's greater reach, to tag a buck crossing a field or peeking at you from the far side of a thinning.

Ironically, most deer-load development in my growing-up years was aimed at increasing the effectiveness of buckshot. Super-heavy charges of hard, plated shot were nested in granulated plastic buffer to cushion them during launch so they stayed round and flew straight. It wasn't until the 1980s, with the advent of rifled shotgun barrels and special-purpose scopes, that slugs began to dominate the "new products" pages of ammo catalogs. Now, if Mike were here, his slug lecture would be much longer. He would note that Foster slugs still account for 60 percent of all slugs sold. But he might say there are better choices. . . .

The New Slugs

"Sabots. *Saybows.* It's French for a projectile within a projectile. Sabot slugs are smaller than the bore, and they come in bore-diameter capsules that fall away leaving the muzzle."

The kid: "What's good about sabots?"

"You can make a sabot slug just about any shape, and you can reduce the weight because you don't have to fill up a big bore. Less weight means faster, flatter flight. Now, you may not care about long shooting. Maybe you want all the weight you can get for close shots in brush. Sabots that weigh as much as Foster slugs are the answer. They penetrate better because they're skinnier and have a higher sectional density. The sabot's main advantage, though, is that plastic capsule. You can fire it in rifled barrels that spin it for better accuracy. You can't shoot hard lead or jacketed slugs in thin-walled rifled shotgun barrels at high speed because the friction could cause too much pressure. The soft lead would strip. With sabots, there's no mark on slug or bullet."

"Bullet?"

"Some sabot slugs are actually jacketed bullets. Nosler's pistol bullet for the .454 Casull goes into Winchester's newest sabot shotgun loads."

The shift to sabot ammunition began when California-based Ballistics Research Institute (BRI) developed an hourglass-shaped slug in a two-piece plastic sleeve split along its length. It was designed for police use but by the mid-1980s had been packaged for deer hunters. I remember shooting prototype rounds through a sleek Benneli autoloader with a rifled barrel. My first three shots went into one hole, and the next two spread the group to just over an inch. Wow! The BRIs promised lots of punch as well: At 445 grains, they more than matched one-ounce 12-gauge Foster slugs in weight, but their reduced diameter gave them a ballistic coefficient of .250—compared to the Foster's .107. The Foster of course, has a bigger frontal area, but it loses speed and energy more quickly. I was mightily impressed with the BRI ammo. So were folks at Winchester, who later bought the company. Federal followed with its own hourglass slug.

What kept the sabot from burying the Foster slug? Well, the Foster shot as accurately, or more accurately, in smoothbore barrels. It flattened readily in whitetail deer, opening a bigger wound channel

Polymag Quik-Shok sabot slugs are among the most effective available to deer hunters.

The Barnes Expander sabot slug is of solid copper. The large hollow cavity ensures expansion at slug velocities. Sabot slugs offer rifle-like accuracy in rifled bores, but perform poorly from smooth bores.

than the slender sabot slugs. Not every shotgun could be fitted with a rifled barrel. And a lot of casual hunters chose not to spend the money for a rifled tube. Besides, sabot slug ammo cost considerably more than shells loaded with Foster slugs.

In 1993 a small New Jersey firm announced the "Lightfield Hybred" sabot, an almost-bore-diameter slug in a thin-walled, two-piece capsule. Designed by Tony Kinchen, the new Lightfield expanded readily and proved deadly on deer-size game. Like the Brenneke Magnum, this slug had an attached wad. Meanwhile, another New Jersey shop, Gun Servicing, continued its experiments with .45 ACP pistol bullets (300-grain Hornadys) in a 12-gauge sabot cup.

That same year, Remington introduced the first commercially successful bullet-shaped slug: the 50-caliber Copper Solid. It had a notched hollow nose to initiate upset, but the light body of a whitetail deer wasn't enough to ensure it. So Remington revamped the slug, using softer metal. By then Barnes had unveiled its Expander MZ bullet, developed for muzzleloading rifles but soon added to the slug market. Federal adopted this 50-bore solid copper hollowpoint for its Premium slug ammunition in, if I recall correctly, 1997. That was the year that Hornady shouldered its way into the slug market with a sabot design. A new Hornady slug, the 300-grain .486 H2K Heavy Mag, appeared two years later.

The turn of the millenium brought a new Brenneke sabot slug into the PMC lineup. This rural Nevada firm loads the one-ounce slug to 1600 fps. Winchester got the spotlight about the same time, with a 385-grain Partition Gold hollowpoint bullet pushed from a 12-bore shotgun at 1900 fps. The Nosler bullet delivers more than a ton and a half of energy at the muzzle, and at 200 yards it still has 1500 foot-pounds: more than a .243 Winchester or a .300 Savage. So much for the image of the slug as pumpkin ball, lethal only as far as you can throw a rock. With the proper sights, a shotgun has the effective reach of a woods carbine.

Of course, ballistic performance depends not only on a slug's design, but on its weight and the powder charge. When I was young, only 12- and 16-gauge shotshells were thought suitable for deer hunting. The 20 (then only a two-and-three-quarter-inch shell) drove a slug as fast as a 12 (to 1600 fps); but a lower ballistic coefficient resulted in fast deceleration, with an attendant loss in energy. At 100 yards, where a 1-ounce 12-gauge Foster slug still carried 1255 foot-pounds of energy, the 20-bore slug managed only 835. The 12 at 100 yards hit like a .30–30 bullet at 125 yards, while the 20-gauge at 100 yards mustered only the snap of a .30–30 at 250. The .410 with its ¼-ounce slug at 1775 fps, turned up just 770 foot-pounds *at the muzzle,* and was down to 435 foot pounds at 50 yards. That's about as much as the .30–30 delivers *at 500 yards.*

But while the .410 is still out of the running as a deer round, some modern 20-bore slug loads make sense. Page 213 has a chart comparing slug types and loads from Federal, which has the most extensive line of slug ammunition. (Remington, Winchester, PMC, and Brenneke offer comparable velocities and energies for given slug types.)

You can glean some useful information from this chart. First, the most sophisticated sabot slugs do not deliver the meanest punch up close. A 1¼-ounce Foster slug from a 3" 12-bore shell generates a ton and a half of smash at the muzzle, 40 percent more than the 3" Barnes Sabot load. Second, there isn't a lot of difference in muzzle velocity across gauges and slug weights. If you ignore the .410 (you might as well) and drop the heaviest 10-gauge slug and slowest 12, you get a real horse race. All the others leave the gate at 1400 to 1680 fps, most of them between 1450 and 1600. Differences in slug weight among gauges translate to energy differences downrange.

Another thing to notice is that for all its machismo at the muzzle, the Foster slug gets tired in a hurry. In fact, that powerful 1¼-ounce magnum 12-bore load can't match the Barnes Sabot for energy at 100 yards.

Lightfield is one of several companies designing and marketing ammunition specifically for whitetail hunters using rifled shotguns.

Over the length of a football field it loses 470 fps to the Sabot's 255, giving up all of its 850 foot-pounds of initial advantage. The Hydra-Shok (hourglass) Sabot slugs can't maintain speed like the Barnes, but they don't decelerate as fast as Foster slugs either.

It's worth noting that the plain 2¾" 20-gauge Foster load outperforms the newer Hydra-Shok sabot in the 3" 20-gauge hull. More speed, more weight, more energy, even at 100 yards. Why would anyone pick the sabot? Better accuracy from a rifled barrel, perhaps.

Among slug types, the trajectory of the Barnes bullet is the flattest. Zeroed at 100 yards, it flies roughly 2 inches high at 50 and 3 inches low at 125. Hydra-Shok slugs arc half an inch higher at midrange and strike almost an inch lower at 125 steps. Foster slugs don't warrant a 100-yard zero, because they would climb too high between 50 and 70 yards. Better to zero closer if you're shooting Foster-style slugs. A 50-yard zero keeps the slug within 2 vertical inches of your aim out to 75 yards. At 100 yards those slugs will land about 5 inches low. If the gun shoots Fosters accurately, it makes more sense to zero a little farther out—say 70 yards. That way you'll be able to hold center on a deer's ribs to 100 yards and be sure of a hit. Foster slugs can kill farther than that, but shot placement becomes a problem.

How much energy is needed for a quick kill? Not as much as is commonly thought. The shotgun slug kills by destroying vital tissue. The more you de-

Shotgun slug comparisons

type, gauge, shell lgth., slug wt. (grs.)		*velocity*			*energy*		
		muzzle	*50*	*100*	*muzzle*	*50*	*100*
Barnes Sabot	12 ga, 3", 438	1525	1390	1270	2260	1870	1560
	12 ga, 2¾", 438	1450	1320	1210	2045	1695	1420
	20 ga, 3", 325	1450	1320	1200	1515	1250	1040
Hydra-Shok Sabot	12 ga, 3", 438	1550	1340	1180	2335	1750	1345
	12 ga 2¾", 438	1450	1260	1120	2045	1545	1215
	20 ga 3", 275	1450	1230	1070	1285	920	705
	20 ga 2¾", 275	1400	1190	1050	1200	860	670
Hydra-Shok HP	12 ga 2¾", 438	1300	1110	1000	1645	1205	865
	12 ga 2¾", 438	1610	1330	1140	2520	1725	1255
Classic Sabot	12 ga 2¾", 438	1450	1260	1120	2045	1545	1215
Classic Foster	10 ga 3½", 766	1280	1080	970	2785	1980	1605
	12 ga 3", 547	1600	1320	1130	3110	2120	1540
	12 ga 2¾", 547	1520	1260	1090	2805	1930	1450
	12 ga 2¾", 438	1610	1330	1140	2520	1725	1255
	16 ga 2¾", 350	1600	1180	990	1990	1075	755
	20 ga 3", 325	1680	1340	1110	2055	1310	890
	20 ga 2¾", 325	1600	1270	1070	1885	1175	835
	.410 2½", 109	1775	1348	1080	770	435	275

Note: Hydra-Shok Sabot and Classic Sabot slugs are of the hourglass type; in Premium loads, they are copper-plated. The Barnes Sabot is a hollowpoint muzzleloading bullet of solid copper. Range is in yards, velocity in feet per second, energy in foot-pounds. One ounce = 437.5 grains.

stroy, the quicker the deer will expire. The main thing is to put that slug where it will do the most damage to the vitals. Dumping a ton of energy in the wrong place can leave you with only a blood trail. A 20-gauge slug sending 700 foot-pounds through a deer's lungs will quickly turn out the lights.

Because slugs are big and heavy, some hunters assume they will hit like a cement truck and literally throw a deer to the ground. That can happen, but generally it's when you shatter the bone structure that supports the deer, or destroy its spine. In my experience, a lightweight, high-speed expanding bullet from a .243 gives more dramatic results than the pile-driver punch of even a 12-gauge slug. One deer I shot through both lungs with a 16-gauge Foster slug showed almost no reaction. It ran off as if unhurt. I followed and found it dead within 60 yards. Other hunters have told me they've seen deer drop as if flung down by the impact of a big slug. Soft slugs that flatten and stay inside the animal can give quicker results. For that reason, and because whitetail deer do not require a deep-penetrating slug, I prefer sabots that expand readily. Foster slugs don't have to expand to plow a big channel, and in my view they're still a great choice for close shooting.

The accuracy of Foster slugs from smoothbore barrels can vary a great deal. I cut the full choke off my Remington 870 16-gauge pump years ago and mounted a 2½X scope. My expectations for rifle-like groups were unfulfilled. At 80 yards the gun would keep Foster slugs inside a 6-inch circle—but it had shot almost that well with the choked barrel. The notion that open chokes yield better accuracy than tight chokes makes sense, but individual guns don't always follow this rule.

Rifled barrels and sabot ammo deliver better and more consistent accuracy. Most slug barrels are rifled 1-in-24 to 1-in-34. Fixed-barrel bolt-action guns by Savage, Mossberg, and Marlin can be expected to outshoot pumps and autoloaders with lightweight, interchangeable barrels. However, I prefer pumps and autoloaders to the bolt-action shotguns, which seem a bit unwieldy. It matters not to me whether a shotgun punches 3-minute groups or makes all the slugs jump through one hole because I'm not going to use a shotgun at long range. While bullets from sabot loads can be deadly at 200 yards and even a bit farther, slugs are at their best in the thickets, where shot ranges are measured in feet, not furlongs. That's where a well-balanced, quick-pointing shotgun (and sometimes a fast repeat shot) helps you most. And it's where hunting whitetails is most fun too!

A lot has been said about shotgun slugs driving undeflected through thickets that would turn a dozer. My tests with 12-gauge Foster slugs showed deflection to be less than that of rifle bullets commonly used for deer (150-grain .280 bullets and even 250-grain .35 Whelen bullets). But thick screens of live sagebrush still tipped and deflected the slugs. Holes in the targets showed that some also flattened and otherwise deformed. While big, heavy projectiles are less susceptible to deflection than small, lightweight bullets, you can't shoot confidently through dense brush or tree limbs. If you're trying to tag a running deer, it's still sound policy to ignore the trees and concentrate on the vitals, just as you block out the alders when you paint a rocketing grouse with shot. It is truly the only way to hit. But you are also smart to pass on shots that put lots of brush between you and your target. If the slug deflects, a crippling hit may result. I count on shooting through the delicate stuff, and on losing my slugs or bullets to any tree that gets in the way. Those half-inch branches are what annoy me. Too small to block the target from view, they won't intercept a slug. But they're big enough to turn it.

Whether you're new to slug shooting or a veteran shotgun deer hunter reconsidering your options, it's a good idea to test various slugs and loads in your gun. To do a good job of that (and to give yourself the best chance on the hunt) you need a proper sight. An aperture sight like the Ashley or Williams with a flat-faced, angled bead up front works better than open sights. Some shooters prefer fiber-optic open sights, but I do not care for them. A low-power scope with a bold reticle is probably best on shotguns with rifled bores. A 2½X such as the Weaver K2.5 is my pick. Mount the scope low, either on a cantilever base or directly to the receiver.

Shooting slugs is like shooting rifle bullets in that practice helps you hit reliably. As with rifles, dry firing helps. So does shooting with light shot loads at "rabbit" targets on sporting clays courses. And if you're shooting Foster slugs through a smooth bore, you'll want to try one of the "light recoil" slug loads now available. They're not as effective as full-power loads in the woods, but they enable you to practice without developing a flinch.

Forty years ago, I listened raptly as Mike explained slugs. He'd have a lot more to talk about now. And I'd still listen.

Handloading, Simply

Many moons ago, when I was young enough to think all girls were sweet, Winchester concluded that it would make more money by re-engineering the Model 70. About that time, Ford came up with the Mustang and actually did make some money. I was thinking about money then too. My Herter's catalog listed a reloading press for $15. Shrewdly I calculated how many cartridges I'd have to load to save such a sum, plus shipping. It was quite a leap for me, writing out that check. What profligacy! I only hoped I'd live long enough to shoot all those bullets.

Well I have. Though it has long since paid for itself, I still use that Herter's press. Alas, the thick Herter's book that flaunted (under the label "Model Perfect") almost every wish in a young man's breast is no more. As is commonly the case, the values were just too good. The company didn't make enough money to survive in its original form. However, reloading—or, more properly, handloading—is still a cheap hobby. It also gives you higher performance and more versatility from your rifle, while teaching you about things that you would never learn from factory ammunition.

Whether you're starting to handload or have been at it awhile, now is a good time to update your catalog stock. There are companies out there whose products you'll want to know about, and they're more than happy to send you wish books at little or no charge. Here's a list of "must have" catalogs:

Cases and primers:

Federal Cartridge Co., 900 Ehlen Dr., Anoka MN 55303; www.federalcartridge.com

Old Western Scrounger, 12924 Hwy A-12, Montague CA 96064; 530–459–5445

Remington Arms Co., 870 Remington Dr., Madison NC 27025; www.remington.com

Winchester, 427 N. Shamrock St., E. Alton IL 62024; www.winchester.com (also powders)

Handloading gives you options not available to hunters who buy factory ammo only. The .338–08 has a lot to recommend it, but you must form the cases and load them yourself.

Bullets:

Barnes Bullets, 750 North 2600 West, American Fork UT 84003; www.barnesbullets.com

Hornady Mfg., POB 1848, Grand Island NE 68803; www.hornady.com (also dies, presses)

Nosler, Inc., POB 671, Bend OR 97709; www.nosler.com

Sierra Bullets, 1400 W. Henry St., Sedalia MO 65301; www.sierrabullets.com

Speer (Blount), 2299 Snake River Ave., Lewiston ID 83501, 208–746–2351 (also CCI primers)

Swift Bullet Co., 201 Main St., POB 27, Quinter KS 67752; 785–754–3959

Powders:

Accurate Arms Co., 5891 Hwy 230 West, McEwen TN 37101; www.accuratepowder.com

Alliant Powder, Rt. 114, POB 6, Radford VA 24141–0096; www.alliantpowder.com

Hodgdon Powder Co., POB 2932, Shawnee Mission KS 66201; www.hodgdon.com

IMR Powder Co., 622 Malone Ridge Rd., Washington PA 15302; 724–228–8949

Vihtavuori (Kaltron-Pettibone), 1241 Ellis St., Bensenville IL 60106; 603–350–1116

Western Powders, POB 158, Miles City MT 59301; www.ramshot.com

Loading equipment:

Dillon Precision, Inc., 8009 E. Dillons Way, Scottsdale AZ 85260; www.dillonprecision.com

Forster Products, Inc., 310 E. Lanark Ave., Lanark IL 61046; www.forsterproducts.com

Graf & Sons, 4050 S. Clark, Mexico MO 65265; www.grafs.com

Lyman Products Corp., 475 Smith St., Middletown CT 06457; www.lymanproducts.com

Lock, Stock & Barrel, Hwy. 20, Drawer B, Valentine NE 69201; 800–228–7925 (also components)

MTM Case-Gard Co., POB 13117, Dayton OH 45413; www.mtmcase-gard.com

MidwayUSA, 5875 W. Van Horn Tavern Rd., Columbia MO 65203; www.midwayusa.com

RCBS (Blount), 605 Oro Dam Blvd., Oroville CA 95965; 916–533–5191

Redding Reloading, 1089 Starr Rd., Cortland NY 13045; www.redding-reloading.com

Sinclair International, 2330 Wayne Haven St., Fort Wayne IN 46803; www.sinclairintl.com

Stoney Point, Inc., 1822 N. Minnesota St., New Ulm MN 56073–0234; www.stoneypoint.com

You don't need a lot of equipment to produce top-quality handloads. Nor do you need tools at the top of the price range. My $15 press has turned out handloads as accurate as any that you might load with a press costing much more. Its one liability: Herter shell-holders are not interchangeable with RCBS-style shell-holders, and those RCBS accessories are lots easier to find these days. Some presses, like the RCBS Rockchucker and Bonanza Co-Ax, have been around for decades. Others, like the Hornady Lock-N-Load and Redding Boss are quite new. The main differences in single-station presses are:

1. size (most will handle .375 H&H-length cartridges; don't buy one that won't)

2. frame shape (of "C", "H", "T," and "O" configurations, the stout "O" is now most popular)

3. linkage (you want a long, palm-filling handle, solidly pinned to a mechanism that maximizes leverage but comprises few moving parts)

If you plan to load lots of ammunition, a progressive or multi-station press makes sense. But it is hardly necessary if, like me, you're assembling a couple of boxes at a time.

Handloading isn't a complicated job, but it does require your attention. It's best to avoid most shortcuts. Anything that promises to be faster or easier may compromise the accuracy of your loads. On the other hand, I'm often pressed for time and find some shortcuts useful. Developing safe handloads is a matter of clear thinking and hewing to proven routines. Making your handloads as powerful and accurate as they can be requires lots of trials at the range. Here are some ways to speed the process:

1. Determine the level of accuracy you want before you start. If you're looking for tiny groups, you'll have to add some procedures that often aren't necessary for inch-and-a-half accuracy at 100 yards. You'll first sort cases by weight, then outside-turn necks to ensure concentricity, and ream flash-holes to uniform diameter, tossing cases with off-center orifices. You may wish to limit bullet selection to bullets with proven accuracy records. If you tighten your accuracy standard after a range session, you won't know which loads merit further trials. Example: Your 74-grain charge of H4831 behind a 180-grain Swift in your .300 Winchester prints 1.5-inch groups. That's middle-of-the-pack. You want subminute clusters, but because you neglected some procedures that might have improved the performance, you can't say whether or not this level of accuracy is as good as you can get. You don't know if this load will respond to benchrest case preparation. Maybe another load that barely stays under 2 inches will respond brilliantly. By changing standards midstream, you force yourself to start over.

2. Use lots of reloading manuals. In this litigious age, they've all become conservative, but a maximum load in one manual can appear as a modest load in another. Unless my rifle has conditions that merit extra caution (a weak lockup, tight neck, short throat) or I'm seating bullets into the lands, I save time by using the highest maximum charge as a benchmark. That doesn't mean I start at the top. If I want a maximum big game load, I generally back off the high "book" load two or three grains, then work up a grain or a grain and a half at a time. Often I'll exceed the book maximum safely, and I get there quickly. Expect a book's starting loads to generate useless velocities. This is not to demean the old caveats about starting well below maximum; and you must pay close attention to the condition of case and primer after firing when you're close to maximum pressures.

Note: If you change any component, back off more to start. Some custom bullets have long bearing surfaces and soft, "sticky" jackets, both of which can jack pressures through the roof. I once blew a primer and froze the bolt with a .338–06 load that seemed reasonable. But I had not used that custom bullet before. It hiked pressures with other loads as well.

3. Study the manuals before you trot to the corner gunshop for components. There are lots of propellants now, and enough bullets for any given bore that you could be testing loads until Congress votes itself a pay cut if you tried them all. A few powders will stand out with several bullet weights. Try those. For a heavy-bullet load in a .300 Weatherby Magnum, for example, your short list should include H1000, IMR 7828, Accurate 8700, and Reloder 22. Others, like Hodgdon and IMR 4831, Accurate 3100 and Reloder 19, deserve a chance too. If you stray far from a preferred burning rate, you'll spend more time eliminating unsatisfactory loads than refining good ones. The same holds true for bullets. Decide what you want in terms of weight, shape, and expansion characteristics; then buy those bullets that hew close to your criteria (splitting boxes with friends helps keep costs down as you work up preliminary loads). When you have, say, half a dozen bullets and as many powders to play with, you have more than enough for first-run combinations. I don't experiment with primers at all for most hunting loads. In my experience, changing primer brands has little effect on hunting rifle accuracy and no discernible effect on performance. My rule of thumb is to use standard primers for charge weights up to 65 grains, and magnum primers for heavier doses of powder.

4. Assemble only four cartridges per load—three for a group and one to either foul the bore or excuse yourself for a bad letoff. Sure, you can load 10 apiece and get more data. But loading and shooting that many cartridges takes time I think is better spent trying other loads. You can get a good feel for accuracy and velocity with three rounds and confirm both later with more shooting. Meanwhile, you'll have a chance to examine a broader selection of loads, keeping only the best for a second round of comparisons or for incremental tuning of the powder charge. Besides, a 4-round batch allows you 10 loads with two boxes of brass. I want data from 10 loads before assembling others. Some shooters point out that you may get an abnormally ragged three-shot group at any time, and could prematurely reject a load based on one group. Well, yes. But most wide groups result from poor shooting, and that's the reason for the extra cartridge. I can throw out a shot if I blunder, but if my shooting is good and the holes are still far apart, I don't want the load. I'll always wonder when that inconsistency will crop up again. Result: no confidence in the load.

5. Do what you're going to do to make the rifle shoot well before you start load development. Floating a barrel, pillar bedding, or even changing guard screw tension may cause a rifle to perform differently with any load. Changing scopes won't have an effect, but if you switch scopes on the suspicion the old one causes wild shooting, you won't have confidence in the earlier load data. Slight changes in how you hold the rifle on the rest may affect group size. You cannot fairly compare the accuracy of loads fired from different rifles or by different people. The rifle and its support comprise a launchpad. To evaluate handloads, you must eliminate variation in the launch.

6. Be realistic in your expectations for velocity. Some barrels are "fast," delivering top listed speeds with no pressure signs. Others, because of throat or

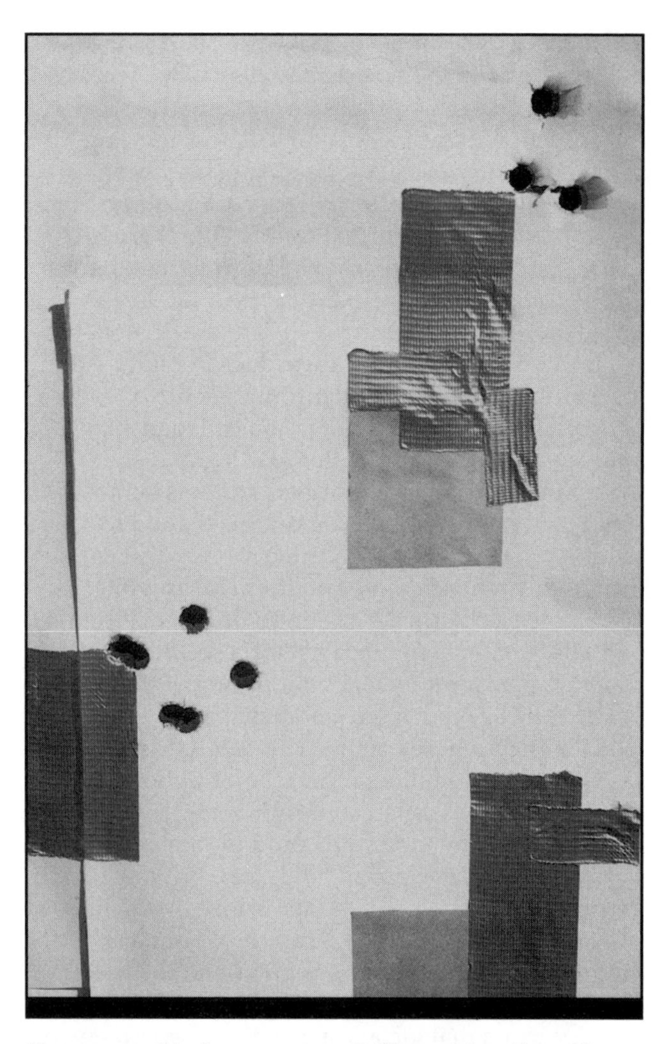

Be reasonable in your expectations of hunting rifles and loads. For big game, groups like this—under 1½ inches at 100 yards—are small enough. Velocity and terminal bullet performance matter too.

bore dimensions, won't let you go as far before warning you with flat primers and sticky bolts not to increase the charge. Remember that a hunting rifle must be 100 percent reliable. Loads that crowd pressure limits may freeze a bolt someday, just when you need another shot. Don't waste time on red-line powder charges. If your preliminary batch of loads includes some that show signs of high pressure, back off a grain or two and try them again. Flat primers and sticky bolts don't show up until you've exceeded safe and useful pressures for the round.

Handloading is a great hobby and a proven path to better ammunition. Shortcuts are not always in order, but when you've no time to play, getting to a favorite load quickly can be as important as having fun at the bench. Make sure you still pay close attention to the details.

PART IV

Terms and Charts

Glossary

Annealing heating metal to soften it. Brass cartridge cases become hard in the neck after repeated resizing. Splits and separations result. You can anneal a case to restore its ductility by heating it a cherry red, then quenching it in cold water. With steel, such an operation would *increase* hardness. Brass is different.

Antimony a metallic element alloyed with lead to increase bullet hardness. In big game bullets, the usual ratio is 97.5 percent lead, 2.5 antimony. A little antimony makes a big difference. Six percent is about the limit. Sierra uses three alloys for rifle bullets, with antimony proportions of 1.5, 3, and 6 percent.

Anvil a rearward-facing part of the primer or primer pocket against which a striker crushes the soft primer cup, pinching the priming compound. The percussive force and friction ignite the primer, which shoots a jet of flame through the flash-hole(s) behind the anvil.

Ball literally, a lead ball used in muzzleloaders, but also a bullet as in "Ball Cartridge" and a (typically double-base) powder whose hard spherical grains roll and slide like tiny shot pellets.

Ballistics the science of projectiles in motion, comprising internal, external, and terminal ballistics—that is, what happens during the launch of a bullet (internal), the flight characteristics of that bullet (external), and the penetration and upset of a bullet in animals (terminal).

Ballistic coefficient the ratio of a bullet's sectional density to its coefficient of form, a measure of the bullet's ability to cleave the air. Ballistic coefficient ("C" in formulas) is reflected in a bullet's rate of velocity loss, and in its vertical drop over distance.

Battery cup a type of primer that comprises anvil and main primer cup inside another cup. Shotshell primers are battery cup primers; modern rifle and pistol primers are not.

Belted case a cartridge case with a thick ring immediately forward of the extractor groove. The ring or belt serves as a headspacing device and has nothing to do with reinforcing the case. Belted cartridges are typically called magnums or belted magnums. But not all magnums are belted.

Berdan primer a type of primer with no integral anvil. The anvil is part of the primer pocket. Common in European cartridges, Berdan primers were named after Hyram Berdan, an American. The primer flame reaches the powder through double flash-holes, one either side of the anvil. Berdan-primed cases cannot be decapped with the central pin in standard handloading dies. The alternative: a special hook or hydraulic pressure.

Black powder a propellant comprising potassium nitrate (saltpeter), charcoal, and sulfur in specified proportion. Black powder substitutes like Pyrodex can be used in guns designed for black powder; they generate lower pressures than smokeless powder. However, Pyrodex and other substitutes vary in bulk density. To match a charge of black powder, you measure by volume, not by weight.

Boat-tail the tapered rear of a bullet designed for long-range shooting (also, "tapered heel"). Standard bullet bases are flat; hence, "flat-base." The ballistic advantage of the more aerodynamic boat-tail bullet is seldom realized at ranges short of 250 or 300 yards.

Bore the inside of a rifle or shotgun barrel. Bore diameter is the measure taken from land to land in a rifled bore, or the inside diameter of the tube before rifling (also, "land diameter," as opposed to "groove diameter," which is greater). "Bore sighting" is a preliminary alignment of the sight with the bore, before zeroing a firearm by shooting at a target from a bench.

Boxer primer the common type of primer used in centerfire rifle and pistol cartridges in the U.S. Named after a British colonel, this primer has an integral anvil. There's one central flash-hole, which makes decapping easy in a die with a central pin.

Brass alloy of copper and zinc used to make cartridge cases (also, a collective term to describe cartridge cases, as in "I have lots of .223 brass").

Bullet a single projectile fired from a cartridge case, usually through a rifled bore (not to be confused with "cartridge," which comprises bullet, case, primer, and powder). Bullet pull is the amount of force needed to extract a seated bullet from the neck of a case. It should be uniform, case to case. Bullet pull that is too light may cause cartridges in a magazine to slip under recoil. An overly tight bullet, due to a tight chamber or crimp, or chemical bonding with the case, can lead to dangerous pressures.

Caliber a measure of a bullet, from the Latin term *qua libra* or "what pound." Initially it was applied to weight, but then exclusively to diameter. A .308 bullet is a 308-caliber bullet. Caliber can be expressed in hundredths or thousandths of an inch (.22 or .224). It is part of cartridge nomenclature but cannot by itself describe a cartridge. There are, for example, many cartridges using .308 bullets—and many of them are designated ".300." The .300 represents bore diameter, the .308 groove diameter or bullet diameter. Case dimensions of the various .300 cartridges are not the same, though all can use the same bullets. You don't load .300 Weatherby Magnum cartridges in a .300 Winchester Magnum rifle. But you could handload the same .308 bullets in either cartridge.

Cannelure circumferential groove around a bullet to identify it and to mark the proper case mouth location for crimping after seating the bullet in the cartridge case.

Canister powder gunpowder ready for retail sales to the public, typically in plastic or cardboard containers or canisters. "Bulk powder" refers to the powders blended to form canister powders in the laboratories and loading rooms of powder manufacturers. Canister powders must meet strict performance requirements, for safety.

Cartridge a unit of ammunition, comprising bullet, powder, primer, and case. "Cartridge" also applies to shotgun ammunition, but "shell" and "shotshell" are by far the more popular terms.

Case the hull or shell of a cartridge, the housing that contains the powder and holds the primer and bullet to form a cartridge. "Case head" is the rear portion that fits against the bolt face of a gun and is gripped by the firearm's extractor.

Centerfire a cartridge whose primer is a removable unit held in a primer pocket at the rear of the case, in the center of the head. Some early cartridge designations included "CF." The ".30 WCF" (Winchester Center Fire) is another name for the .30–30.

Chamber in a rifle, pistol, or shotgun, the rear portion of the bore reamed to precise dimensions to accept a cartridge ready for firing. A chamber is big enough to allow easy insertion of the appropriate cartridge, but small enough to prevent undue case stretch when the cartridge is fired. A "chamber cast" is sometimes taken of relic guns to determine chamber dimensions.

Charge amount of powder, usually by weight, specified in a load. Charge also includes the type of powder, as in 90 grains FFg or 56.5 grains H-4350. A compressed charge is one that fills the case to a point above the normal seating depth of the bullet, so that the bullet presses the powder down during seating. Black powder charges are always compressed by the ball or bullet, which must be tightly seated to avoid an air space and a pressure wave that spikes when it hits the immobile projectile.

Choke restriction in a shotgun barrel to bring the pellets into a tighter column, thereby extending the effective range of the gun. Discovered by market hunter Fred Kimble shortly after the Civil War, choke boring is tailored to the shotgun's use. "Full" choke has maximum constriction and gives tight patterns for long shooting at waterfowl. "Modified" is more open, an all-purpose choke. "Improved-Cylinder" helps you with fast-moving birds up close because it allows the shot to spread quickly. "Cylinder bore," best with slugs, means there's no choke at all. There are some intermediate chokes, like "Improved-Modified." Choke dimensions and shotshell type determine pattern densities. The way to measure choke is to shoot at a white paper at 40 yards, draw a 30-inch circle around the pattern's center and count the number of holes in that circle. You assign choke based on the number of pellet holes as a percentage of the shot pellets to start with. A full-choke barrel may put 80 percent of its shot in that circle, a modified 65 percent. But the percentages vary by shot size and shell type. Modern shotshells with plated pellets and protective plastic filler to keep pellets round can throw very tight patterns from open chokes. Pattern uniformity is important but not measured in this test. There's no way to tell by measuring a shotgun's muzzle what kind of pattern it will throw, though you can generally expect tighter patterns from more constriction. At first, choke was swaged into the end of the barrel, but now interchangeable chokes are popular. These tubes thread into the barrel and do not affect the looks or balance of the gun—unlike the post-war "Poly-Choke" with its thick collar squeezing the barrel (split into fingers) down to the desired constriction.

Chronograph an instrument for measuring bullet speed. Most chronographs record bullet passage over two points by registering the bullet's shadow with two electric eyes. The distance between the points is known, and when the chronograph comes up with the time interval (hence, "chrono") between crossings, speed can be calculated. For a long time, chronographs were laboratory instruments. Beginning in the 1970s, Dr. Ken Oehler made them available to shooters with portable, inexpensive models—much like Bill Weaver gave riflemen an affordable, functional rifle scope in the 1930s.

Collimator an optical device used in bore-sighting rifles and pistols. A collimator attaches to the muzzle. Its screen has a grid you use to align your sight with the bore's axis. It's important to shoot at a distance to establish a zero after using a collimator. This instrument merely helps ensure that your first bullets land close to point of aim.

Corrosive primer a primer whose residue is hygroscopic (attracts moisture), causing rust in the bore. The corrosive agent in many early primers was potassium chlorate. Corrosive priming was discontinued shortly after World War II, when lead styphnate became the primary ingredient in military and commercial small arms primers.

Crimp an inward bending of the case mouth to bite into the cannelure of a bullet. Crimping increases bullet pull and helps secure heavy bullets with short shanks, such as big-bore pistol bullets. Crimping is also used on dangerous-game ammunition to prevent the bullets in a loaded magazine from backing out under recoil and seizing the magazine. Ammunition for tubular magazines and revolvers likewise warrants a crimp, particularly if recoil is severe. Because there's some bullet deformation, crimped ammo may not give you the best accuracy. A lot of factory ammunition is crimped, however, and can be expected to deliver good hunting accuracy. Bullets without cannelures should not be crimped.

CUP copper units of pressure, a measure of breech pressure obtained by measuring a copper pellet of specified starting dimensions after it has been crushed by a piston thrust outward through a hole in the barrel of a test gun. Copper crushers gauge the pressure of high-performance rifle and pistol ammunition. Lead units of pressure are generally used for shotguns and low-pressure pistol loads.

Drift lateral movement of a bullet away from the boreline during flight. Drift is the horizontal equivalent of drop, which results from gravity's pull on the bullet. Unlike gravity, the air movement that causes drift is not always present. And when it is, its speed and direction cannot be assumed. Good "wind-dopers" know that wind action (and drift) may differ at different points along the bullet's path.

Energy the amount of "work" that can be done by a moving bullet at a given distance from launch. Muzzle energy is a common standard, calculated by squaring the velocity of the bullet (fps), multiplying that figure by the bullet's weight (grains) and dividing the product by 450,240.

Erosion wearing of the bore caused by the friction and heat of firing. Erosion occurs whether or not you clean the bore and no matter what type of powder, primer, or bullet you use. It accelerates as bullet velocity increases. The most severe erosion occurs in the throat, just ahead of the chamber, where temperatures and pressures are highest.

Small-bore, high-speed cartridges with slow-burning powders generate the most erosion. Erosion can be kept to a minimum by letting the barrel cool between shots.

Expansion ratio interior case volume divided by bore volume. Cartridges with very high expansion ratios are said to be "overbore capacity." That is, they have big powder chambers relative to the bore size. Slow powders and long barrels are necessary to get the most from these rounds.

Extrusion the result of material flow under pressure. Bullet jackets and cores are formed by extrusion in dies. So is tubular powder, which is then diced into short kernels. An extruded primer, in which case the striker dimple has a raised perimeter, can be caused by too much pressure, excess headspace or an oversize striker hole in the bolt face.

Fps feet per second, a measure of a bullet's speed, like miles per hour for an automobile.

Fire-form shaping a case to new dimensions by firing it in a chamber of those new dimensions. This is a common practice with non-standard or "wildcat" cartridges but is safe only if the proper headspace has already been established. A full-power load should not be fired to establish a new headspace dimension.

Firing pin a spring-loaded rod that strikes the primer of a cartridge when you pull the trigger of a gun. Also called the striker in bolt-action rifles, a firing pin can be long or short. Shapes vary. In early revolvers, a nipple on the hammer nose served the function.

Flash-hole the tunnel connecting the primer pocket to the powder chamber of the case through the web. Boxer-primed cases have one flash-hole; Berdan-primed cases have two. The flash-hole conducts the primer's flame to the powder charge, just as a touch-hole introduced spark in early muzzleloaders.

Foot-pound a unit of energy commonly used for bullets, the force required to raise a one-pound weight a foot against the resistance of gravity.

Forcing cone the beveled forward edge of a chamber that brings chamber diameter down to bore diameter. Most commonly, this term applies to shotguns, which have relatively long forcing cones.

Form factor a multiplier determined by the shape of a bullet and used, with sectional density, to determine its ballistic coefficient (C).

Freebore an unrifled section of bore immediately in front of the chamber. Freebore can include a parallel section or be cut so as to slope gradually from chamber mouth to full land height. It is this variability that causes some confusion. Freebore is loosely used as a substitute term for throat, the section of bore between chamber mouth and full land diameter. But a steep, short throat is not freebored. Freebore becomes, then, an unnecessary term. Freebore is a generic, unspecific way to describe long throats, which allow the bullet to start moving without interference, reducing upward slope of the pressure curve. Some rifles, notably Weatherbys, are noted for freebore, which allows more aggressive powder charges than do rifles without freebore. Higher velocity results. Also, the freebore lets you seat bullets shallowly, boosting case capacity and, again, velocity. However, increasing the "free" travel of a bullet does nothing for accuracy.

Full metal jacket a bullet type designed to maintain its shape during penetration, as opposed to a softpoint or hollowpoint designed to expand and create a large wound channel. Full-jacket (FMJ) bullets are typically used for target shooting (because the jacketed nose is a good airfoil and does not easily deform in handling) or for hunting large African game (because the jacket up front protects the lead core from deformation during entry, ensuring minimal resistance to penetration and a straight wound channel). Full-jacket bullets are made with the jacket opening to the bullet's base, where the core may be exposed.

Gilding metal an alloy of copper and zinc commonly used for bullet jackets. Zinc comprises 5 to 10 percent of this alloy, copper the rest.

Grain in shooting, a unit of weight, not a description of a particle. That is, 54 grains of powder means that charge weighs 54 grains. It may have hundreds of *kernels* of powder. There are 7000 grains to a pound, 437.5 to an ounce. Bullets and powder are thus weighed by the same measure.

Grooves spiral channels cut or ironed into the bore of a barrel by single cutter, broach button, or high-pressure hammers. Grooves and lands (the uncut sections between grooves) comprise rifling, which spins the bullet around its axis, making it stable in flight and more accurate than a bullet from a smooth bore. A bullet's unfired diameter is, ideally, groove diameter. The lands cut into the bullet, grabbing and spinning it as it is shoved forward on firing.

Hangfire a delayed ignition of the main charge of powder after the striker hits the primer. Hangfires can be caused by a large air space in the case, a weak primer, powder that is hard to ignite. They are ruinous to accuracy because the bullet leaves late, when your sights are no longer perfectly aligned. A hangfire that delays more than a fraction of a second is rare, but misfires should be treated as hangfires for safety's sake, the rifle kept pointed downrange for 30 seconds before you open the bolt.

Headspace the measure between the bolt face and that part of the chamber that acts as a cartridge stop when the round is fully chambered. Headspace on rimmed cases is measured to the front of the rim, on belted cases to the front of the belt. Rimless bottleneck cartridges headspace on a datum line on the case shoulder. A few pistol cartridges like the .45 Automatic headspace on the case mouth. Headspace in any firearm is measured with steel "go" and "no go" gauges, precisely machined to minimum and maximum dimensions. A rifle's bolt should close on a "go" gauge but not on a "no go" gauge. Too much headspace allows the cartridge to move forward when the striker hits the primer. When the expanding gas irons the pliable front of the case tight to the chamber wall, there is nothing to keep the thicker rear of the case (the head) from moving rearward, stretching the brass forward of the case web, sometimes to the point of separation. Released into the chamber through a crack in the case wall, high-pressure gas can damage a rifle and maim the shooter.

Improved a case design fashioned by "blowing out" a standard cartridge in a chamber of more generous dimensions, typically with less body taper and a sharper shoulder. Result: more powder capacity. Improved cartridges generally take the name of their parent, e.g., the .257 Improved, .280 Improved. Headspace is not changed. If the new chamber does have a different headspace measurement, the factory cartridge cannot safely be fire-formed. It must first be reshaped to establish proper headspace.

IMR Improved Military Rifle, a powder designation of E.I. DuPont de Nemours, which replaced the old Military Rifle line of propellants with IMR powders in the 1920s, when four-digit numbers supplanted the two-digit MR designations. In 1986, DuPont sold its powder business to EXPRO, and the IMR Powder Company was established as a testing and marketing arm of that firm. IMR powders are still manufactured for handloaders; only the corporate umbrellas have changed.

Ingalls Tables ballistics tables computed by Colonel James Ingalls and first published in 1916. These tables have since served American ammunition makers and ballisticians as the basis for calculating ballistic coefficients and bullet flight characteristics. As with French ballisticians in his day, Ingalls' work followed that of Isaac Newton, Galileo, Benjamin Robins (who developed the ballistic pendulum), the Krupp factory in Germany (which, with other agencies, conducted firing tests to determine ballistic coefficients), and a 19th-century Russian named Mayevski (who constructed a mathematical model of the standard drag deceleration of the Krupp bullet). In France, the Gavre Commission had found a flaw in the assumption that drag on a bullet was proportional to some power of the velocity within a range of velocities. At high speed (above 6000 fps) there was a sharp rise in retardation. The Gavre Commission developed tables to show this—really, the first ballistics tables ever. British ballisticians came up with a better one in 1909, another in 1929. The Ingalls tables were produced using a bullet much like the one-pound, one-inch-diameter British projectile with its 2-caliber ogive. They are valid to velocities of about 3600 fps, at which point the British 1909 tables must be used.

Jacket a metal covering or envelope that protects a bullet's lead core from the heat of friction produced in its travel down the bore. Unjacketed lead bullets were sufficient in early muzzleloaders and black powder cartridge guns that kept velocities under 1800 fps. Higher bullet speeds stripped lead from the bullets, ruining accuracy and fouling the bore. Bullet jackets are typically copper or gilding metal (copper alloyed with zinc), but steel jackets have been used. Jacket material and thickness can affect breech pressures. They, with jacket design, also have a lot to do with how an expanding bullet opens in a game animal.

Keyhole the perforation made by a bullet entering a target sideways. Keyholing is the mark of a tumbling bullet that is both inaccurate and ballistically inefficient. Tumbling can be caused by insufficient rifling twist, a defective bullet, or bullet contact with a twig or other obstruction.

Lands the raised sections of rifling with a bore. Lands lie between the grooves and engrave the bullet as it is forced down the bore by powder gas. Land diameter is less than groove or bullet diameter, so the lands are really what spin the bullet, making it stable in flight like a well-thrown football. Some caliber designations reflect land diameter, some groove diameter. The .300 Savage and .308 Winchester both use .308 bullets in bores with land diameters of .300.

Leade same as "throat," that section of the bore between the chamber mouth and full land diameter.

Loading density in ammunition, the ratio of the volume of the powder charge, expressed in grains weight, to the volume of the case, also in grains weight.

Lock time in a firearm, the interval between release of the sear by the trigger and detonation of the primer.

LUP lead units of pressure, a measure of breech pressure obtained by measuring a lead pellet of specified starting dimensions after it has been crushed by a piston thrust outward through a hole in the barrel of a test gun. Lead units of pressure are generally used for shotguns and low-pressure pistol loads. Copper crushers measure the pressure of high-performance rifle and pistol ammunition.

Magnum a cartridge (or, by extension, a firearm) of unusually high performance. A magnum designation may indicate the existence of a standard cartridge of the same bore dimensions but lesser power (the .270 Weatherby Magnum drives a .270 bullet faster than the .270 Winchester because it has a bigger case and more powder). But some magnums, like the .264 Winchester Magnum, have no standard counterpart. Magnum rifle cartridges are generally belted; not so the .357 and .44 Magnum handgun cartridges. The term "magnum" may be overused, because there are several commercial .300 magnums, all of different dimensions and ballistic potential.

Maximum ordinate the point at which a bullet reaches its greatest vertical distance above line of sight, typically just over half the zero distance for ordinary hunting bullets in rifles zeroed at normal ranges. "Max ord" can move farther downrange, relative to midpoint, as zero range is increased. That is because a bullet's arc is parabolic. So the term "midrange trajectory" is really a misnomer for maximum ordinate.

Meplat the diameter of the flat nose of a bullet.

Mercuric primer an old primer used successfully with black powder arms because the bulky black powder residue absorbed the primer's residue. But with clean-burning smokeless rounds, the mercury fulminate was left to attack the brass cartridge case, weakening it. Non-mercuric primers arrived when potassium chlorate replaced mercury fulminate as the primary ingredient. Non-mercuric, non-corrosive primers followed in the 1940s.

Metal fouling deposits of bullet jacket left in the rifling.

Minute of angle a term describing shot dispersion. A circle of 100 yards radius has 360 degrees of roughly 60 inches per degree on its perimeter. Each degree can be divided into 60 minutes of about an inch on the perimeter (totalling 21,600 minutes). A 1-minute group is a series of shots whose greatest dispersion to the centers of the outside holes in the target measures an inch (really, a minute is 1.047 inch) at 100 yards. A one-inch group is the same size. But a one-inch group at *200* yards is a *half-minute* group. A one-minute group at 200 yards measures 2 inches; at 300 it measures 3 inches. Divergence from group center increases with distance, though the rate of divergence remains the same (or may even decrease as the rotating bullet "goes to sleep" at long range). A benchrest rifle is expected to shoot groups as small as a quarter-minute; a rifle for prairie dogs should shoot well under a minute. A big game rifle, however, needn't be so accurate, because a deer's chest is a big target. Three-minute accuracy will allow you to keep all your shots in the deer's vitals out to 300 yards, a very long shot. Most modern rifles and loads are capable of better accuracy, however, and it's not too much to expect 1½-inch groups from your hunting rifle.

Mushroom shape of a bullet after expansion in game. Softpoint bullets designed for big game are made to mushroom, delivering energy as the bullet slows and plowing a wide wound channel. Frangible bullets for small animals like coyotes and prairie dogs, where penetration is not an issue, are made to disintegrate for a lightning-like kill.

Ogive the curved portion of a bullet between nose and shank. "Secant ogive" and "tangent ogive" refer to the placement of a compass used to scribe the arc that determines the nose profile. The radius of the curve is typically expressed in calibers. A 2-caliber ogive would be a curve with a radius twice bore diameter.

Overbore a condition or cartridge characterized by a high expansion ratio; that is, the cartridge case is big in relation to bore diameter. Overbore cartridges have produced the highest bullet velocities, but they don't operate efficiently in short barrels, and they typically require large charges of very slow-burning powder. Because of high throat temperatures and the exit of considerable unburned powder from the case during firing, throat erosion proceeds more rapidly in rifles chambered to overbore cartridges.

Patched ball a ball or bullet, usually used in black powder rifles and most commonly in muzzleloaders, that has a cloth or paper patch protecting it from powder gas and sealing that gas behind. Patches also reduce fouling.

Point-blank range any distance at which you can hit your target without aiming high or low to correct for the trajectory of the bullet. Maximum point-blank range is the farthest distance at which you can hold in the middle and not hit too low. Point-blank ranges vary, depending on the load, the target, and the acceptable deviation from center. The less deviation you tolerate, the shorter will be your maximum point-blank range. If you don't mind your bullet hitting a couple of inches high or low with a center hold (that's not too much deviation for most big game hunters), you'll extend your point-blank range farther than if you insisted on hitting no more than half an inch high at mid-range, and half an inch low beyond zero. If you accept a 3-inch deviation, you can stretch than point-blank range even farther. Zero too far out, however, and your mid-range trajectory will carry bullets too high, causing a miss or a crippling hit. If you're shooting at targets, point-blank range is the range at which your target is fixed, the range at which you've zeroed, because precision matters. On the other hand, point-blank range is often used colloquially to describe very short-range shooting, where sights aren't used at all.

Port pressure in a gas-operated autoloading firearm, the gas pressure at the port, typically some distance down the barrel.

Powder in this book, gunpowder, which includes black powder, semi-smokeless and smokeless powder, as well as black powder substitutes. Powder is a granular fuel (not really a "powder") that burns very fast. It does not detonate from impact as will a primer. Gas formed from burning powder expands to push a bullet or shot charge down the barrel. Powder granules are of various shapes and sizes, depending on their intended use. Coarse, slow-burning powder is used in large-volume cases behind fast bullets. Small cases with big mouths call for powders of faster burn rates. Single-base powders are mainly nitrocellulose, while double-base powders have a significant amount of nitroglycerine. Progressive-burning powders are either shaped or treated to release energy in a controlled way, increasing gas production over time rather than "burning down" from a high initial energy release. Powder charges are measured in grains weight (437.5 grains to the ounce). Black powder substitutes do not all have the same bulk density. They're formulated to be measured in black-powder measures, bulk for bulk. Labels on shotshell boxes show powder charge in "drams equivalent," a designation held over from when the propellant was black powder. A dram is a unit of weight; 16 drams equal one ounce. When smokeless powder supplanted black powder at the turn of the century, it was of a type known as "bulk powder" and could be loaded in place of black powder "bulk for bulk" (not by weight). "Dense" smokeless powders came later. They took up less space in the shell, so neither bulk or weight measures of earlier loads applied. A "3¼-dram equivalent" charge is a smokeless charge that approximates the performance of a 3¼-dram black-powder charge. It has nothing to do with the amount of smokeless powder actually loaded.

Primer a small metal cup containing a sensitive detonating compound which, when crushed by the blow of a firearm's striker against an internal anvil, throws a spark. The anvil may be part of the primer (Boxer) or part of the primer pocket in the case (Berdan). The spark is directed through a flash-hole (or flash holes) in the case web to ignite the main charge of powder. You can buy standard and magnum primers (the magnums give you a spark of longer duration) and primers for "large rifle," "small rifle," "large pistol," and "small pistol" cases. "Battery cup" primers for shotshells are encased in a larger cup that adds support in the thin shotshell head.

Pyrodex a black powder substitute developed by Dan Pawlak, who died in a powder fire at his factory in Washington state. Hodgdon Powder Company bought the rights to manufacture and market Pyrodex and has now introduced Pyrodex pellets for quicker recharging of muzzleloading rifles.

Rifling in the bore of a rifle or pistol (and now shotguns designed to shoot slugs), lands and grooves that spin a bullet. Rifling gives any projectile (even a patched ball) greater accuracy and, by keeping bullets nose-first, greater range. Rifling twist is the rate at which a bullet is spun, expressed as the distance it travels while making one complete revolution. A 1-in-14 twist means the bullet turns over one time for every 14 inches of forward travel. The proper rate of twist varies with bullet profile, weight, and even speed. But in 1879 Briton Sir Alfred George Greenhill came up with a formula that works for most bullets most of the time: The required twist, in calibers, is 150 divided by the length of the bullet in calibers. So if you have a 180-grain .30-caliber bullet 1.35 inches long, you first divide 1.35 by .30 to get the length in calibers (4.5). Then you divide 150 by 4.5 and get a fraction over 33. That's in calibers, so to bring it into inches of linear measure, you multiply it by .30. The final number is very close to 10, which is a useful rate of spin for most popular 30-caliber hunting cartridges, from the .308 to the .300 Weatherby Magnum.

Rim in cartridges, the edge of the case head upon which the extractor seizes to pull the case from the chamber. Rimless cases have a rim behind a deep extractor groove, but it's the same diameter as the body of the case and does not protrude beyond as does a rimmed or semi-rimless case. A rimmed case headspaces on the rim and does not need an extractor groove. A rebated rim is one that is smaller in diameter than the case body (e.g., the .284 Winchester). British rimmed cartridges are typically said to be "flanged."

Sabot a lightweight hull or envelope, typically of groove diameter, that carries a smaller, more ballistically efficient projectile out the barrel, then falls away. The central projectile (commonly a shotgun slug or a small-diameter bullet), is spun by the sabot but bears none of the rifling marks. It benefits from the sabot's large-diameter base during launch but is not burdened with excess weight or diameter in flight. The sabot idea was first tested in French artillery.

Seating the act of inserting a bullet in a case neck. Seating depth is a critical element in getting the most accuracy from a firearm, and can affect pressures as well.

Sectional density a bullet's weight in pounds divided by the square of its diameter in inches. Sectional density and the bullet's profile or form combine to yield ballistic coefficient, a measure of the bullet's effectiveness in battling drag in flight.

Spitzer a pointed bullet, derived from a German term that described the first 8mm German military bullets of aerodynamic shape. Spitzers have a more streamlined form and, thus, higher ballistic coefficients than flatnose or roundnose bullets of the same weight.

Throat the unrifled section of bore between the case mouth and full land diameter (also known as leade and, not quite correctly, as freebore).

Web the solid portion of a cartridge case between its base and the powder chamber. The primer pocket is at the rear of the web, and the flash-hole is drilled or punched through the web.

Wildcat A cartridge that is not loaded commercially. Wildcat cartridges are designed by handloaders who want something different. They're made by reshaping parent cases in special dies so they headspace properly. Fire-forming completes the forming operation. Then the cases can be sized and handloaded as you might handload for any commercial cartridge.

Zero in shooting, the range at which the sightline crosses the bullet's path the second time, farthest from the gun. The bullet begins to drop as soon as it leaves the muzzle, but the sightline is at a slight angle to the bore and meets the descending bullet arc a few yards (typically 25 to 35) from the muzzle. The angle of the sightline is such that it crosses the bullet's path there and continues forward in a straight but descending line, eventually to meet the trajectory again as the bullet's rate of drop accelerates with distance. That final crossing is the zero range. Zeroing at 200 yards gives riflemen a useful "point-blank" range, in which the bullet stays within a negligible vertical distance from sightline. A 200-yard zero puts most bullets from modern hunting cartridges between 2 and 3 inches high at mid-range or maximum ordinate. They drop 3 inches low somewhere around 250 yards. Bullets with steeper trajectories must be zeroed at shorter range. Foster-style shotgun slugs might call for a 75-yard zero, and traditional handgun bullets given an even closer zero.

Comprehensive Ballistics Tables for Currently Manufactured Sporting Rifle Cartridges

Note: Barrel length affects velocity at various rates depending on the load. As a rule, figure 50 fps per inch of barrel, plus or minus, if your barrel is longer or shorter than 22 inches.

Bullets are given by make, weight (in grains) and type. Most abbreviations are self-explanatory: BT=Boat-Tail, FMJ=Full Metal Jacket, HP=Hollow Point, SP=Soft Point—except in Hornady listings, where SP is the firm's Spire Point. TNT and TXP are trademarked designations of Speer and Norma. XLC identifies a coated Barnes X-Bullet. HE indicates a Federal High Energy load, similar to the Hornady LM (Light Magnum) and HM (Heavy Magnum) cartridges.

Arc (trajectory) is based on a zero range published by the manufacturer, from 100 to 300 yards. If a zero does not fall in a yardage column, it lies halfway between—at 150 yards, for example, if the bullet's strike is "+" at 100 yards and "−" at 200.

cartridge ***bullet***	***range, yards:***	***0***	***100***	***200***	***300***	***400***
.17 Remington						
Rem. 25 HP Power-Lokt	velocity, fps:	4040	3284	2644	2086	1606
	energy, ft-lb:	906	599	388	242	143
	arc, inches:		+1.8	0	−3.3	−16.6
.218 Bee						
Win. 46 Hollow Point	velocity, fps:	2760	2102	1550	1155	961
	energy, ft-lb:	778	451	245	136	94
	arc, inches:		0	−7.2	−29.4	
.22 Hornet						
Hornady 35 V-Max	velocity, fps:	3100	2278	1601	1135	929
	energy, ft-lb:	747	403	199	100	67
	arc, inches:		+2.8	0	−16.9	−60.4
Rem. 45 Pointed Soft Point	velocity, fps:	2690	2042	1502	1128	948
	energy, ft-lb:	723	417	225	127	90
	arc, inches:		0	−7.1	−30.0	
Rem. 45 Hollow Point	velocity, fps:	2690	2042	1502	1128	948
	energy, ft-lb:	723	417	225	127	90
	arc, inches:		0	−7.1	−30.0	
Win. 34 Jacketed HP	velocity, fps:	3050	2132	1415	1017	852
	energy, ft-lb:	700	343	151	78	55
	arc, inches:		0	−6.6	−29.9	
Win. 45 Soft Point	velocity, fps:	2690	2042	1502	1128	948
	energy, ft-lb:	723	417	225	127	90
	arc, inches:		0	−7.7	−31.3	

Win. 46 Hollow Point	velocity, fps:	2690	2042	1502	1128	948
	energy, ft-lb:	739	426	230	130	92
	arc, inches:		0	−7.7	−31.3	
.222 Remington						
Federal 50 Hi-Shok	velocity, fps:	3140	2600	2120	1700	1350
	energy, ft-lb:	1095	750	500	320	200
	arc, inches:		+1.9	0	−9.7	−31.6
Federal 55 FMJ boat-tail	velocity, fps:	3020	2740	2480	2230	1990
	energy, ft-lb:	1115	915	750	610	484
	arc, inches:		+1.6	0	−7.3	−21.5
Hornady 40 V-Max	velocity, fps:	3600	3117	2673	2269	1911
	energy, ft-lb:	1151	863	634	457	324
	arc, inches:		+1.1	0	−6.1	−18.9
Hornady 50 V-Max	velocity, fps:	3140	2729	2352	2008	1710
	energy, ft-lb:	1094	827	614	448	325
	arc, inches:		+1.7	0	−7.9	−24.4
Norma 50 Soft Point	velocity, fps:	3199	2667	2193	1771	
	energy, ft-lb:	1136	790	534	348	
	arc, inches:		+1.7	0	−9.1	
Norma 50 FMJ	velocity, fps:	2789	2326	1910	1547	
	energy, ft-lb:	864	601	405	266	
	arc, inches:		+2.5	0	−12.2	
Norma 62 Soft Point	velocity, fps:	2887	2457	2067	1716	
	energy, ft-lb:	1148	831	588	405	
	arc, inches:		+2.1	0	−10.4	
PMC 50 Pointed Soft Point	velocity, fps:	3044	2727	2354	2012	1651
	energy, ft-lb:	1131	908	677	494	333
	arc, inches:		+1.6	0	−7.9	−24.5
Rem. 50 Pointed Soft Point	velocity, fps:	3140	2602	2123	1700	1350
	energy, ft-lb:	1094	752	500	321	202
	arc, inches:		+1.9	0	−9.7	−31.7
Rem. 50 HP Power-Lokt	velocity, fps:	3140	2635	2182	1777	1432
	energy, ft-lb:	1094	771	529	351	228
	arc, inches:		+1.8	0	−9.2	−29.6
Rem. 50 V-Max boat-tail	velocity, fps:	3140	2744	2380	2045	1740
	energy, ft-lb:	1094	836	629	464	336
	arc, inches:		+1.6	0	−7.8	−23.9
Win. 40 Ballistic Silvertip	velocity, fps:	3370	2915	2503	2127	1786
	energy, ft-lb:	1009	755	556	402	283
	arc, inches:		+1.3	0	−6.9	−21.5
Win. 50 Pointed Soft Point	velocity, fps:	3140	2602	2123	1700	1350
	energy, ft-lb:	1094	752	500	321	202
	arc, inches:		+2.2	0	−10.0	−32.3
.223 Remington						
Federal 50 Jacketed HP	velocity, fps:	3400	2910	2460	2060	1700
	energy, ft-lb:	1285	940	675	470	320
	arc, inches:		+1.3	0	−7.1	−22.7
Federal 50 Speer TNT HP	velocity, fps:	3300	2860	2450	2080	1750
	energy, ft-lb:	1210	905	670	480	340
	arc, inches:		+1.4	0	−7.3	−22.6
Federal 52 Sierra MatchKing BTHP	velocity, fps:	3300	2860	2460	2090	1760
	energy, ft-lb:	1255	945	700	505	360
	arc, inches:		+1.4	0	−7.2	−22.4
Federal 55 Hi-Shok	velocity, fps:	3240	2750	2300	1910	1550
	energy, ft-lb:	1280	920	650	445	295
	arc, inches:		+1.6	0	−8.2	−26.1
Federal 55 FMJ boat-tail	velocity, fps:	3240	2950	2670	2410	2170
	energy, ft-lb:	1280	1060	875	710	575
	arc, inches:		+1.3	0	−6.1	−18.3
Federal 55 Sierra GameKing BTHP	velocity, fps:	3240	2770	2340	1950	1610
	energy, ft-lb:	1280	935	670	465	315
	arc, inches:		+1.5	0	−8.0	−25.3
Federal 55 Trophy Bonded	velocity, fps:	3100	2630	2210	1830	1500
	energy, ft-lb:	1175	845	595	410	275
	arc, inches:		+1.8	0	−8.9	−28.7

Federal 55 Nosler Bal. Tip	velocity, fps:	3240	2870	2530	2220	1920
	energy, ft-lb:	1280	1005	780	600	450
	arc, inches:		+1.4	0	−6.8	−20.8
Federal 55 Sierra BlitzKing	velocity, fps:	3240	2870	2520	2200	1910
	energy, ft-lb:	1280	1005	775	590	445
	arc, inches:		+1.4	0	−6.9	−20.9
Federal 62 FMJ	velocity, fps:	3020	2650	2310	2000	1710
	energy, ft-lb:	1225	970	735	550	405
	arc, inches:		+1.7	0	−8.4	−25.5
Federal 69 Sierra MatchKing BTHP	velocity, fps:	3000	2720	2460	2210	1980
	energy, ft-lb:	1380	1135	925	750	600
	arc, inches:		+1.6	0	−7.4	−21.9
Hornady 40 V-Max	velocity, fps:	3800	3305	2845	2424	2044
	energy, ft-lb:	1282	970	719	522	371
	arc, inches:		+0.8	0	−5.3	−16.6
Hornady 53 Hollow Point	velocity, fps:	3330	2882	2477	2106	1710
	energy, ft-lb:	1305	978	722	522	369
	arc, inches:		+1.7	0	−7.4	−22.7
Hornady 55 V-Max	velocity, fps:	3240	2859	2507	2181	1891
	energy, ft-lb:	1282	998	767	581	437
	arc, inches:		+1.4	0	−7.1	−21.4
Hornady 55 Urban Tactical	velocity, fps:	2970	2626	2307	2011	1739
	energy, ft-lb:	1077	842	650	494	369
	arc, inches:		+1.5	0	−8.1	−24.9
Hornady 60 Soft Point	velocity, fps:	3150	2782	2442	2127	1837
	energy, ft-lb:	1322	1031	795	603	450
	arc, inches:		+1.6	0	−7.5	−22.5
Hornady 60 Urban Tactical	velocity, fps:	2950	2619	2312	2025	1762
	energy, ft-lb:	1160	914	712	546	413
	arc, inches:		+1.6	0	−8.1	−24.7
Hornady 75 BTHP Match	velocity, fps:	2790	2554	2330	2119	1926
	energy, ft-lb:	1296	1086	904	747	617
	arc, inches:		+2.4	0	−8.8	−25.1
Hornady 75 BTHP Tactical	velocity, fps:	2630	2409	2199	2000	1814
	energy, ft-lb:	1152	966	805	666	548
	arc, inches:		−2.0	0	−9.2	−25.9
PMC 55 HP boat-tail	velocity, fps:	3240	2717	2250	1832	1473
	energy, ft-lb:	1282	901	618	410	265
	arc, inches:		+1.6	0	−8.6	−27.7
PMC 55 FMJ boat-tail	velocity, fps:	3195	2882	2525	2169	1843
	energy, ft-lb:	1246	1014	779	574	415
	arc, inches:		+1.4	0	−6.8	−21.1
PMC 55 Pointed Soft Point	velocity, fps:	3112	2767	2421	2100	1806
	energy, ft-lb:	1182	935	715	539	398
	arc, inches:		+1.5	0	−7.5	−22.9
PMC 64 Pointed Soft Point	velocity, fps:	2775	2511	2261	2026	1806
	energy, ft-lb:	1094	896	726	583	464
	arc, inches:		+2.0	0	−8.8	−26.1
PMC 69 BTHP Match	velocity, fps:					
	energy, ft-lb:					
	arc, inches:					
Rem. 50 V-Max, boat-tail	velocity, fps:	3300	2889	2514	2168	1851
	energy, ft-lb:	1209	927	701	522	380
	arc, inches:		+1.4	0	−6.9	−21.2
Rem. 55 Pointed Soft Point	velocity, fps:	3240	2747	2304	1905	1554
	energy, ft-lb:	1282	921	648	443	295
	arc, inches:		+1.6	0	−8.2	−26.2
Rem. 55 HP Power-Lokt	velocity, fps:	3240	2773	2352	1969	1627
	energy, ft-lb:	1282	939	675	473	323
	arc, inches:		+1.5	0	−7.9	−24.8
Rem. 55 Metal Case	velocity, fps:	3240	2759	2326	1933	1587
	energy, ft-lb:	1282	929	660	456	307
	arc, inches:		+1.6	0	−8.1	−25.5
Rem. 62 HP Match	velocity, fps:	3025	2572	2162	1792	1471
	energy, ft-lb:	1260	911	643	442	298
	arc, inches:		+1.9	0	−9.4	−29.9

Win. 40 Ballistic Silvertip	velocity, fps:	3700	3166	2693	2265	1879
	energy, ft-lb:	1216	891	644	456	314
	arc, inches:		+1.0	0	−5.8	−18.4
Win. 50 Ballistic Silvertip	velocity, fps:	3410	2982	2593	2235	1907
	energy, ft-lb:	1291	987	746	555	404
	arc, inches:		+1.2	0	−6.4	−19.8
Win. 53 Hollow Point	velocity, fps:	3330	2882	2477	2106	1770
	energy, ft-lb:	1305	978	722	522	369
	arc, inches:		+1.7	0	−7.4	−22.7
Win. 55 Pointed Soft Point	velocity, fps:	3240	2747	2304	1905	155
	energy, ft-lb:	1282	921	648	443	295
	arc, inches:		+1.9	0	−8.5	−26.7
Win. 55 Super Clean NT	velocity, fps:	3150	2520	1970	1505	1165
	energy, ft-lb:	1212	776	474	277	166
	arc, inches:		+2.8	0	−11.9	−38.9
Win. 64 Power-Point	velocity, fps:	3020	2656	2320	2009	1724
	energy, ft-lb:	1296	1003	765	574	423
	arc, inches:		+1.7	0	−8.2	−25.1
Win. 64 Power-Point Plus	velocity, fps:	3090	2684	2312	1971	1664
	energy, ft-lb:	1357	1024	760	552	393
	arc, inches:		+1.7	0	−8.2	−25.4
.5.6 x 52 R						
Norma 71 Soft Point	velocity, fps:	2789	2446	2128	1835	
	energy, ft-lb:	1227	944	714	531	
	arc, inches:		+2.1	0	−9.9	
.22 PPC						
A-Square 52 Berger	velocity, fps:	3300	2952	2629	2329	2049
	energy, ft-lb:	1257	1006	798	626	485
	arc, inches:		+1.3	0	−6.3	−19.1
.225 Winchester						
Win. 55 Pointed Soft Point	velocity, fps:	3570	3066	2616	2208	1838
	energy, ft-lb:	1556	1148	836	595	412
	arc, inches:		+2.4	+2.0	−3.5	−16.3
.224 Weatherby Mag.						
Wby. 55 Pointed Expanding	velocity, fps:	3650	3192	2780	2403	2056
	energy, ft-lb:	1627	1244	944	705	516
	arc, inches:		+2.8	+3.7	0	−9.8
.22–250 Remington						
Federal 40 Sierra Varminter	velocity, fps:	4000	3320	2720	2200	1740
	energy, ft-lb:	1420	980	660	430	265
	arc, inches:		+0.8	0	−5.6	−18.4
Federal 55 Hi-Shok	velocity, fps:	3680	3140	2660	2220	1830
	energy, ft-lb:	1655	1200	860	605	410
	arc, inches:		+1.0	0	−6.0	−19.1
Federal 55 Sierra BlitzKing	velocity, fps:	3680	3270	2890	2540	2220
	energy, ft-lb:	1655	1300	1020	790	605
	arc, inches:		+0.9	0	−5.1	−15.6
Federal 55 Sierra GameKing	velocity, fps:	3680	3280	2920	2590	2280
BTHP	energy, ft-lb:	1655	1315	1040	815	630
	arc, inches:		+0.9	0	−5.0	−15.1
Federal 55 Trophy Bonded	velocity, fps:	3600	3080	2610	2190	1810
	energy, ft-lb:	1585	1155	835	590	400
	arc, inches:		+1.1	0	−6.2	−19.8
Hornady 40 V-Max	velocity, fps:	4150	3631	3147	2699	2293
	energy, ft-lb:	1529	1171	879	647	467
	arc, inches:		+0.5	0	−4.2	−13.3
Hornady 50 V-Max	velocity, fps:	3800	3349	2925	2535	2178
	energy, ft-lb:	1603	1245	950	713	527
	arc, inches:		+0.8	0	−5.0	−15.6
Hornady 53 Hollow Point	velocity, fps:	3680	3185	2743	2341	1974
	energy, ft-lb:	1594	1194	886	645	459
	arc, inches:		+1.0	0	−5.7	−17.8
Hornady 55 V-Max	velocity, fps:	3680	3265	2876	2517	2183
	energy, ft-lb:	1654	1302	1010	772	582
	arc, inches:		+0.9	0	−5.3	−16.1

Hornady 60 Soft Point	velocity, fps:	3600	3195	2826	2485	2169
	energy, ft-lb:	1727	1360	1064	823	627
	arc, inches:		+1.0	0	−5.4	−16.3
Norma 53 Soft Point	velocity, fps:	3707	3234	2809	1716	
	energy, ft-lb:	1618	1231	928	690	
	arc, inches:		+0.9	0	−5.3	
PMC 55 HP boat-tail	velocity, fps:	3680	3104	2596	2141	1737
	energy, ft-lb:	1654	1176	823	560	368
	arc, inches:		+1.1	0	−6.3	−20.2
PMC 55 Pointed Soft Point	velocity, fps:	3586	3203	2852	2505	2178
	energy, ft-lb:	1570	1253	993	766	579
	arc, inches:		+1.0	0	−5.2	−16.0
Rem 45 JHP (UMC)	velocity, fps:	4000	3340	2770	2267	1820
	energy, ft-lbs:	1598	1114	767	513	331
	arc, inches:		+1.6	+1.6	−3.0	−14.5
Rem. 50 V-Max boat-tail (also in EtronX)	velocity, fps:	3725	3272	2864	2491	2147
	energy, ft-lb:	1540	1188	910	689	512
	arc, inches:		+1.7	+1.6	−2.8	−12.8
Rem. 55 Pointed Soft Point	velocity, fps:	3680	3137	2656	2222	1832
	energy, ft-lb:	1654	1201	861	603	410
	arc, inches:		+1.9	+1.8	−3.3	−15.5
Rem. 55 HP Power-Lokt	velocity, fps:	3680	3209	2785	2400	2046
	energy, ft-lb:	1654	1257	947	703	511
	arc, inches:		+1.8	+1.7	−3.0	−13.7
Rem. 60 Nosler Part. (also in EtronX)	velocity, fps:	3500	3045	2634	2258	1914
	energy, ft-lb:	1632	1235	924	679	488
	arc, inches:		+0.4	−1.5	−8.4	−22.2
Win. 40 Ballistic Silvertip	velocity, fps:	4150	3591	3099	2658	2257
	energy, ft-lb:	1530	1146	853	628	453
	arc, inches:		+0.6	0	−4.2	−13.4
Win. 50 Ballistic Silvertip	velocity, fps:	3810	3341	2919	2536	2182
	energy, ft-lb:	1611	1239	946	714	529
	arc, inches:		+0.8	0	−4.9	−15.2
Win. 55 Pointed Soft Point	velocity, fps:	3680	3137	2656	2222	1832
	energy, ft-lb:	1654	1201	861	603	410
	arc, inches:		+2.3	+1.9	−3.4	−15.9
.220 Swift						
Federal 52 Sierra MatchKing BTHP	velocity, fps:	3830	3370	2960	2600	2230
	energy, ft-lb:	1690	1310	1010	770	575
	arc, inches:		+0.8	0	−4.8	−14.9
Federal 55 Sierra BlitzKing	velocity, fps:	3800	3370	2990	2630	2310
	energy, ft-lb:	1765	1390	1090	850	650
	arc, inches:		+0.8	0	−4.7	−14.4
Federal 55 Trophy Bonded	velocity, fps:	3700	3170	2690	2270	1880
	energy, ft-lb:	1670	1225	885	625	430
	arc, inches:		+1.0	0	−5.8	−18.5
Hornady 40 V-Max	velocity, fps:	4200	3678	3190	2739	2329
	energy, ft-lb:	1566	1201	904	666	482
	arc, inches:		+0.5	0	−4.0	−12.9
Hornady 50 V-Max	velocity, fps:	3850	3396	2970	2576	2215
	energy, ft-lb:	1645	1280	979	736	545
	arc, inches:		+0.7	0	−4.8	−15.1
Hornady 50 SP	velocity, fps:	3850	3327	2862	2442	2060
	energy, ft-lb:	1645	1228	909	662	471
	arc, inches:		+0.8	0	−5.1	−16.1
Hornady 55 V-Max	velocity, fps:	3680	3265	2876	2517	2183
	energy, ft-lb:	1654	1302	1010	772	582
	arc, inches:		+0.9	0	−5.3	−16.1
Hornady 60 Hollow Point	velocity, fps:	3600	3199	2824	2475	2156
	energy, ft-lb:	1727	1364	1063	816	619
	arc, inches:		+1.0	0	−5.4	−16.3
Norma 50 Soft Point	velocity, fps:	4019	3380	2826	2335	
	energy, ft-lb:	1794	1268	887	605	
	arc, inches:		+0.7	0	−5.1	
Rem. 50 Pointed Soft Point	velocity, fps:	3780	3158	2617	2135	1710
	energy, ft-lb:	1586	1107	760	506	325
	arc, inches:		+0.3	−1.4	−8.2	

Rem. 50 V-Max boat-tail (also in EtronX)	velocity, fps:	3780	3321	2908	2532	2185
	energy, ft-lb:	1586	1224	939	711	530
	arc, inches:		+0.8	0	−5.0	−15.4
Win. 40 Ballistic Silvertip	velocity, fps:	4050	3518	3048	2624	2238
	energy, ft-lb:	1457	1099	825	611	445
	arc, inches:		+0.7	0	−4.4	−13.9
Win. 50 Pointed Soft Point	velocity, fps:	3870	3310	2816	2373	1972
	energy, ft-lb:	1663	1226	881	625	432
	arc, inches:		+0.8	0	−5.2	−16.7
6mm PPC						
A-Square 68 Berger	velocity, fps:	3100	2751	2428	2128	1850
	energy, ft-lb:	1451	1143	890	684	516
	arc, inches:		+1.5	0	−7.5	−22.6
.243 Winchester						
Federal 70 Nosler Bal. Tip	velocity, fps:	3400	3070	2760	2470	2200
	energy, ft-lb:	1795	1465	1185	950	755
	arc, inches:		+1.1	0	−5.7	−17.1
Federal 70 Speer TNT HP	velocity, fps:	3400	3040	2700	2390	2100
	energy, ft-lb:	1795	1435	1135	890	685
	arc, inches:		+1.1	0	−5.9	−18.0
Federal 80 Sierra Pro-Hunter	velocity, fps:	3350	2960	2590	2260	1950
	energy, ft-lb:	1995	1550	1195	905	675
	arc, inches:		+1.3	0	−6.4	−19.7
Federal 85 Sierra GameKing BTHP	velocity, fps:	3320	3070	2830	2600	2380
	energy, ft-lb:	2080	1770	1510	1280	1070
	arc, inches:		+1.1	0	−5.5	−16.1
Federal 90 Trophy Bonded	velocity, fps:	3100	2850	2610	2380	2160
	energy, ft-lb:	1920	1620	1360	1130	935
	arc, inches:		+1.4	0	−6.1	−19.2
Federal 100 Hi-Shok	velocity, fps:	2960	2700	2450	2220	1990
	energy, ft-lb:	1945	1615	1330	1090	880
	arc, inches:		+1.6	0	−7.5	−22.0
Federal 100 Sierra GameKing BTSP	velocity, fps:	2960	2760	2570	2380	2210
	energy, ft-lb:	1950	1690	1460	1260	1080
	arc, inches:		+1.5	0	−6.8	−19.8
Federal 100 Nosler Partition	velocity, fps:	2960	2730	2510	2300	2100
	energy, ft-lb:	1945	1650	1395	1170	975
	arc, inches:		+1.6	0	−7.1	−20.9
Hornady 58 V-Max	velocity, fps:	3750	3319	2913	2539	2195
	energy, ft-lb:	1811	1418	1093	830	620
	arc, inches:		+1.2	0	−5.5	−16.4
Hornady 75 Hollow Point	velocity, fps:	3400	2970	2578	2219	1890
	energy, ft-lb:	1926	1469	1107	820	595
	arc, inches:		+1.2	0	−6.5	−20.3
Hornady 100 BTSP	velocity, fps:	2960	2728	2508	2299	2099
	energy, ft-lb:	1945	1653	1397	1174	979
	arc, inches:		+1.6	0	−7.2	−21.0
Hornady 100 BTSP LM	velocity, fps:	3100	2839	2592	2358	2138
	energy, ft-lb:	2133	1790	1491	1235	1014
	arc, inches:		+1.5	0	−6.8	−19.8
Norma 100 FMJ	velocity, fps:	3018	2747	2493	2252	
	energy, ft-lb:	2023	1677	1380	1126	
	arc, inches:		+1.5	0	−7.1	
Norma 100 Soft Point	velocity, fps:	3018	2748	2493	2252	
	energy, ft-lb:	2023	1677	1380	1126	
	arc, inches:		+1.5	0	−7.1	
PMC 80 Pointed Soft Point	velocity, fps:	2940	2684	2444	2215	1999
	energy, ft-lb:	1535	1280	1060	871	709
	arc, inches:		+1.7	0	−7.5	−22.1
PMC 85 HP boat-tail	velocity, fps:	3275	2922	2596	2292	2009
	energy, ft-lb:	2024	1611	1272	991	761
	arc, inches:		+1.3	0	−6.5	−19.7
PMC 100 Pointed Soft Point	velocity, fps:	2743	2507	2283	2070	1869
	energy, ft-lb:	1670	1395	1157	951	776
	arc, inches:		+2.0	0	−8.7	−25.5
PMC 100 SP boat-tail	velocity, fps:	2960	2742	2534	2335	2144
	energy, ft-lb:	1945	1669	1425	1210	1021
	arc, inches:		+1.6	0	−7.0	−20.5

Rem. 75 V-Max boat-tail	velocity, fps:	3375	3065	2775	2504	2248
	energy, ft-lb:	1897	1564	1282	1044	842
	arc, inches:		+2.0	+1.8	−3.0	−13.3
Rem. 80 Pointed Soft Point	velocity, fps:	3350	2955	2593	2259	1951
	energy, ft-lb:	1993	1551	1194	906	676
	arc, inches:		+2.2	+2.0	−3.5	−15.8
Rem. 80 HP Power-Lokt	velocity, fps:	3350	2955	2593	2259	1951
	energy, ft-lb:	1993	1551	1194	906	676
	arc, inches:		+2.2	+2.0	−3.5	−15.8
Rem. 90 Nosler Bal. Tip	velocity, fps:	3120	2871	2635	2411	2199
(also in EtronX)	energy, ft-lb:	1946	1647	1388	1162	966
	arc, inches:		+1.4	0	−6.4	−18.8
Rem. 100 PSP Core-Lokt	velocity, fps:	2960	2697	2449	2215	1993
(also in EtronX)	energy, ft-lb:	1945	1615	1332	1089	882
	arc, inches:		+1.6	0	−7.5	−22.1
Rem. 100 PSP boat-tail	velocity, fps:	2960	2720	2492	2275	2069
	energy, ft-lb:	1945	1642	1378	1149	950
	arc, inches:		+2.8	+2.3	−3.8	−16.6
Speer 100 Grand Slam	velocity, fps:	2950	2684	2434	2197	
	energy, ft-lb:	1932	1600	1315	1072	
	arc, inches:		+1.7	0	−7.6	−22.4
Win. 55 Ballistic Silvertip	velocity, fps:	4025	3597	3209	2853	2525
	energy, ft-lb:	1978	1579	1257	994	779
	arc, inches:		+0.6	0	−4.0	−12.2
Win. 80 Pointed Soft Point	velocity, fps:	3350	2955	2593	2259	1951
	energy, ft-lb:	1993	1551	1194	906	676
	arc, inches:		+2.6	+2.1	−3.6	−16.2
Win. 95 Ballistic Silvertip	velocity, fps:	3100	2854	2626	2410	2203
	energy, ft-lb:	2021	1719	1455	1225	1024
	arc, inches:		+1.4	0	−6.4	−18.9
Win. 100 Power-Point	velocity, fps:	2960	2697	2449	2215	1993
	energy, ft-lb:	1945	1615	1332	1089	882
	arc, inches:		+1.9	0	−7.8	−22.6
Win. 100 Power-Point Plus	velocity, fps:	3090	2818	2562	2321	2092
	energy, ft-lb:	2121	1764	1458	1196	972
	arc, inches:		+1.4	0	−6.7	−20.0
6mm Remington						
Federal 80 Sierra Pro-Hunter	velocity, fps:	3470	3060	2690	2350	2040
	energy, ft-lb:	2140	1665	1290	980	735
	arc, inches:		+1.1	0	−5.9	−18.2
Federal 100 Hi-Shok	velocity, fps:	3100	2830	2570	2330	2100
	energy, ft-lb:	2135	1775	1470	1205	985
	arc, inches:		+1.4	0	−6.7	−19.8
Federal 100 Nosler Partition	velocity, fps:	3100	2860	2640	2420	2220
	energy, ft-lb:	2135	1820	1545	1300	1090
	arc, inches:		+1.4	0	−6.3	−18.7
Hornady 100 SP boat-tail	velocity, fps:	3100	2861	2634	2419	2231
	energy, ft-lb:	2134	1818	1541	1300	1088
	arc, inches:		+1.3	0	−6.5	−18.9
Hornady 100 SPBT LM	velocity, fps:	3250	2997	2756	2528	2311
	energy, ft-lb:	2345	1995	1687	1418	1186
	arc, inches:		+1.6	0	−6.3	−18.2
Rem. 75 V-Max boat-tail	velocity, fps:	3400	3088	2797	2524	2267
	energy, ft-lb:	1925	1587	1303	1061	856
	arc, inches:		+1.9	+1.7	−3.0	−13.1
Rem. 100 PSP Core-Lokt	velocity, fps:	3100	2829	2573	2332	2104
	energy, ft-lb:	2133	1777	1470	1207	983
	arc, inches:		+1.4	0	−6.7	−19.8
Rem. 100 PSP boat-tail	velocity, fps:	3100	2852	2617	2394	2183
	energy, ft-lb:	2134	1806	1521	1273	1058
	arc, inches:		+1.4	0	−6.5	−19.1
Win. 100 Power-Point	velocity, fps:	3100	2829	2573	2332	2104
	energy, ft-lb:	2133	1777	1470	1207	983
	arc, inches:		+1.7	0	−7.0	−20.4
.240 Weatherby Mag.						
Wby. 87 Pointed Expanding	velocity, fps:	3523	3199	2898	2617	2352
	energy, ft-lb:	2397	1977	1622	1323	1069
	arc, inches:		+2.7	+3.4	0	−8.4

Wby. 90 Barnes-X	velocity, fps:	3500	3222	2962	2717	2484
	energy, ft-lb:	2448	2075	1753	1475	1233
	arc, inches:		+2.6	+3.3	0	−8.0
Wby. 95 Nosler Bal. Tip	velocity, fps:	3420	3146	2888	2645	2414
	energy, ft-lb:	2467	2087	1759	1475	1229
	arc, inches:		+2.7	+3.5	0	−8.4
Wby. 100 Pointed Expanding	velocity, fps:	3406	3134	2878	2637	2408
	energy, ft-lb:	2576	2180	1839	1544	1287
	arc, inches:		+2.8	+3.5	0	−8.4
Wby. 100 Partition	velocity, fps:	3406	3136	2882	2642	2415
	energy, ft-lb:	2576	2183	1844	1550	1294
	arc, inches:		+2.8	+3.5	0	−8.4
.25–20 Winchester						
Rem. 86 Soft Point	velocity, fps:	1460	1194	1030	931	858
	energy, ft-lb:	407	272	203	165	141
	arc, inches:		0	−22.9	−78.9	−173.0
Win. 86 Soft Point	velocity, fps:	1460	1194	1030	931	858
	energy, ft-lb:	407	272	203	165	141
	arc, inches:		0	−23.5	−79.6	−175.9
.25–35 Winchester						
Win. 117 Soft Point	velocity, fps:	2230	1866	1545	1282	1097
	energy, ft-lb:	1292	904	620	427	313
	arc, inches:		+2.1	−5.1	−27.0	−70.1
.250 Savage						
Rem. 100 Pointed SP	velocity, fps:	2820	2504	2210	1936	1684
	energy, ft-lb:	1765	1392	1084	832	630
	arc, inches:		+2.0	0	−9.2	−27.7
Win. 100 Silvertip	velocity, fps:	2820	2467	2140	1839	1569
	energy, ft-lb:	1765	1351	1017	751	547
	arc, inches:		+2.4	0	−10.1	−30.5
.257 Roberts						
Federal 120 Nosler Partition	velocity, fps:	2780	2560	2360	2160	1970
	energy, ft-lb:	2060	1750	1480	1240	1030
	arc, inches:		+1.9	0	−8.2	−24.0
Hornady 117 SP boat-tail	velocity, fps:	2780	2550	2331	2122	1925
	energy, ft-lb:	2007	1689	1411	1170	963
	arc, inches:		+1.9	0	−8.3	−24.4
Hornady 117 SP boat-tail LM	velocity, fps:	2940	2694	2460	2240	2031
	energy, ft-lb:	2245	1885	1572	1303	1071
	arc, inches:		+1.7	0	−7.6	−21.8
Rem. 117 SP Core-Lokt	velocity, fps:	2650	2291	1961	1663	1404
	energy, ft-lb:	1824	1363	999	718	512
	arc, inches:		+2.6	0	−11.7	−36.1
Win. 117 Power-Point	velocity, fps:	2780	2411	2071	1761	1488
	energy, ft-lb:	2009	1511	1115	806	576
	arc, inches:		+2.6	0	−10.8	−33.0
.25–06 Remington						
Federal 90 Sierra Varminter	velocity, fps:	3440	3040	2680	2340	2030
	energy, ft-lb:	2365	1850	1435	1100	825
	arc, inches:		+1.1	0	−6.0	−18.3
Federal 100 Barnes XLC	velocity, fps:	3210	2970	2750	2540	2330
	energy, ft-lb:	2290	1965	1680	1430	1205
	arc, inches:		+1.2	0	−5.8	−17.0
Federal 100 Nosler Bal. Tip	velocity, fps:	3210	2960	2720	2490	2280
	energy, ft-lb:	2290	1940	1640	1380	1150
	arc, inches:		+1.2	0	−6.0	−17.5
Federal 115 Nosler Partition	velocity, fps:	2990	2750	2520	2300	2100
	energy, ft-lb:	2285	1930	1620	1350	1120
	arc, inches:		+1.6	0	−7.0	−20.8
Federal 115 Trophy Bonded	velocity, fps:	2990	2740	2500	2270	2050
	energy, ft-lb:	2285	1910	1590	1310	1075
	arc, inches:		+1.6	0	−7.2	−21.1
Federal 117 Sierra Pro Hunt.	velocity, fps:	2990	2730	2480	2250	2030
	energy, ft-lb:	2320	1985	1645	1350	1100
	arc, inches:		+1.6	0	−7.2	−21.4
Federal 117 Sierra GameKing BTSP	velocity, fps:	2990	2770	2570	2370	2190
	energy, ft-lb:	2320	2000	1715	1465	1240
	arc, inches:		+1.5	0	−6.8	−19.9

Hornady 117 SP boat-tail	velocity, fps:	2990	2749	2520	2302	2096
	energy, ft-lb:	2322	1962	1649	1377	1141
	arc, inches:		+1.6	0	−7.0	−20.7
Hornady 117 SP boat-tail LM	velocity, fps:	3110	2855	2613	2384	2168
	energy, ft-lb:	2512	2117	1774	1476	1220
	arc, inches:		+1.8	0	−7.1	−20.3
PMC 117 PSP	velocity, fps:					
	energy, ft-lb:					
	arc, inches:					
Rem. 100 PSP Core-Lokt	velocity, fps:	3230	2893	2580	2287	2014
	energy, ft-lb:	2316	1858	1478	1161	901
	arc, inches:		+1.3	0	−6.6	−19.8
Rem. 120 PSP Core-Lokt	velocity, fps:	2990	2730	2484	2252	2032
	energy, ft-lb:	2382	1985	1644	1351	1100
	arc, inches:		+1.6	0	−7.2	−21.4
Speer 120 Grand Slam	velocity, fps:	3130	2835	2558	2298	
	energy, ft-lb:	2610	2141	1743	1407	
	arc, inches:		+1.4	0	−6.8	−20.1
Win. 90 Pos. Exp. Point	velocity, fps:	3440	3043	2680	2344	2034
	energy, ft-lb:	2364	1850	1435	1098	827
	arc, inches:		+2.4	+2.0	−3.4	−15.0
Win. 115 Ballistic Silvertip	velocity, fps:	3060	2825	2603	2390	2188
	energy, ft-lb:	2391	2038	1729	1459	1223
	arc, inches:		+1.4	0	−6.6	−19.2
.257 Weatherby Mag.						
Federal 115 Nosler Partition	velocity, fps:	3150	2900	2660	2440	2220
	energy, ft-lb:	2535	2145	1810	1515	1260
	arc, inches:		+1.3	0	−6.2	−18.4
Federal 115 Trophy Bonded	velocity, fps:	3150	2890	2640	2400	2180
	energy, ft-lb:	2535	2125	1775	1470	1210
	arc, inches:		+1.4	0	−6.3	−18.8
Wby. 87 Pointed Expanding	velocity, fps:	3825	3472	3147	2845	2563
	energy, ft-lb:	2826	2328	1913	1563	1269
	arc, inches:		+2.1	+2.8	0	−7.1
Wby. 100 Pointed Expanding	velocity, fps:	3602	3298	3016	2750	2500
	energy, ft-lb:	2881	2416	2019	1680	1388
	arc, inches:		+2.4	+3.1	0	−7.7
Wby. 115 Nosler Bal. Tip	velocity, fps:	3400	3170	2952	2745	2547
	energy, ft-lb:	2952	2566	2226	1924	1656
	arc, inches:		+3.0	+3.5	0	−7.9
Wby. 115 Barnes X	velocity, fps:	3400	3158	2929	2711	2504
	energy, ft-lb:	2952	2546	2190	1877	1601
	arc, inches:		+2.7	+3.4	0	−8.1
Wby. 117 RN Expanding	velocity, fps:	3402	2984	2595	2240	1921
	energy, ft-lb:	3007	2320	1742	1302	956
	arc, inches:		+3.4	+4.31	0	−11.1
Wby. 120 Nosler Partition	velocity, fps:	3305	3046	2801	2570	2350
	energy, ft-lb:	2910	2472	2091	1760	1471
	arc, inches:		+3.0	+3.7	0	−8.9
6.53 (.257) Scramjet						
Lazzeroni 85 Nosler Bal. Tip	velocity, fps:	3960	3652	3365	3096	2844
	energy, ft-lb:	2961	2517	2137	1810	1526
	arc, inches:		+1.7	+2.4	0	−6.0
Lazzeroni 100 Nosler Part.	velocity, fps:	3740	3465	3208	2965	2735
	energy, ft-lb:	3106	2667	2285	1953	1661
	arc, inches:		+2.1	+2.7	0	−6.7
6.5x50 Japanese						
Norma 156 Alaska	velocity, fps:	2067	1832	1615	1423	
	energy, ft-lb:	1480	1162	904	701	
	arc, inches:		+4.4	0	−17.8	
6.5x52 Carcano						
Norma 156 Alaska	velocity, fps:	2428	2169	1926	1702	
	energy, ft-lb:	2043	1630	1286	1004	
	arc, inches:		+2.9	0	−12.3	
6.5x55 Swedish						
Federal 140 Hi-Shok	velocity, fps:	2600	2400	2220	2040	1860
	energy, ft-lb:	2100	1795	1525	1285	1080
	arc, inches:		+2.3	0	−9.4	−27.2

Federal 140 Trophy Bonded	velocity, fps:	2550	2350	2160	1980	1810
	energy, ft-lb:	2020	1720	1450	1220	1015
	arc, inches:		+2.4	0	−9.8	−28.4
Federal 140 Sierra MatchKg. BTHP	velocity, fps:	2630	2460	2300	2140	2000
	energy, ft-lb:	2140	1880	1640	1430	1235
	arc, inches:		+16.4	+28.8	+33.9	+31.8
Hornady 129 SP LM	velocity, fps:	2770	2561	2361	2171	1994
	energy, ft-lb:	2197	1878	1597	1350	1138
	arc, inches:		+2.0	0	−8.2	−23.2
Hornady140 SP LM	velocity, fps:	2740	2541	2351	2169	1999
	energy, ft-lb:	2333	2006	1717	1463	1242
	arc, inches:		+2.4	0	−8.7	−24.0
Norma 139 Vulkan	velocity, fps:	2854	2569	2302	2051	
	energy, ft-lb:	2515	2038	1636	1298	
	arc, inches:		+1.8	0	−8.4	
Norma 140 Nosler Partition	velocity, fps:	2789	2592	2403	2223	
	energy, ft-lb:	2419	2089	1796	1536	
	arc, inches:		+1.8	0	−7.8	
Norma 156 TXP Swift A-Fr.	velocity, fps:	2526	2276	2040	1818	
	energy, ft-lb:	2196	1782	1432	1138	
	arc, inches:		+2.6	0	−10.9	
Norma 156 Alaska	velocity, fps:	2559	2245	1953	1687	
	energy, ft-lb:	2269	1746	1322	986	
	arc, inches:		+2.7	0	−11.9	
Norma 156 Vulkan	velocity, fps:	2644	2395	2159	1937	
	energy, ft-lb:	2422	1987	1616	1301	
	arc, inches:		+2.2	0	−9.7	
Norma 156 Oryx	velocity, fps:	2559	2308	2070	1848	
	energy, ft-lb:	2269	1845	1485	1183	
	arc, inches:		+2.5	0	−10.6	
PMC 139 Pointed Soft Point	velocity, fps:	2850	2560	2290	2030	1790
	energy, ft-lb:	2515	2025	1615	1270	985
	arc, inches:		+2.2	0	−8.9	−26.3
PMC 140 HP boat-tail	velocity, fps:	2560	2398	2243	2093	1949
	energy, ft-lb:	2037	1788	1563	1361	1181
	arc, inches:		+2.3	0	−9.2	−26.4
PMC 140 SP boat-tail	velocity, fps:	2560	2386	2218	2057	1903
	energy, ft-lb:	2037	1769	1529	1315	1126
	arc, inches:		+2.3	0	−9.4	−27.1
PMC 144 FMJ	velocity, fps:	2650	2370	2110	1870	1650
	energy, ft-lb:	2425	1950	1550	1215	945
	arc, inches:		+2.7	0	−10.5	−30.9
Rem. 140 PSP Core-Lokt	velocity, fps:	2550	2353	2164	1984	1814
	energy, ft-lb:	2021	1720	1456	1224	1023
	arc, inches:		+2.4	0	−9.8	−27.0
Speer 140 Grand Slam	velocity, fps:	2550	2318	2099	1892	
	energy, ft-lb:	2021	1670	1369	1112	
	arc, inches:		+2.5	0	−10.4	−30.6
Win. 140 Soft Point	velocity, fps:	2550	2359	2176	2002	1836
	energy, ft-lb:	2022	1731	1473	1246	1048
	arc, inches:		+2.4	0	−9.7	−28.1
.260 Remington						
Federal 140 Sierra GameKing BTSP	velocity, fps:	2750	2570	2390	2220	2060
	energy, ft-lb:	2350	2045	1775	1535	1315
	arc, inches:		+1.9	0	−8.0	−23.1
Federal 140 Trophy Bonded	velocity, fps:	2750	2540	2340	2150	1970
	energy, ft-lb:	2350	2010	1705	1440	1210
	arc, inches:		+1.9	0	−8.4	−24.1
Rem. 120 Nosler Bal. Tip	velocity, fps:	2890	2688	2494	2309	2131
	energy, ft-lb:	2226	1924	1657	1420	1210
	arc, inches:		+1.7	0	−7.3	−21.1
Rem. 125 Nosler Partition	velocity, fps:	2875	2669	2473	2285	2105
	energy, ft-lb:	2294	1977	1697	1449	1230
	arc, inches:		+1.71	0	−7.4	−21.4
Rem. 140 PSP Core-Lokt	velocity, fps:	2750	2544	2347	2158	1979
	energy, ft-lb:	2351	2011	1712	1448	1217
	arc, inches:		+1.9	0	−8.3	−24.0

Speer 140 Grand Slam	velocity, fps:	2750	2518	2297	2087	
	energy, ft-lb:	2351	1970	1640	1354	
	arc, inches:		+2.3	0	−8.9	−25.8
.264 Winchester Mag.						
Rem. 140 PSP Core-Lokt	velocity, fps:	3030	2782	2548	2326	2114
	energy, ft-lb:	2854	2406	2018	1682	1389
	arc, inches:		+1.5	0	−6.9	−20.2
Win. 140 Power-Point	velocity, fps:	3030	2782	2548	2326	2114
	energy, ft-lb:	2854	2406	2018	1682	1389
	arc, inches:		+1.8	0	−7.2	−20.8
.270 Winchester						
Federal 130 Hi-Shok	velocity, fps:	3060	2800	2560	2330	2110
	energy, ft-lb:	2700	2265	1890	1565	1285
	arc, inches:		+1.5	0	−6.8	−20.0
Federal 130 Sierra Pro-Hunt.	velocity, fps:	3060	2830	2600	2390	2190
	energy, ft-lb:	2705	2305	1960	1655	1390
	arc, inches:		+1.4	0	−6.4	−19.0
Federal 130 Sierra GameKing	velocity, fps:	3060	2830	2620	2410	2220
	energy, ft-lb:	2700	2320	1980	1680	1420
	arc, inches:		+1.4	0	−6.5	−19.0
Federal 130 Nosler Bal. Tip	velocity, fps:	3060	2840	2630	2430	2230
	energy, ft-lb:	2700	2325	1990	1700	1440
	arc, inches:		+1.4	0	−6.5	−18.8
Federal 130 Barnes XLC	velocity, fps:	3060	2840	2620	2420	2220
	energy, ft-lb:	2705	2320	1985	1690	1425
	arc, inches:		+1.4	0	−6.4	−18.9
Federal 130 Trophy Bonded	velocity, fps:	3060	2810	2570	2340	2130
	energy, ft-lb:	2705	2275	1905	1585	1310
	arc, inches:		+1.5	0	−6.7	−19.8
Federal 140 Trophy Bonded	velocity, fps:	2940	2700	2480	2260	2060
	energy, ft-lb:	2685	2270	1905	1590	1315
	arc, inches:		+1.6	0	−7.3	−21.5
Federal 140 Tr. Bonded HE	velocity, fps:	3100	2860	2620	2400	2200
	energy, ft-lb:	2990	2535	2140	1795	1500
	arc, inches:		+1.4	0	−6.4	−18.9
Federal 150 Hi-Shok RN	velocity, fps:	2850	2500	2180	1890	1620
	energy, ft-lb:	2705	2085	1585	1185	870
	arc, inches:		+2.0	0	−9.4	−28.6
Federal 150 Sierra GameKing	velocity, fps:	2850	2660	2480	2300	2130
	energy, ft-lb:	2705	2355	2040	1760	1510
	arc, inches:		+1.7	0	−7.4	−21.4
Federal 150 Sierra GameKing HE	velocity, fps:	3000	2800	2620	2430	2260
	energy, ft-lb:	2995	2615	2275	1975	1700
	arc, inches:		+1.5	0	−6.5	−18.9
Federal 150 Nosler Partition	velocity, fps:	2850	2590	2340	2100	1880
	energy, ft-lb:	2705	2225	1815	1470	1175
	arc, inches:		+1.9	0	−8.3	−24.4
Hornady 130 SP	velocity, fps:	3060	2800	2560	2330	2110
	energy, ft-lb:	2700	2265	1890	1565	1285
	arc, inches:		+1.8	0	−7.1	−20.6
Hornady 130 SST LM	velocity, fps:	3215	2998	2790	2590	2400
	energy, ft-lb:	2983	2594	2246	1936	1662
	arc, inches:		+1.2	0	−5.8	−17.0
Hornady 140 SP boat-tail	velocity, fps:	2940	2747	2562	2385	2214
	energy, ft-lb:	2688	2346	2041	1769	1524
	arc, inches:		+1.6	0	−7.0	−20.2
Hornady 140 SP boat-tail LM	velocity, fps:	3100	2894	2697	2508	2327
	energy, ft-lb:	2987	2604	2261	1955	1684
	arc, inches:		+1.4	0	6.3	−18.3
Hornady 150 SP	velocity, fps:	2800	2684	2478	2284	2100
	energy, ft-lb:	2802	2400	2046	1737	1469
	arc, inches:		+1.7	0	−7.4	−21.6
Norma 130 SP	velocity, fps:	3140	2862	2601	2354	
	energy, ft-lb:	2847	2365	1953	1600	
	arc, inches:	0	+1.3	0	−6.5	
Norma 150 SP	velocity, fps:	2799	2555	2323	2104	
	energy, ft-lb:	2610	2175	1798	1475	
	arc, inches:	0	+1.9	0	−8.3	

PMC 130 Barnes X	velocity, fps:	2910	2717	2533	2356	2186
	energy, ft-lb:	2444	2131	1852	1602	1379
	arc, inches:		+1.6	0	−7.1	−20.4
PMC 130 SP boat-tail	velocity, fps:	3050	2830	2620	2421	2229
	energy, ft-lb:	2685	2312	1982	1691	1435
	arc, inches:		+1.5	0	−6.5	−19.0
PMC 130 Pointed Soft Point	velocity, fps:	2816	2593	2381	2179	1987
	energy, ft-lb:	2288	1941	1636	1370	1139
	arc, inches:		+1.8	0	−8.0	−23.2
PMC 150 Barnes X	velocity, fps:	2700	2541	2387	2238	2095
	energy, ft-lb:	2428	2150	1897	1668	1461
	arc, inches:		+2.0	0	−8.1	−23.1
PMC 150 SP boat-tail	velocity, fps:	2850	2660	2477	2302	2134
	energy, ft-lb:	2705	2355	2043	1765	1516
	arc, inches:		+1.7	0	−7.4	−21.4
PMC 150 Pointed Soft Point	velocity, fps:	2547	2368	2197	2032	1875
	energy, ft-lb:	2160	1868	1607	1375	1171
	arc, inches:		+2.4	0	−9.5	−27.5
Rem. 100 Pointed Soft Point	velocity, fps:	3320	2924	2561	2225	1916
	energy, ft-lb:	2448	1898	1456	1099	815
	arc, inches:		+2.3	+2.0	−3.6	−16.2
Rem. 130 PSP Core-Lokt	velocity, fps:	3060	2776	2510	2259	2022
	energy, ft-lb:	2702	2225	1818	1472	1180
	arc, inches:		+1.5	0	−7.0	−20.9
Rem. 130 Bronze Point	velocity, fps:	3060	2802	2559	2329	2110
	energy, ft-lb:	2702	2267	1890	1565	1285
	arc, inches:		+1.5	0	−6.8	−20.0
Rem. 130 Swift Scirocco	velocity, fps:	3060	2838	2627	2425	2232
	energy, ft-lb:	2702	2325	1991	1697	1438
	arc, inches:		+2.4	+2.0	−3.4	−14.8
Rem. 140 Swift A-Frame	velocity, fps:	2925	2652	2394	2152	1923
	energy, ft-lb:	2659	2186	1782	1439	1150
	arc, inches:		+1.7	0	−7.8	−23.2
Rem. 140 PSP boat-tail	velocity, fps:	2960	2749	2548	2355	2171
	energy, ft-lb:	2723	2349	2018	1724	1465
	arc, inches:		+1.6	0	−6.9	−20.1
Rem. 140 Nosler Bal. Tip	velocity, fps:	2960	2754	2557	2366	2187
	energy, ft-lb:	2724	2358	2032	1743	1487
	arc, inches:		+1.6	0	−6.9	−20.0
Rem. 150 SP Core-Lokt	velocity, fps:	2850	2504	2183	1886	1618
	energy, ft-lb:	2705	2087	1587	1185	872
	arc, inches:		+2.0	0	−9.4	−28.6
Rem. 150 Nosler Partition	velocity, fps:	2850	2652	2463	2282	2108
	energy, ft-lb:	2705	2343	2021	1734	1480
	arc, inches:		+1.7	0	−7.5	−21.6
Speer 130 Grand Slam	velocity, fps:	3050	2774	2514	2269	
	energy, ft-lb:	2685	2221	1824	1485	
	arc, inches:		+1.5	0	−7.0	−20.9
Speer 150 Grand Slam	velocity, fps:	2830	2594	2369	2156	
	energy, ft-lb:	2667	2240	1869	1548	
	arc, inches:		+1.8	0	−8.1	−23.6
Win. 130 Power-Point	velocity, fps:	3060	2802	2559	2329	2110
	energy, ft-lb:	2702	2267	1890	1565	1285
	arc, inches:		+1.8	0	−7.1	−20.6
Win. 130 Power-Point Plus	velocity, fps:	3150	2881	2628	2388	2161
	energy, ft-lb:	2865	2396	1993	1646	1348
	arc, inches:		+1.3	0	−6.4	−18.9
Win. 130 Silvertip	velocity, fps:	3060	2776	2510	2259	2022
	energy, ft-lb:	2702	2225	1818	1472	1180
	arc, inches:		+1.8	0	−7.4	−21.6
Win. 130 Ballistic Silvertip	velocity, fps:	3050	2828	2618	2416	2224
	energy, ft-lb:	2685	2309	1978	1685	1428
	arc, inches:		+1.4	0	−6.5	−18.9
Win. 140 Fail Safe	velocity, fps:	2920	2671	2435	2211	1999
	energy, ft-lb:	2651	2218	1843	1519	1242
	arc, inches:		+1.7	0	−7.6	−22.3

Win. 150 Power-Point	velocity, fps:	2850	2585	2336	2100	1879
	energy, ft-lb:	2705	2226	1817	1468	1175
	arc, inches:		+2.2	0	−8.6	−25.0
Win. 150 Power-Point Plus	velocity, fps:	2950	2679	2425	2184	1957
	energy, ft-lb:	2900	2391	1959	1589	1276
	arc, inches:		+1.7	0	−7.6	−22.6
Win. 150 Partition Gold	velocity, fps:	2930	2693	2468	2254	2051
	energy, ft-lb:	2860	2416	2030	1693	1402
	arc, inches:		+1.7	0	−7.4	−21.6
.270 Weatherby Mag.						
Federal 130 Nosler Partition	velocity, fps:	3200	2960	2740	2520	2320
	energy, ft-lb:	2955	2530	2160	1835	1550
	arc, inches:		+1.2	0	−5.9	−17.3
Federal 130 Sierra GameKing BTSP	velocity, fps:	3200	2980	2780	2580	2400
	energy, ft-lb:	2955	2570	2230	1925	1655
	arc, inches:		+1.2	0	−5.7	−16.6
Federal 140 Trophy Bonded	velocity, fps:	3100	2840	2600	2370	2150
	energy, ft-lb:	2990	2510	2100	1745	1440
	arc, inches:		+1.4	0	−6.6	−19.3
Wby. 100 Pointed Expanding	velocity, fps:	3760	3396	3061	2751	2462
	energy, ft-lb:	3139	2560	2081	1681	1346
	arc, inches:		+2.3	+3.0	0	−7.6
Wby. 130 Pointed Expanding	velocity, fps:	3375	3123	2885	2659	2444
	energy, ft-lb:	3288	2815	2402	2041	1724
	arc, inches:		+2.8	+3.5	0	−8.4
Wby. 130 Nosler Partition	velocity, fps:	3375	3127	2892	2670	2458
	energy, ft-lb:	3288	2822	2415	2058	1744
	arc, inches:		+2.8	+3.5	0	−8.3
Wby. 140 Nosler Bal. Tip	velocity, fps:	3300	3077	2865	2663	2470
	energy, ft-lb:	3385	2943	2551	2204	1896
	arc, inches:		+2.9	+3.6	0	−8.4
Wby. 140 Barnes X	velocity, fps:	3250	3032	2825	2628	2438
	energy, ft-lb:	3283	2858	2481	2146	1848
	arc, inches:		+3.0	+3.7	0	−8.7
Wby. 150 Pointed Expanding	velocity, fps:	3245	3028	2821	2623	2434
	energy, ft-lb:	3507	3053	2650	2292	1973
	arc, inches:		+3.0	+3.7	0	−8.7
Wby. 150 Nosler Partition	velocity, fps:	3245	3029	2823	2627	2439
	energy, ft-lb:	3507	3055	2655	2298	1981
	arc, inches:		+3.0	+3.7	0	−8.
7–30 Waters						
Federal 120 Sierra GameKing BTSP	velocity, fps:	2700	2300	1930	1600	1330
	energy, ft-lb:	1940	1405	990	685	470
	arc, inches:		+2.6	0	−12.0	−37.6
7mm Mauser (7x57)						
Federal 140 Sierra Pro-Hunt.	velocity, fps:	2660	2450	2260	2070	1890
	energy, ft-lb:	2200	1865	1585	1330	1110
	arc, inches:		+2.1	0	−9.0	−26.1
Federal 140 Nosler Partition	velocity, fps:	2660	2450	2260	2070	1890
	energy, ft-lb:	2200	1865	1585	1330	1110
	arc, inches:		+2.1	0	−9.0	−26.1
Federal 175 Hi-Shok RN	velocity, fps:	2440	2140	1860	1600	1380
	energy, ft-lb:	2315	1775	1340	1000	740
	arc, inches:		+3.1	0	−13.3	−40.1
Hornady 139 SP boat-tail	velocity, fps:	2700	2504	2316	2137	1965
	energy, ft-lb:	2251	1936	1656	1410	1192
	arc, inches:		+2.0	0	−8.5	−24.9
Hornady 139 SP boat-tail LM	velocity, fps:	2830	2620	2450	2250	2070
	energy, ft-lb:	2475	2135	1835	1565	1330
	arc, inches:		+1.8	0	−7.6	−22.1
Hornady 139 SP LM	velocity, fps:	2950	2736	2532	2337	2152
	energy, ft-lb:	2686	2310	1978	1686	1429
	arc, inches:		+2.0	0	−7.6	−21.5
Norma 150 Soft Point	velocity, fps:	2690	2479	2278	2087	
	energy, ft-lb:	2411	2048	1729	1450	
	arc, inches:		+2.0	0	−8.8	

PMC 140 Pointed Soft Point	velocity, fps:	2660	2450	2260	2070	1890
	energy, ft-lb:	2200	1865	1585	1330	1110
	arc, inches:		+2.4	0	−9.6	−27.3
PMC 175 Soft Point	velocity, fps:	2440	2140	1860	1600	1380
	energy, ft-lb:	2315	1775	1340	1000	740
	arc, inches:		+1.5	−3.6	−18.6	−46.8
Rem. 140 PSP Core-Lokt	velocity, fps:	2660	2435	2221	2018	1827
	energy, ft-lb:	2199	1843	1533	1266	1037
	arc, inches:		+2.2	0	−9.2	−27.4
Win. 145 Power-Point	velocity, fps:	2660	2413	2180	1959	1754
	energy, ft-lb:	2279	1875	1530	1236	990
	arc, inches:		+1.1	−2.8	−14.1	−34.4
7x57 R						
Norma 150 FMJ	velocity, fps:	2690	2489	2296	2112	
	energy, ft-lb:	2411	2063	1756	1486	
	arc, inches:		+2.0	0	−8.6	
Norma 154 Soft Point	velocity, fps:	2625	2417	2219	2030	
	energy, ft-lb:	2357	1999	1684	1410	
	arc, inches:		+2.2	0	−9.3	
7mm-08 Remington						
Federal 140 Nosler Partition	velocity, fps:	2800	2590	2390	2200	2020
	energy, ft-lb:	2435	2085	1775	1500	1265
	arc, inches:		+1.8	0	−8.0	−23.1
Federal 140 Nosler Bal. Tip	velocity, fps:	2800	2610	2430	2260	2100
	energy, ft-lb:	2440	2135	1840	1590	1360
	arc, inches:		+1.8	0	−7.7	−22.3
Federal 140 Tr. Bonded HE	velocity, fps:	2950	2660	2390	2140	1900
	energy, ft-lb:	2705	2205	1780	1420	1120
	arc, inches:		+1.7	0	−7.9	−23.2
Federal 150 Sierra Pro-Hunt.	velocity, fps:	2650	2440	2230	2040	1860
	energy, ft-lb:	2340	1980	1660	1390	1150
	arc, inches:		+2.2	0	−9.2	−26.7
Hornady 139 SP boat-tail LM	velocity, fps:	3000	2790	2590	2399	2216
	energy, ft-lb:	2777	2403	2071	1776	1515
	arc, inches:		+1.5	0	−6.7	−19.4
PMC 140 PSP	velocity, fps:					
	energy, ft-lb:					
	arc, inches:					
Rem. 120 Hollow Point	velocity, fps:	3000	2725	2467	2223	1992
	energy, ft-lb:	2398	1979	1621	1316	1058
	arc, inches:		+1.6	0	−7.3	−21.7
Rem. 140 PSP Core-Lokt	velocity, fps:	2860	2625	2402	2189	1988
	energy, ft-lb:	2542	2142	1793	1490	1228
	arc, inches:		+1.8	0	−7.8	−22.9
Rem. 140 PSP boat-tail	velocity, fps:	2860	2656	2460	2273	2094
	energy, ft-lb:	2542	2192	1881	1606	1363
	arc, inches:		+1.7	0	−7.5	−21.7
Rem. 140 Nosler Bal. Tip	velocity, fps:	2860	2670	2488	2313	2145
	energy, ft-lb:	2543	2217	1925	1663	1431
	arc, inches:		+1.7	0	−7.3	−21.2
Rem. 140 Nosler Part.	velocity, fps:	2860	2648	2446	2253	2068
	energy, ft-lb:	2542	2180	1860	1577	1330
	arc, inches:		+1.7	0	−7.6	−22.0
Speer 145 Grand Slam	velocity, fps:	2845	2567	2305	2059	
	energy, ft-lb:	2606	2121	1711	1365	
	arc, inches:		+1.9	0	−8.4	−25.5
Win. 140 Power-Point	velocity, fps:	2800	2523	2268	2027	1802
	energy, ft-lb:	2429	1980	1599	1277	1010
	arc, inches:		+2.0	0	−8.8	−26.0
Win. 140 Power-Point Plus	velocity, fps:	2875	2597	2336	2090	1859
	energy, ft-lb:	2570	1997	1697	1358	1075
	arc, inches:		+2.0	0	−8.8	26.0
Win. 140 Fail Safe	velocity, fps:	2760	2506	2271	2048	1839
	energy, ft-lb:	2360	1953	1603	1304	1051
	arc, inches:		+2.0	0	−8.8	−25.9
Win. 140 Ballistic Silvertip	velocity, fps:	2770	2572	2382	2200	2026
	energy, ft-lb:	2386	2056	1764	1504	1276
	arc, inches:		+1.9	0	−8.0	−23.8

7x64 Brenneke						
Federal 160 Nosler Partition	velocity, fps:	2650	2480	2310	2150	2000
	energy, ft-lb:	2495	2180	1895	1640	1415
	arc, inches:		+2.1	0	−8.7	−24.9
Norma 154 Soft Point	velocity, fps:	2821	2605	2399	2203	
	energy, ft-lb:	2722	2321	1969	1660	
	arc, inches:		+1.8	0	−7.8	
Norma 170 Vulkan	velocity, fps:	2756	2501	2259	2031	
	energy, ft-lb:	2868	2361	1927	1558	
	arc, inches:		+2.0	0	−8.8	
Norma 170 Oryx	velocity, fps:	2756	2481	2222	1979	
	energy, ft-lb:	2868	2324	1864	1478	
	arc, inches:		+2.1	0	−9.2	
Norma 170 Plastic Point	velocity, fps:	2756	2519	2294	2081	
	energy, ft-lb:	2868	2396	1987	1635	
	arc, inches:		+2.0	0	−8.6	
Rem. 175 PSP Core-Lokt	velocity, fps:	2650	2445	2248	2061	1883
	energy, ft-lb:	2728	2322	1964	1650	1378
	arc, inches:		+2.2	0	−9.1	−26.4
Speer 160 Grand Slam	velocity, fps:	2600	2376	2164	1962	
	energy, ft-lb:	2401	2006	1663	1368	
	arc, inches:		+2.3	0	−9.8	−28.6
Speer 175 Grand Slam	velocity, fps:	2650	2461	2280	2106	
	energy, ft-lb:	2728	2353	2019	1723	
	arc, inches:		+2.4	0	−9.2	−26.2
7x65 R						
Norma 170 Plastic Point	velocity, fps:	2625	2390	2167	1956	
	energy, ft-lb:	2602	2157	1773	1445	
	arc, inches:		+2.3	0	−9.7	
Norma 170 Vulkan	velocity, fps:	2657	2392	2143	1909	
	energy, ft-lb:	2666	2161	1734	1377	
	arc, inches:		+2.3	0	−9.9	
Norma 170 Oryx	velocity, fps:	2657	2378	2115	1871	
	energy, ft-lb:	2666	2135	1690	1321	
	arc, inches:		+2.3	0	−10.1	
.284 Winchester						
Win. 150 Power-Point	velocity, fps:	2860	2595	2344	2108	1886
	energy, ft-lb:	2724	2243	1830	1480	1185
	arc, inches:		+2.1	0	−8.5	−24.8
.280 Remington						
Federal 140 Sierra Pro-Hunt.	velocity, fps:	2990	2740	2500	2270	2060
	energy, ft-lb:	2770	2325	1940	1605	1320
	arc, inches:		+1.6	0	−7.0	−20.8
Federal 140 Trophy Bonded	velocity, fps:	2990	2630	2310	2040	1730
	energy, ft-lb:	2770	2155	1655	1250	925
	arc, inches:		+1.6	0	−8.4	−25.4
Federal 140 Tr. Bonded HE	velocity, fps:	3150	2850	2570	2300	2050
	energy, ft-lb:	3085	2520	2050	1650	1310
	arc, inches:		+1.4	0	−6.7	−20.0
Federal 150 Hi-Shok	velocity, fps:	2890	2670	2460	2260	2060
	energy, ft-lb:	2780	2370	2015	1695	1420
	arc, inches:		+1.7	0	−7.5	−21.8
Federal 150 Nosler Partition	velocity, fps:	2890	2690	2490	2310	2130
	energy, ft-lb:	2780	2405	2070	1770	1510
	arc, inches:		+1.7	0	−7.2	−21.1
Federal 160 Trophy Bonded	velocity, fps:	2800	2570	2350	2140	1940
	energy, ft-lb:	2785	2345	1960	1625	1340
	arc, inches:		+1.9	0	−8.3	−24.0
Hornady 139 SPBT LMmoly	velocity, fps:	3110	2888	2675	2473	2280
	energy, ft-lb:	2985	2573	2209	1887	1604
	arc, inches:		+1.4	0	−6.5	−18.6
Norma 170 Vulkan	velocity, fps:	2592	2346	2113	1894	
	energy, ft-lb:	2537	2078	1686	1354	
	arc, inches:		+2.4	0	−10.2	
Norma 170 Oryx	velocity, fps:	2690	2416	2159	1918	
	energy, ft-lb:	2732	2204	1760	1389	
	arc, inches:		+2.2	0	−9.7	

Norma 170 Plastic Point	velocity, fps:	2707	2468	2241	2026	
	energy, ft-lb:	2767	2299	1896	1550	
	arc, inches:		+2.1	0	−9.1	
Rem. 140 PSP Core-Lokt	velocity, fps:	3000	2758	2528	2309	2102
	energy, ft-lb:	2797	2363	1986	1657	1373
	arc, inches:		+1.5	0	−7.0	−20.5
Rem. 140 PSP boat-tail	velocity, fps:	2860	2656	2460	2273	2094
	energy, ft-lb:	2542	2192	1881	1606	1363
	arc, inches:		+1.7	0	−7.5	−21.7
Rem. 140 Nosler Bal. Tip	velocity, fps:	3000	2804	2616	2436	2263
	energy, ft-lb:	2799	2445	2128	1848	1593
	arc, inches:		+1.5	0	−6.8	−19.0
Rem. 150 PSP Core-Lokt	velocity, fps:	2890	2624	2373	2135	1912
	energy, ft-lb:	2781	2293	1875	1518	1217
	arc, inches:		+1.8	0	−8.0	−23.6
Rem. 165 SP Core-Lokt	velocity, fps:	2820	2510	2220	1950	1701
	energy, ft-lb:	2913	2308	1805	1393	1060
	arc, inches:		+2.0	0	−9.1	−27.4
Speer 145 Grand Slam	velocity, fps:	2900	2619	2354	2105	
	energy, ft-lb:	2707	2207	1784	1426	
	arc, inches:		+2.1	0	−8.4	−24.7
Speer 160 Grand Slam	velocity, fps:	2890	2652	2425	2210	
	energy, ft-lb:	2967	2497	2089	1735	
	arc, inches:		+1.7	0	−7.7	−22.4
Win. 140 Fail Safe	velocity, fps:	3050	2756	2480	2221	1977
	energy, ft-lb:	2893	2362	1913	1533	1216
	arc, inches:		+1.5	0	−7.2	−21.5
Win. 140 Ballistic Silvertip	velocity, fps:	3040	2842	2653	2471	2297
	energy, ft-lb:	2872	2511	2187	1898	1640
	arc, inches:		+1.4	0	−6.3	−18.4
7mm Remington Mag.						
A-Square 175 Monolithic Solid	velocity, fps:	2860	2557	2273	2008	1771
	energy, ft-lb:	3178	2540	2008	1567	1219
	arc, inches:		+1.92	0	−8.7	−25.9
Federal 140 Nosler Partition	velocity, fps:	3150	2930	2710	2510	2320
	energy, ft-lb:	3085	2660	2290	1960	1670
	arc, inches:		+1.3	0	−6.0	−17.5
Federal 140 Trophy Bonded	velocity, fps:	3150	2910	2680	2460	2250
	energy, ft-lb:	3085	2630	2230	1880	1575
	arc, inches:		+1.3	0	−6.1	−18.1
Federal 150 Hi-Shok	velocity, fps:	3110	2830	2570	2320	2090
	energy, ft-lb:	3220	2670	2200	1790	1450
	arc, inches:		+1.4	0	−6.7	−19.9
Federal 150 Sierra GameKing BTSP	velocity, fps:	3110	2920	2750	2580	2410
	energy, ft-lb:	3220	2850	2510	2210	1930
	arc, inches:		+1.3	0	−5.9	−17.0
Federal 150 Nosler Bal. Tip	velocity, fps:	3110	2910	2720	2540	2370
	energy, ft-lb:	3220	2825	2470	2150	1865
	arc, inches:		+1.3	0	−6.0	−17.4
Federal 160 Sierra Pro-Hunt.	velocity, fps:	2940	2730	2520	2320	2140
	energy, ft-lb:	3070	2640	2260	1920	1620
	arc, inches:		+1.6	0	−7.1	−20.6
Federal 160 Nosler Partition	velocity, fps:	2950	2770	2590	2420	2250
	energy, ft-lb:	3090	2715	2375	2075	1800
	arc, inches:		+1.5	0	−6.7	−19.4
Federal 160 Trophy Bonded	velocity, fps:	2940	2660	2390	2140	1900
	energy, ft-lb:	3070	2505	2025	1620	1280
	arc, inches:		+1.7	0	−7.9	−23.3
Federal 165 Sierra GameKing BTSP	velocity, fps:	2950	2800	2650	2510	2370
	energy, ft-lb:	3190	2865	2570	2300	2050
	arc, inches:		+1.5	0	−6.4	−18.4
Federal 175 Hi-Shok	velocity, fps:	2860	2650	2440	2240	2060
	energy, ft-lb:	3180	2720	2310	1960	1640
	arc, inches:		+1.7	0	−7.6	−22.1
Federal 175 Trophy Bonded	velocity, fps:	2860	2600	2350	2120	1900
	energy, ft-lb:	3180	2625	2150	1745	1400
	arc, inches:		+1.8	0	−8.2	−24.0

Hornady 139 SPBT	velocity, fps:	3150	2933	2727	2530	2341
	energy, ft-lb:	3063	2656	2296	1976	1692
	arc, inches:		+1.2	0	−6.1	−17.7
Hornady 139 SPBT HMmoly	velocity, fps:	3250	3041	2822	2613	2413
	energy, ft-lb:	3300	2854	2458	2106	1797
	arc, inches:		+1.1	0	−5.7	−16.6
Hornady 154 Soft Point	velocity, fps:	3035	2814	2604	2404	2212
	energy, ft-lb:	3151	2708	2319	1977	1674
	arc, inches:		+1.3	0	−6.7	−19.3
Hornady 162 SP boat-tail	velocity, fps:	2940	2757	2582	2413	2251
	energy, ft-lb:	3110	2735	2399	2095	1823
	arc, inches:		+1.6	0	−6.7	−19.7
Hornady 175 SP	velocity, fps:	2860	2650	2440	2240	2060
	energy, ft-lb:	3180	2720	2310	1960	1640
	arc, inches:		+2.0	0	−7.9	−22.7
Norma 170 Vulkan	velocity, fps:	3018	2747	2493	2252	
	energy, ft-lb:	3439	2850	2346	1914	
	arc, inches:		+1.5	0	−2.8	
Norma 170 Oryx	velocity, fps:	2887	2601	2333	2080	
	energy, ft-lb:	3147	2555	2055	1634	
	arc, inches:		+1.8	0	−8.2	
Norma 170 Plastic Point	velocity, fps:	3018	2762	2519	2290	
	energy, ft-lb:	3439	2880	2394	1980	
	arc, inches:		+1.5	0	−7.0	
PMC 140 Barnes X	velocity, fps:	3000	2808	2624	2448	2279
	energy, ft-lb:	2797	2451	2141	1863	1614
	arc, inches:		+1.5	0	−6.6	18.9
PMC 140 Pointed Soft Point	velocity, fps:	3099	2878	2668	2469	2279
	energy, ft-lb:	2984	2574	2212	1895	1614
	arc, inches:		+1.4	0	−6.2	−18.1
PMC 140 SP boat-tail	velocity, fps:	3125	2891	2669	2457	2255
	energy, ft-lb:	3035	2597	2213	1877	1580
	arc, inches:		+1.4	0	−6.3	−18.4
PMC 160 Barnes X	velocity, fps:	2800	2639	2484	2334	2189
	energy, ft-lb:	2785	2474	2192	1935	1703
	arc, inches:		+1.8	0	−7.4	−21.2
PMC 160 Pointed Soft Point	velocity, fps:	2914	2748	2586	2428	2276
	energy, ft-lb:	3016	2682	2375	2095	1840
	arc, inches:		+1.6	0	−6.7	−19.4
PMC 160 SP boat-tail	velocity, fps:	2900	2696	2501	2314	2135
	energy, ft-lb:	2987	2582	2222	1903	1620
	arc, inches:		+1.7	0	−7.2	−21.0
PMC 175 Pointed Soft Point	velocity, fps:	2860	2645	2442	2244	2957
	energy, ft-lb:	3178	2718	2313	1956	1644
	arc, inches:		+2.0	0	−7.9	−22.7
Rem. 140 PSP Core-Lokt	velocity, fps:	3175	2923	2684	2458	2243
	energy, ft-lb:	3133	2655	2240	1878	1564
	arc, inches:		+2.2	+1.9	−3.2	−14.2
Rem. 140 PSP boat-tail	velocity, fps:	3175	2956	2747	2547	2356
	energy, ft-lb:	3133	2715	2345	2017	1726
	arc, inches:		+2.2	+1.6	−3.1	−13.4
Rem. 150 PSP Core-Lokt	velocity, fps:	3110	2830	2568	2320	2085
	energy, ft-lb:	3221	2667	2196	1792	1448
	arc, inches:		+1.3	0	−6.6	−20.2
Rem. 150 Nosler Bal. Tip	velocity, fps:	3110	2912	2723	2542	2367
	energy, ft-lb:	3222	2825	2470	2152	1867
	arc, inches:		+1.2	0	−5.9	−17.3
Rem. 150 Swift Scirocco	velocity, fps:	3100	2927	2751	2582	2419
	energy, ft-lb:	3221	2852	2520	2220	1948
	arc, inches:		+2.2	+1.9	−3.1	−13.3
Rem. 160 Swift A-Frame	velocity, fps:	2900	2659	2430	2212	2006
	energy, ft-lb:	2987	2511	2097	1739	1430
	arc, inches:		+1.7	0	−7.6	−22.4
Rem. 160 Nosler Partition	velocity, fps:	2950	2752	2563	2381	2207
	energy, ft-lb:	3091	2690	2333	2014	1730
	arc, inches:		+0.6	−1.9	−9.6	−23.6
Rem. 175 PSP Core-Lokt	velocity, fps:	2860	2645	2440	2244	2057
	energy, ft-lb:	3178	2718	2313	1956	1644
	arc, inches:		+1.7	0	−7.6	−22.1

Speer 145 Grand Slam	velocity, fps:	3140	2843	2565	2304	
	energy, ft-lb:	3174	2602	2118	1708	
	arc, inches:		+1.4	0	−6.7	
Speer 175 Grand Slam	velocity, fps:	2850	2653	2463	2282	
	energy, ft-lb:	3156	2734	2358	2023	
	arc, inches:		+1.7	0	−7.5	−21.7
Win. 140 Fail Safe	velocity, fps:	3150	2861	2589	2333	2092
	energy, ft-lb:	3085	2544	2085	1693	1361
	arc, inches:		+1.4	0	−6.6	−19.5
Win. 140 Ballistic Silvertip	velocity, fps:	3100	2889	2687	2494	2310
	energy, ft-lb:	2988	2595	2245	1934	1659
	arc, inches:		+1.3	0	−6.2	−17.9
Win. 150 Power-Point	velocity, fps:	3090	2812	2551	2304	2071
	energy, ft-lb:	3181	2634	2167	1768	1429
	arc, inches:		+1.5	0	−6.8	−20.2
Win. 150 Power-Point Plus	velocity, fps:	3130	2849	2586	2337	2102
	energy, ft-lb:	3264	2705	2227	1819	1472
	arc, inches:		+1.4	0	−6.6	−19.6
Win. 150 Ballistic Silvertip	velocity, fps:	3100	2903	2714	2533	2359
	energy, ft-lb:	3200	2806	2453	2136	1853
	arc, inches:		+1.3	0	−6.0	−17.5
Win. 160 Partition Gold	velocity, fps:	2950	2743	2546	2357	2176
	energy, ft-lb:	3093	2674	2303	1974	1682
	arc, inches:		+1.6	0	−6.9	−20.1
Win. 160 Fail Safe	velocity, fps:	2920	2678	2449	2331	2025
	energy, ft-lb:	3030	2549	2131	1769	1457
	arc, inches:		+1.7	0	−7.5	−22.0
Win. 175 Power-Point	velocity, fps:	2860	2645	2440	2244	2057
	energy, ft-lb:	3178	2718	2313	1956	1644
	arc, inches:		+2.0	0	−7.9	−22.7
7mm Weatherby Mag.						
Federal 160 Nosler Partition	velocity, fps:	3050	2850	2650	2470	2290
	energy, ft-lb:	3305	2880	2505	2165	1865
	arc, inches:		+1.4	0	−6.3	−18.4
Federal 160 Sierra GameKing BTSP	velocity, fps:	3050	2880	2710	2560	2400
	energy, ft-lb:	3305	2945	2615	2320	2050
	arc, inches:		+1.4	0	−6.1	−17.4
Federal 160 Trophy Bonded	velocity, fps:	3050	2730	2420	2140	1880
	energy, ft-lb:	3305	2640	2085	1630	1255
	arc, inches:		+1.6	0	−7.6	−22.7
Hornady 154 Soft Point	velocity, fps:	3200	2971	2753	2546	2348
	energy, ft-lb:	3501	3017	2592	2216	1885
	arc, inches:		+1.2	0	−5.8	−17.0
Hornady 175 Soft Point	velocity, fps:	2910	2709	2516	2331	2154
	energy, ft-lb:	3290	2850	2459	2111	1803
	arc, inches:		+1.6	0	−7.1	−20.6
Wby. 139 Pointed Expanding	velocity, fps:	3340	3079	2834	2601	2380
	energy, ft-lb:	3443	2926	2478	2088	1748
	arc, inches:		+2.9	+3.6	0	−8.7
Wby. 140 Nosler Partition	velocity, fps:	3303	3069	2847	2636	2434
	energy, ft-lb:	3391	2927	2519	2159	1841
	arc, inches:		+2.9	+3.6	0	−8.5
Wby. 150 Nosler Bal. Tip	velocity, fps:	3300	3093	2896	2708	2527
	energy, ft-lb:	3627	3187	2793	2442	2127
	arc, inches:		+2.8	+3.5	0	−8.2
Wby. 150 Barnes X	velociy, fps:	3100	2901	2710	2527	2352
	energy, ft-lb:	3200	2802	2446	2127	1842
	arc, inches:		+3.3	+4.0	0	−9.4
Wby. 154 Pointed Expanding	velocity, fps:	3260	3028	2807	2597	2397
	energy, ft-lb:	3634	3134	2694	2307	1964
	arc, inches:		+3.0	+3.7	0	−8.8
Wby. 160 Nosler Partition	velocity, fps:	3200	2991	2791	2600	2417
	energy, ft-lb:	3638	3177	2767	2401	2075
	arc, inches:		+3.1	+3.8	0	−8.9
Wby. 175 Pointed Expanding	velocity, fps:	3070	2861	2662	2471	2288
	energy, ft-lb:	3662	3181	2753	2373	2034
	arc, inches:		+3.5	+4.2	0	−9.9

7mm Dakota						
Dakota 140 Barnes X	velocity, fps:	3500	3253	3019	2798	2587
	energy, ft-lb:	3807	3288	2833	2433	2081
	arc, inches:		+2.0	+2.1	−1.5	−9.6
Dakota 160 Barnes X	velocity, fps:	3200	3001	2811	2630	2455
	energy, ft-lb:	3637	3200	2808	2456	2140
	arc, inches:		+2.1	+1.9	−2.8	−12.5
7mm STW						
A-Square 140 Nos. Bal. Tip	velocity, fps:	3450	3254	3067	2888	2715
	energy, ft-lb:	3700	3291	2924	2592	2292
	arc, inches:		+2.2	+3.0	0	−7.3
A-Square 160 Nosler Part.	velocity, fps:	3250	3071	2900	2735	2576
	energy, ft-lb:	3752	3351	2987	2657	2357
	arc, inches:		+2.8	+3.5	0	−8.2
A-Square 160 SP boat-tail	velocity, fps:	3250	3087	2930	2778	2631
	energy, ft-lb:	3752	3385	3049	2741	2460
	arc, inches:		+2.8	+3.4	0	−8.0
Federal 140 Trophy Bonded	velocity, fps:	3330	3080	2850	2630	2420
	energy, ft-lb:	3435	2950	2520	2145	1815
	arc, inches:		+1.1	0	−5.4	−15.8
Federal 150 Trophy Bonded	velocity, fps:	3250	3010	2770	2560	2350
	energy, ft-lb:	3520	3010	2565	2175	1830
	arc, inches:		+1.2	0	−5.7	−16.7
Federal 160 Sierra GameKing BTSP	velocity, fps:	3200	3020	2850	2670	2530
	energy, ft-lb:	3640	3245	2890	2570	2275
	arc, inches:		+1.1	0	−5.5	−15.7
Rem. 140 PSP Core-Lokt	velocity, fps:	3325	3064	2818	2585	2364
	energy, ft-lb:	3436	2918	2468	2077	1737
	arc, inches:		+2.0	+1.7	−2.9	−12.8
Rem. 140 Swift A-Frame	velocity, fps:	3325	3020	2735	2467	2215
	energy, ft-lb:	3436	2834	2324	1892	1525
	arc, inches:		+2.1	+1.8	−3.1	−13.8
Speer 145 Grand Slam	velocity, fps:	3300	2992	2075	2435	
	energy, ft-lb:	3506	2882	2355	1909	
	arc, inches:		+1.2	0	−6.0	−17.8
Win. 140 Ballistic Silvertip	velocity, fps:	3320	3100	2890	2690	2499
	energy, ft-lb:	3427	2982	2597	2250	1941
	arc, inches:		+1.1	0	−5.2	−15.2
Win. 150 Power-Point	velocity, fps:	3250	2957	2683	2424	2181
	energy, ft-lb:	3519	2913	2398	1958	1584
	arc, inches:		+1.2	0	−6.1	−18.1
Win. 160 Fail Safe	velocity, fps:	3150	2894	2652	2422	2204
	energy, ft-lb:	3526	2976	2499	2085	1727
	arc, inches:		+1.3	0	−6.3	−18.5
7mm Remington Ultra Mag						
Rem. 140 Nosler Part.	velocity, fps:	3475	3231	3001	2782	2573
	energy, ft-lb:	3753	3245	2798	2405	2058
	arc, inches:		+1.7	+1.5	−2.5	−11.1
Rem. 140 PSP Core-Lokt	velocity, fps:	3475	3205	2951	2711	2483
	energy, ft-lb:	3753	3192	2706	2284	1917
	arc, inches:		+1.7	+1.5	−2.6	−11.6
7.21 (.284) Firehawk						
Lazzeroni 140 Nosler Part.	velocity, fps:	3580	3349	3130	2923	2724
	energy, ft-lb:	3985	3488	3048	2656	2308
	arc, inches:		+2.2	+2.9	0	−7.0
Lazzeroni 160 Swift A-Fr.	velocity, fps:	3385	3167	2961	2763	2574
	energy, ft-lb:	4072	3565	3115	2713	2354
	arc, inches:		+2.6	+3.3	0	−7.8
7.5x55 Swiss						
Norma 180 Soft Point	velocity, fps:	2651	2432	2223	2025	
	energy, ft-lb:	2810	2364	1976	1639	
	arc, inches:		+2.2	0	−9.3	
7.62x39 Russian						
Federal 123 Hi-Shok	velocity, fps:	2300	2030	1780	1550	1350
	energy, ft-lb:	1445	1125	860	655	500
	arc, inches:		0	−7.0	−25.1	

Federal 124 FMJ	velocity, fps:	2300	2030	1780	1560	1360
	energy, ft-lb:	1455	1135	875	670	510
	arc, inches:		+3.5	0	−14.6	−43.5
Norma 150 Soft Point	velocity, fps:	2953	2622	2314	2028	
	energy, ft-lb:	2905	2291	1784	1370	
	arc, inches:		+1.8	0	−8.3	
Norma 180 Soft Point	velocity, fps:	2575	2360	2154	1960	
	energy, ft-lb:	2651	2226	1856	1536	
	arc, inches:		+2.4	0	−9.9	
PMC 123 FMJ	velocity, fps:	2350	2072	1817	1583	1368
	energy, ft-lb:	1495	1162	894	678	507
	arc, inches:		0	−5.0	−26.4	−67.8
PMC 125 Pointed Soft Point	velocity, fps:	2320	2046	1794	1563	1350
	energy, ft-lb:	1493	1161	893	678	505
	arc, inches:		0	−5.2	−27.5	−70.6
Rem. 125 Pointed Soft Point	velocity, fps:	2365	2062	1783	1533	1320
	energy, ft-lb:	1552	1180	882	652	483
	arc, inches:		0	−6.7	−24.5	
Win. 123 Soft Point	velocity, fps:	2365	2033	1731	1465	1248
	energy, ft-lb:	1527	1129	818	586	425
	arc, inches:		+3.8	0	−15.4	−46.3
.30 Carbine						
Federal 110 Hi-Shok RN	velocity, fps:	1990	1570	1240	1040	920
	energy, ft-lb:	965	600	375	260	210
	arc, inches:		0	−12.8	−46.9	
Federal 110 FMJ	velocity, fps:	1990	1570	1240	1040	920
	energy, ft-lb:	965	600	375	260	210
	arc, inches:		0	−12.8	−46.9	
PMC 110 FMJ	velocity, fps:	1927	1548	1248		
	energy, ft-lb:	906	585	380		
	arc, inches:		0	−14.2		
PMC 110 RNSP	velocity, fps:					
	energy, ft-lb:					
	arc, inches:					
Rem. 110 Soft Point	velocity, fps:	1990	1567	1236	1035	923
	energy, ft-lb:	967	600	373	262	208
	arc, inches:		0	−12.9	−48.6	
Win. 110 Hollow Soft Point	velocity, fps:	1990	1567	1236	1035	923
	energy, ft-lb:	967	600	373	262	208
	arc, inches:		0	−13.5	−49.9	
.30–30 Winchester						
Federal 125 Hi-Shok HP	velocity, fps:	2570	2090	1660	1320	1080
	energy, ft-lb:	1830	1210	770	480	320
	arc, inches:		+3.3	0	−16.0	−50.9
Federal 150 Hi-Shok FN	velocity, fps:	2390	2020	1680	1400	1180
	energy, ft-lb:	1900	1355	945	650	460
	arc, inches:		+3.6	0	−15.9	−49.1
Federal 170 Hi-Shok RN	velocity, fps:	2200	1900	1620	1380	1190
	energy, ft-lb:	1830	1355	990	720	535
	arc, inches:		+4.1	0	−17.4	−52.4
Federal 170 Sierra Pro-Hunt.	velocity, fps:	2200	1820	1500	1240	1060
	energy, ft-lb:	1830	1255	845	575	425
	arc, inches:		+4.5	0	−20.0	−63.5
Federal 170 Nosler Partition	velocity, fps:	2200	1900	1620	1380	1190
	energy, ft-lb:	1830	1355	990	720	535
	arc, inches:		+4.1	0	−17.4	−52.4
Hornady 150 Round Nose	velocity, fps:	2390	1973	1605	1303	1095
	energy, ft-lb:	1902	1296	858	565	399
	arc, inches:		0	−8.2	−30.0	
Hornady 170 Flat Point	velocity, fps:	2200	1895	1619	1381	1191
	energy, ft-lb:	1827	1355	989	720	535
	arc, inches:		0	−8.9	−31.1	
Norma 150 Soft Point	velocity, fps:	2329	2008	1716	1459	
	energy, ft-lb:	1807	1344	981	709	
	arc, inches:		+3.6	0	−15.5	
PMC 150 Starfire HP	velocity, fps:	2100	1769	1478		
	energy, ft-lb:	1469	1042	728		
	arc, inches:		0	−10.8		

PMC 150 Flat Nose	velocity, fps:	2159	1819	1554		
	energy, ft-lb:	1552	1102	804		
	arc, inches:		0	−9.0		
PMC 170 Flat Nose	velocity, fps:	1965	1680	1480		
	energy, ft-lb:	1457	1065	827		
	arc, inches:		0	−10.7		
Rem. 55 PSP (sabot) "Accelerator"	velocity, fps:	3400	2693	2085	1570	1187
	energy, ft-lb:	1412	886	521	301	172
	arc, inches:		+1.7	0	−9.9	−34.3
Rem. 150 SP Core-Lokt	velocity, fps:	2390	1973	1605	1303	1095
	energy, ft-lb:	1902	1296	858	565	399
	arc, inches:		0	−7.6	−28.8	
Rem. 170 SP Core-Lokt	velocity, fps:	2200	1895	1619	1381	1191
	energy, ft-lb:	1827	1355	989	720	535
	arc, inches:		0	−8.3	−29.9	
Rem. 170 HP Core-Lokt	velocity, fps:	2200	1895	1619	1381	1191
	energy, ft-lb:	1827	1355	989	720	535
	arc, inches:		0	−8.3	−29.9	
Speer 150 Flat Nose	velocity, fps:	2370	2067	1788	1538	
	energy, ft-lb:	1870	1423	1065	788	
	arc, inches:		+3.3	0	−14.4	−43.7
Win. 150 Hollow Point	velocity, fps:	2390	2018	1684	1398	1177
	energy, ft-lb:	1902	1356	944	651	461
	arc, inches:		0	−7.7	−27.9	
Win. 150 Power-Point	velocity, fps:	2390	2018	1684	1398	1177
	energy, ft-lb:	1902	1356	944	651	461
	arc, inches:		0	−7.7	−27.9	
Win. 150 Silvertip	velocity,fps:	2390	2018	1684	1398	1177
	energy, ft-lb:	1902	1356	944	651	461
	arc, inches:		0	−7.7	−27.9	
Win. 150 Power-Point Plus	velocity, fps:	2480	2095	1747	1446	1209
	energy, ft-lb:	2049	1462	1017	697	487
	arc, inches:		0	−6.5	−24.5	
Win. 170 Power-Point	velocity, fps:	2200	1895	1619	1381	1191
	energy, ft-lb:	1827	1355	989	720	535
	arc, inches:		0	−8.9	−31.1	
Win. 170 Silvertip	velocity, fps:	2200	1895	1619	1381	1191
	energy, ft-lb:	1827	1355	989	720	535
	arc, inches:		0	−8.9	−31.1	
.300 Savage						
Federal 150 Hi-Shok	velocity, fps:	2630	2350	2100	1850	1630
	energy, ft-lb:	2305	1845	1460	1145	885
	arc, inches:		+2.4	0	−10.4	−30.9
Federal 180 Hi-Shok	velocity, fps:	2350	2140	1940	1750	1570
	energy, ft-lb:	2205	1825	1495	1215	985
	arc, inches:		+3.1	0	−12.4	−36.1
Rem. 150 PSP Core-Lokt	velocity, fps:	2630	2354	2095	1853	1631
	energy, ft-lb:	2303	1845	1462	1143	806
	arc, inches:		+2.4	0	−10.4	−30.9
Rem. 180 SP Core-Lokt	velocity, fps:	2350	2025	1728	1467	1252
	energy, ft-lb:	2207	1639	1193	860	626
	arc, inches:		0	−7.1	−25.9	
Win. 150 Power-Point	velocity, fps:	2630	2311	2015	1743	1500
	energy, ft-lb:	2303	1779	1352	1012	749
	arc, inches:		+2.8	0	−11.5	−34.4
.307 Winchester						
Win. 180 Power-Point	velocity, fps:	2510	2179	1874	1599	1362
	energy, ft-lb:	2519	1898	1404	1022	742
	arc, inches:		+1.5	−3.6	−18.6	−47.1
.30–40 Krag						
Rem. 180 PSP Core-Lokt	velocity, fps:	2430	2213	2007	1813	1632
	energy, ft-lb:	2360	1957	1610	1314	1064
	arc, inches:		0	−5.6	−18.6	
Win. 180 Power-Point	velocity, fps:	2430	2099	1795	1525	1298
	energy, ft-lb:	2360	1761	1288	929	673
	arc, inches:		0	−7.1	−25.0	

.308 Winchester						
Federal 150 Hi-Shok	velocity, fps:	2820	2530	2260	2010	1770
	energy, ft-lb:	2650	2140	1705	1345	1050
	arc, inches:		+2.0	0	−8.8	−26.3
Federal 150 Nosler Bal. Tip.	velocity, fps:	2820	2610	2410	2220	2040
	energy, ft-lb:	2650	2270	1935	1640	1380
	arc, inches:		+1.8	0	−7.8	−22.7
Federal 150 FMJ boat-tail	velocity, fps:	2820	2620	2430	2250	2070
	energy, ft-lb:	2650	2285	1965	1680	1430
	arc, inches:		+1.8	0	−7.7	−22.4
Federal 150 Barnes XLC	velocity, fps:	2820	2610	2400	2210	2030
	energy, ft-lb:	2650	2265	1925	1630	1370
	arc, inches:		+1.8	0	−7.8	−22.9
Federal 155 Sierra MatchKg. BTHP	velocity, fps:	2950	2740	2540	2350	2170
	energy, ft-lb:	2995	2585	2225	1905	1620
	arc, inches:		+13.2	+23.3	+28.1	+26.5
Federal 165 Sierra GameKing BTSP	velocity, fps:	2700	2520	2330	2160	1990
	energy, ft-lb:	2670	2310	1990	1700	1450
	arc, inches:		+2.0	0	−8.4	−24.3
Federal 165 Trophy Bonded	velocity, fps:	2700	2440	2200	1970	1760
	energy, ft-lb:	2670	2185	1775	1425	1135
	arc, inches:		+2.2	0	−9.4	−27.7
Federal 165 Tr. Bonded HE	velocity, fps:	2870	2600	2350	2120	1890
	energy, ft-lb:	3020	2485	2030	1640	1310
	arc, inches:		+1.8	0	−8.2	−24.0
Federal 168 Sierra MatchKg. BTHP	velocity, fps:	2600	2410	2230	2060	1890
	energy, ft-lb:	2520	2170	1855	1580	1340
	arc, inches:		+17.7	+31.0	+37.2	+35.4
Federal 180 Hi-Shok	velocity, fps:	2620	2390	2180	1970	1780
	energy, ft-lb:	2745	2290	1895	1555	1270
	arc, inches:		+2.3	0	−9.7	−28.3
Federal 180 Sierra Pro-Hunt.	velocity, fps:	2620	2410	2200	2010	1820
	energy, ft-lb:	2745	2315	1940	1610	1330
	arc, inches:		+2.3	0	−9.3	−27.1
Federal 180 Nosler Partition	velocity, fps:	2620	2430	2240	2060	1890
	energy, ft-lb:	2745	2355	2005	1700	1430
	arc, inches:		+2.2	0	−9.2	−26.5
Federal 180 Nosler Part. HE	velocity, fps:	2740	2550	2370	2200	2030
	energy, ft-lb:	3000	2600	2245	1925	1645
	arc, inches:		+1.9	0	−8.2	−23.5
Hornady 110 Urban Tactical	velocity, fps:	3170	2825	2504	2206	1937
	energy, ft-lb:	2454	1950	1532	1189	916
	arc, inches:		+1.5	0	−7.2	−21.2
Hornady 150 SP boat-tail	velocity, fps:	2820	2560	2315	2084	1866
	energy, ft-lb:	2648	2183	1785	1447	1160
	arc, inches:		+2.0	0	−8.5	−25.2
Hornady 150 SP LM	velocity, fps:	2980	2703	2442	2195	1964
	energy, ft-lb:	2959	2433	1986	1606	1285
	arc, inches:		+1.6	0	−7.5	−22.2
Hornady 155 A-Max	velocity, fps:	2815	2610	2415	2229	2051
	energy, ft-lb:	2727	2345	2007	1709	1448
	arc, inches:		+1.9	0	−7.9	−22.6
Hornady 165 SP boat-tail	velocity, fps:	2700	2496	2301	2115	1937
	energy, ft-lb:	2670	2283	1940	1639	1375
	arc, inches:		+2.0	0	−8.7	−25.2
Hornady 165 SPBT LM	velocity, fps:	2870	2658	2456	2283	2078
	energy, ft-lb:	3019	2589	2211	1877	1583
	arc, inches:		+1.7	0	−7.5	−21.8
Hornady 168 BTHP Match	velocity, fps:	2700	2524	2354	2191	2035
	energy, ft-lb:	2720	2377	2068	1791	1545
	arc, inches:		+2.0	0	−8.4	−23.9
Hornady 168 BTHP Match LM	velocity, fps:	2640	2630	2429	2238	2056
	energy, ft-lb:	3008	2579	2201	1868	1577
	arc, inches:		+1.8	0	−7.8	−22.4
Hornady 168 A-Max Match	velocity fps:	2620	2446	2280	2120	1972
	energy, ft-lb:	2560	2232	1939	1677	1450
	arc, inches:		+2.6	0	−9.2	−25.6

Hornady 168 A-Max	velocity, fps:	2700	2491	2292	2102	1921
	energy, ft-lb:	2719	2315	1959	1648	1377
	arc, inches:		+2.4	0	−9.0	−25.9
Hornady 178 A-Max	velocity, fps:	2965	2778	2598	2425	2259
	energy, ft-lb:	3474	3049	2666	2323	2017
	arc, inches:		+1.6	0	−6.9	−19.8
Hornady 180 A-Max Match	velocity, fps:	2550	2397	2249	2106	1974
	energy, ft-lb:	2598	2295	2021	1773	1557
	arc, inches:		+2.7	0	−9.5	−26.2
Norma 150 Soft Point	velocity, fps:	2861	2537	2235	1954	
	energy, ft-lb:	2727	2144	1664	1272	
	arc, inches:		+2.0	0	−9.0	
Norma 165 TXP Swift A-Fr.	velocity, fps:	2700	2459	2231	2015	
	energy, ft-lb:	2672	2216	1824	1488	
	arc, inches:		+2.1	0	−9.1	
Norma 180 Plastic Point	velocity, fps:	2612	2365	2131	1911	
	energy, ft-lb:	2728	2235	1815	1460	
	arc, inches:		+2.4	0	−10.1	
Norma 180 Nosler Partition	velocity, fps:	2612	2414	2225	2044	
	energy, ft-lb:	2728	2330	1979	1670	
	arc, inches:		+2.2	0	−9.3	
Norma 180 Alaska	velocity, fps:	2612	2269	1953	1667	
	energy, ft-lb:	2728	2059	1526	1111	
	arc, inches:		+2.7	0	−11.9	
Norma 180 Vulkan	velocity, fps:	2612	2325	2056	1806	
	energy, ft-lb:	2728	2161	1690	1304	
	arc, inches:		+2.5	0	−10.8	
Norma 180 Oryx	velocity, fps:	2612	2305	2019	1755	
	energy, ft-lb:	2728	2124	1629	1232	
	arc, inches:		+2.5	0	−11.1	
Norma 200 Vulkan	velocity, fps:	2461	2215	1983	1767	
	energy, ft-lb:	2690	2179	1747	1387	
	arc, inches:		+2.8	0	−11.7	
PMC 147 FMJ boat-tail	velocity, fps:	2751	2473	2257	2052	1859
	energy, ft-lb:	2428	2037	1697	1403	1150
	arc, inches:		+2.3	0	−9.3	−27.3
PMC 150 Barnes X	velocity, fps:	2700	2504	2316	2135	1964
	energy, ft-lb:	2428	2087	1786	1518	1284
	arc, inches:		+2.0	0	−8.6	−24.7
PMC 150 Pointed Soft Point	velocity, fps:	2643	2417	2203	1999	1807
	energy, ft-lb:	2326	1946	1615	1331	1088
	arc, inches:		+2.2	0	−9.4	−27.5
PMC 150 SP boat-tail	velocity, fps:	2820	2581	2354	2139	1935
	energy, ft•lb:	2648	2218	1846	1523	1247
	arc, inches:		+1.9	0	−8.2	−24.0
PMC 165 Barnes X	velocity, fps:	2600	2425	2256	2095	1940
	energy, ft-lb:	2476	2154	1865	1608	1379
	arc, inches:		+2.2	0	−9.0	−26.0
PMC 168 HP boat-tail	velocity, fps:	2650	2460	2278	2103	1936
	energy, ft-lb:	2619	2257	1935	1649	1399
	arc, inches:		+2.1	0	−8.8	−25.6
PMC 180 Pointed Soft Point	velocity, fps:	2410	2223	2044	1874	1714
	energy, ft-lb:	2320	1975	1670	1404	1174
	arc, inches:		+2.8	0	−11.1	−32.0
PMC 180 SP boat-tail	velocity, fps:	2620	2446	2278	2117	1962
	energy, ft-lb:	2743	2391	2074	1790	1538
	arc, inches:		+2.2	0	−8.9	−25.4
Rem. 150 PSP Core-Lokt	velocity, fps:	2820	2533	2263	2009	1774
	energy, ft-lb:	2648	2137	1705	1344	1048
	arc, inches:		+2.0	0	−8.8	−26.2
Rem. 150 Swift Scirocco	velocity, fps:	2820	2611	2410	2219	2037
	energy, ft-lb:	2648	2269	1935	1640	1381
	arc, inches:		+1.8	0	−7.8	−22.7
Rem. 165 PSP boat-tail	velocity, fps:	2700	2497	2303	2117	1941
	energy, ft-lb:	2670	2284	1942	1642	1379
	arc, inches:		+2.0	0	−8.6	−25.0

Rem. 165 Nosler Bal. Tip	velocity, fps:	2700	2613	2333	2161	1996
	energy, ft-lb:	2672	2314	1995	1711	1460
	arc, inches:		+2.0	0	−8.4	−24.3
Rem. 165 Swift Scirocco	velocity, fps:	2700	2513	2333	2161	1996
	energy, ft-lb:	2670	2313	1994	1711	1459
	arc, inches:		+2.0	0	−8.4	−24.3
Rem. 168 HPBT Match	velocity, fps:	2680	2493	2314	2143	1979
	energy, ft-lb:	2678	2318	1998	1713	1460
	arc, inches:		+2.1	0	−8.6	−24.7
Rem. 180 SP Core-Lokt	velocity, fps:	2620	2274	1955	1666	1414
	energy, ft-lb:	2743	2066	1527	1109	799
	arc, inches:		+2.6	0	−11.8	−36.3
Rem. 180 PSP Core-Lokt	velocity, fps:	2620	2393	2178	1974	1782
	energy, ft-lb:	2743	2288	1896	1557	1269
	arc, inches:		+2.3	0	−9.7	−28.3
Rem. 180 Nosler Partition	velocity, fps:	2620	2436	2259	2089	1927
	energy, ft-lb:	2743	2371	2039	1774	1485
	arc, inches:		+2.2	0	−9.0	−26.0
Speer 150 Grand Slam	velocity, fps:	2900	2599	2317	2053	
	energy, ft-lb:	2800	2249	1788	1404	
	arc, inches:		+2.1	0	−8.6	−24.8
Speer 165 Grand Slam	velocity, fps:	2700	2475	2261	2057	
	energy, ft-lb:	2670	2243	1872	1550	
	arc, inches:		+2.1	0	−8.9	−25.9
Speer 180 Grand Slam	velocity, fps:	2620	2420	2229	2046	
	energy, ft-lb:	2743	2340	1985	1674	
	arc, inches:		+2.2	0	−9.2	−26.6
Win. 150 Power-Point	velocity, fps:	2820	2488	2179	1893	1633
	energy, ft-lb:	2648	2061	1581	1193	888
	arc, inches:		+2.4	0	−9.8	−29.3
Win. 150 Power-Point Plus	velocity, fps:	2900	2558	2241	1946	1678
	energy, ft-lb:	2802	2180	1672	1262	938
	arc, inches:		+1.9	0	−8.9	−27.0
Win. 150 Partition Gold	velocity, fps:	2900	2645	2405	2177	1962
	energy, ft-lb:	2802	2332	1927	1579	1282
	arc, inches:		+1.7	0	−7.8	−22.9
Win. 150 Ballistic Silvertip	velocity, fps:	2810	2601	2401	2211	2028
	energy, ft-lb:	2629	2253	1920	1627	1370
	arc, inches:		+1.8	0	−7.8	−22.8
Win. 150 Fail Safe	velocity, fps:	2820	2533	2263	2010	1775
	energy, ft-lb:	2649	2137	1706	1346	1049
	arc, inches:		+2.0	0	−8.8	−26.2
Win. 168 Ballistic Silvertip	velocity, fps:	2670	2484	2306	2134	1971
	energy, ft-lb:	2659	2301	1983	1699	1449
	arc, inches:		+2.1	0	−8.6	−24.8
Win. 168 HP boat-tail Match	velocity, fps:	2680	2485	2297	2118	1948
	energy, ft-lb:	2680	2303	1970	1674	1415
	arc, inches:		+2.1	0	−8.7	−25.1
Win. 180 Power-Point	velocity, fps:	2620	2274	1955	1666	1414
	energy, ft-lb:	2743	2066	1527	1109	799
	arc, inches:		+2.9	0	−12.1	−36.9
Win. 180 Silvertip	velocity, fps:	2620	2393	2178	1974	1782
	energy, ft-lb:	2743	2288	1896	1557	1269
	arc, inches:		+2.6	0	−9.9	−28.9
.30-06 Springfield						
A-Square 180 M & D-T	velocity, fps:	2700	2365	2054	1769	1524
	energy, ft-lb:	2913	2235	1687	1251	928
	arc, inches:		+2.4	0	−10.6	−32.4
A-Square 220 Monolythic Solid	velocity, fps:	2380	2108	1854	1623	1424
	energy, ft-lb:	2767	2171	1679	1287	990
	arc, inches:		+3.1	0	−13.6	−39.9
Federal 125 Sierra Pro-Hunt.	velocity, fps:	3140	2780	2450	2140	1850
	energy, ft-lb:	2735	2145	1660	1270	955
	arc, inches:		+1.5	0	−7.3	−22.3
Federal 150 Hi-Shok	velocity, fps:	2910	2620	2340	2080	1840
	energy, ft-lb:	2820	2280	1825	1445	1130
	arc, inches:		+1.8	0	−8.2	−24.4

Federal 150 Sierra Pro-Hunt.	velocity, fps:	2910	2640	2380	2130	1900
	energy, ft-lb:	2820	2315	1880	1515	1205
	arc, inches:		+1.7	0	−7.9	−23.3
Federal 150 Sierra GameKing BTSP	velocity, fps:	2910	2690	2480	2270	2070
	energy, ft-lb:	2820	2420	2040	1710	1430
	arc, inches:		+1.7	0	−7.4	−21.5
Federal 150 Nosler Bal. Tip	velocity, fps:	2910	2700	2490	2300	2110
	energy, ft-lb:	2820	2420	2070	1760	1485
	arc, inches:		+1.6	0	−7.3	−21.1
Federal 150 FMJ boat-tail	velocity, fps:	2910	2710	2510	2320	2150
	energy, ft-lb:	2820	2440	2100	1800	1535
	arc, inches:		+1.6	0	−7.1	−20.8
Federal 165 Sierra Pro-Hunt.	velocity, fps:	2800	2560	2340	2130	1920
	energy, ft-lb:	2875	2410	2005	1655	1360
	arc, inches:		+1.9	0	−8.3	−24.3
Federal 165 Sierra GameKing BTSP	velocity, fps:	2800	2610	2420	2240	2070
	energy, ft-lb:	2870	2490	2150	1840	1580
	arc, inches:		+1.8	0	−7.8	−22.4
Federal 165 Sierra GameKing HE	velocity, fps:	3140	2900	2670	2450	2240
	energy, ft-lb:	3610	3075	2610	2200	1845
	arc, inches:		+1.5	0	−6.9	−20.4
Federal 165 Nosler Bal. Tip	velocity, fps:	2800	2610	2430	2250	2080
	energy, ft-lb:	2870	2495	2155	1855	1585
	arc, inches:		+1.8	0	−7.7	−22.3
Federal 165 Trophy Bonded	velocity, fps:	2800	2540	2290	2050	1830
	energy, ft-lb:	2870	2360	1915	1545	1230
	arc, inches:		+2.0	0	−8.7	−25.4
Federal 165 Tr. Bonded HE	velocity, fps:	3140	2860	2590	2340	2100
	energy, ft-lb:	3610	2990	2460	2010	1625
	arc, inches:		+1.6	0	−7.4	−21.9
Federal 168 Sierra MatchKg. BTHP	velocity, fps:	2700	2510	2320	2150	1980
	energy, ft-lb:	2720	2350	2010	1720	1460
	arc, inches:		+16.2	+28.4	+34.1	+32.3
Federal 180 Hi-Shok	velocity, fps:	2700	2470	2250	2040	1850
	energy, ft-lb:	2915	2435	2025	1665	1360
	arc, inches:		+2.1	0	−9.0	−26.4
Federal 180 Sierra Pro-Hunt. RN	velocity, fps:	2700	2350	2020	1730	1470
	energy, ft-lb:	2915	2200	1630	1190	860
	arc, inches:		+2.4	0	−11.0	−33.6
Federal 180 Nosler Partition	velocity, fps:	2700	2500	2320	2140	1970
	energy, ft-lb:	2915	2510	2150	1830	1550
	arc, inches:		+2.0	0	−8.6	−24.6
Federal 180 Nosler Part. HE	velocity, fps:	2880	2690	2500	2320	2150
	energy, ft-lb:	3315	2880	2495	2150	1845
	arc, inches:		+1.7	0	−7.2	−21.0
Federal 180 Sierra GameKing BTSP	velocity, fps:	2700	2540	2380	2220	2080
	energy, ft-lb:	2915	2570	2260	1975	1720
	arc, inches:		+1.9	0	−8.1	−23.1
Federal 180 Barnes XLC	velocity, fps:	2700	2530	2360	2200	2040
	energy, ft-lb:	2915	2550	2220	1930	1670
	arc, inches:		+2.0	0	−8.3	−23.8
Federal 180 Trophy Bonded	velocity, fps:	2700	2460	2220	2000	1800
	energy, ft-lb:	2915	2410	1975	1605	1290
	arc, inches:		+2.2	0	−9.2	−27.0
Federal 180 Tr. Bonded HE	velocity, fps:	2880	2630	2380	2160	1940
	energy, ft-lb:	3315	2755	2270	1855	1505
	arc, inches:		+1.8	0	−8.0	−23.3
Federal 220 Sierra Pro-Hunt. RN	velocity, fps:	2410	2130	1870	1630	1420
	energy, ft-lb:	2835	2215	1705	1300	985
	arc, inches:		+3.1	0	−13.1	−39.3
Hornady 150 SP	velocity, fps:	2910	2617	2342	2083	1843
	energy, ft-lb:	2820	2281	1827	1445	1131
	arc, inches:		+2.1	0	−8.5	−25.0
Hornady 150 SP LM	velocity, fps:	3100	2815	2548	2295	2058
	energy, ft-lb:	3200	2639	2161	1755	1410
	arc, inches:		+1.4	0	−6.8	−20.3

Hornady 150 SP boat-tail	velocity, fps:	2910	2683	2467	2262	2066
	energy, ft-lb:	2820	2397	2027	1706	1421
	arc, inches:		+2.0	0	−7.7	−22.2
Hornady 165 SP boat-tail	velocity, fps:	2800	2591	2392	2202	2020
	energy, ft-lb:	2873	2460	2097	1777	1495
	arc, inches:		+1.8	0	−8.0	−23.3
Hornady 165 SPBT LM	velocity, fps:	3015	2790	2575	2370	2176
	energy, ft-lb:	3330	2850	2428	2058	1734
	arc, inches:		+1.6	0	−7.0	−20.1
Hornady 168 HPBT Match	velocity, fps:	2790	2620	2447	2280	2120
	energy, ft-lb:	2925	2561	2234	1940	1677
	arc, inches:		+1.7	0	−7.7	−22.2
Hornady 180 SP	velocity, fps:	2700	2469	2258	2042	1846
	energy, ft-lb:	2913	2436	2023	1666	1362
	arc, inches:		+2.4	0	−9.3	−27.0
Hornady 180 SPBT LM	velocity, fps:	2880	2676	2480	2293	2114
	energy, ft-lb:	3316	2862	2459	2102	1786
	arc, inches:		+1.7	0	−7.3	−21.3
Norma 150 Soft Point	velocity, fps:	2972	2640	2331	2043	
	energy, ft-lb:	2943	2321	1810	1390	
	arc, inches:		+1.8	0	−8.2	
Norma 180 Alaska	velocity, fps:	2700	2351	2028	1734	
	energy, ft-lb:	2914	2209	1645	1202	
	arc, inches:		+2.4	0	−11.0	
Norma 180 Nosler Partition	velocity, fps:	2700	2494	2297	2108	
	energy, ft-lb:	2914	2486	2108	1777	
	arc, inches:		+2.1	0	−8.7	
Norma 180 Plastic Point	velocity, fps:	2700	2455	2222	2003	
	energy, ft-lb:	2914	2409	1974	1603	
	arc, inches:		+2.1	0	−9.2	
Norma 180 Vulkan	velocity, fps:	2700	2416	2150	1901	
	energy, ft-lb:	2914	2334	1848	1445	
	arc, inches:		+2.2	0	−9.8	
Norma 180 Oryx	velocity, fps:	2700	2387	2095	1825	
	energy, ft-lb:	2914	2278	1755	1332	
	arc, inches:		+2.3	0	−10.2	
Norma 180 TXP Swift A-Fr.	velocity, fps:	2700	2479	2268	2067	
	energy, ft-lb:	2914	2456	2056	1708	
	arc, inches:		+2.0	0	−8.8	
Norma 200 Vulkan	velocity, fps:	2641	2385	2143	1916	
	energy, ft-lb:	3098	2527	2040	1631	
	arc, inches:		+2.3	0	−9.9	
Norma 200 Oryx	velocity, fps:	2625	2362	2115	1883	
	energy, ft-lb:	3061	2479	1987	1575	
	arc, inches:		+2.3	0	−10.1	
PMC 150 X-Bullet	velocity, fps:	2750	2552	2361	2179	2005
	energy, ft-lb:	2518	2168	1857	1582	1339
	arc, inches:		+2.0	0	−8.2	−23.7
PMC 150 Pointed Soft Point	velocity, fps:	2773	2542	2322	2113	1916
	energy, ft-lb:	2560	2152	1796	1487	1222
	arc, inches:		+1.9	0	−8.4	−24.6
PMC 150 SP boat-tail	velocity, fps:	2900	2657	2427	2208	2000
	energy, ft-lb:	2801	2351	1961	1623	1332
	arc, inches:		+1.7	0	−7.7	−22.5
PMC 150 FMJ	velocity, fps:	2773	2542	2322	2113	1916
	energy, ft-lb:	2560	2152	1796	1487	1222
	arc, inches:		+1.9	0	−8.4	−24.6
PMC 165 Barnes X	velocity, fps:	2750	2569	2395	2228	2067
	energy, ft-lb:	2770	2418	2101	1818	1565
	arc, inches:		+1.9	0	−8.0	−23.0
PMC 180 Barnes X	velocity, fps:	2650	2487	2331	2179	2034
	energy, ft-lb:	2806	2472	2171	1898	1652
	arc, inches:		+2.1	0	−8.5	−24.3
PMC 180 Pointed Soft Point	velocity, fps:	2550	2357	2172	1996	1829
	energy, ft-lb:	2598	2220	1886	1592	1336
	arc, inches:		+2.4	0	−9.7	−28.2

PMC 180 SP boat-tail	velocity, fps:	2700	2523	2352	2188	2030
	energy, ft-lb:	2913	2543	2210	1913	1646
	arc, inches:		+2.0	0	−8.3	−23.9
Rem. 55 PSP (sabot)	velocity, fps:	4080	3484	2964	2499	2080
"Accelerator"	energy, ft-lb:	2033	1482	1073	763	528
	arc, inches:		+1.4	+1.4	−2.6	−12.2
Rem. 125 Pointed Soft Point	velocity, fps:	3140	2780	2447	2138	1853
	energy, ft-lb:	2736	2145	1662	1269	953
	arc, inches:		+1.5	0	−7.4	−22.4
Rem. 150 PSP Core-Lokt	velocity, fps:	2910	2617	2342	2083	1843
	energy, ft-lb:	2820	2281	1827	1445	1131
	arc, inches:		+1.8	0	−8.2	−24.4
Rem. 150 Bronze Point	velocity, fps:	2910	2656	2416	2189	1974
	energy, ft-lb:	2820	2349	1944	1596	1298
	arc, inches:		+1.7	0	−7.7	−22.7
Rem. 150 Nosler Bal. Tip	velocity, fps:	2910	2696	2492	2298	2112
	energy, ft-lb:	2821	2422	2070	1769	1485
	arc, inches:		+1.6	0	−7.3	−21.1
Rem. 150 Swift Scirocco	velocity, fps:	2910	2696	2492	2298	2111
	energy, ft-lb:	2820	2421	2069	1758	1485
	arc, inches:		+1.6	0	−7.3	−21.1
Rem. 165 PSP Core-Lokt	velocity, fps:	2800	2534	2283	2047	1825
	energy, ft-lb:	2872	2352	1909	1534	1220
	arc, inches:		+2.0	0	−8.7	−25.9
Rem. 165 PSP boat-tail	velocity, fps:	2800	2592	2394	2204	2023
	energy, ft-lb:	2872	2462	2100	1780	1500
	arc, inches:		+1.8	0	−7.9	−23.0
Rem. 165 Nosler Bal. Tip	velocity, fps:	2800	2609	2426	2249	2080
	energy, ft-lb:	2873	2494	2155	1854	1588
	arc, inches:		+1.8	0	−7.7	−22.3
Rem. 180 SP Core-Lokt	velocity, fps:	2700	2348	2023	1727	1466
	energy, ft-lb:	2913	2203	1635	1192	859
	arc, inches:		+2.4	0	−11.0	−33.8
Rem. 180 PSP Core-Lokt	velocity, fps:	2700	2469	2250	2042	1846
	energy, ft-lb:	2913	2436	2023	1666	1362
	arc, inches:		+2.1	0	−9.0	−26.3
Rem. 180 Bronze Point	velocity, fps:	2700	2485	2280	2084	1899
	energy, ft-lb:	2913	2468	2077	1736	1441
	arc, inches:		+2.1	0	−8.8	−25.5
Rem. 180 Swift A-Frame	velocity, fps:	2700	2465	2243	2032	1833
	energy, ft-lb:	2913	2429	2010	1650	1343
	arc, inches:		+2.1	0	−9.1	−26.6
Rem. 180 Nosler Partition	velocity, fps:	2700	2512	2332	2160	1995
	energy, ft-lb:	2913	2522	2174	1864	1590
	arc, inches:		+2.0	0	−8.4	−24.3
Rem. 220 SP Core-Lokt	velocity, fps:	2410	2130	1870	1632	1422
	energy, ft-lb:	2837	2216	1708	1301	988
	arc, inches, s:		0	−6.2	−22.4	
Speer 150 Grand Slam	velocity, fps:	2975	2669	2383	2114	
	energy, ft-lb:	2947	2372	1891	1489	
	arc, inches:		+2.0	0	−8.1	−24.1
Speer 165 Grand Slam	velocity, fps:	2790	2560	2342	2134	
	energy, ft-lb:	2851	2401	2009	1669	
	arc, inches:		+1.9	0	−8.3	−24.1
Speer 180 Grand Slam	velocity, fps:	2690	2487	2293	2108	
	energy, ft-lb:	2892	2472	2101	1775	
	arc, inches:		+2.1	0	−8.8	−25.1
Win. 125 Pointed Soft Point	velocity, fps:	3140	2780	2447	2138	1853
	energy, ft-lb:	2736	2145	1662	1269	953
	arc, inches:		+1.8	0	−7.7	−23.0
Win. 150 Power-Point	velocity, fps:	2920	2580	2265	1972	1704
	energy, ft-lb:	2839	2217	1708	1295	967
	arc, inches:		+2.2	0	−9.0	−27.0
Win. 150 Power-Point Plus	velocity, fps:	3050	2685	2352	2043	1760
	energy, ft-lb:	3089	2402	1843	1391	1032
	arc, inches:		+1.7	0	−8.0	−24.3

Win. 150 Silvertip	velocity, fps:	2910	2617	2342	2083	1843
	energy, ft-lb:	2820	2281	1827	1445	1131
	arc, inches:		+2.1	0	−8.5	−25.0
Win. 150 Partition Gold	velocity, fps:	2960	2705	2464	2235	2019
	energy, ft-lb:	2919	2437	2022	1664	1358
	arc, inches:		+1.6	0	−7.4	−21.7
Win. 150 Ballistic Silvertip	velocity, fps:	2900	2687	2483	2289	2103
	energy, ft-lb:	2801	2404	2054	1745	1473
	arc, inches:		+1.7	0	−7.3	−21.2
Win. 150 Fail Safe	velocity, fps:	2920	2625	2349	2089	1848
	energy, ft-lb:	2841	2296	1838	1455	1137
	arc, inches:		+1.8	0	−8.1	−24.3
Win. 165 Pointed Soft Point	velocity, fps:	2800	2573	2357	2151	1956
	energy, ft-lb:	2873	2426	2036	1696	1402
	arc, inches:		+2.2	0	−8.4	−24.4
Win. 165 Fail Safe	velocity, fps:	2800	2540	2295	2063	1846
	energy, ft-lb:	2873	2365	1930	1560	1249
	arc, inches:		+2.0	0	−8.6	−25.3
Win. 168 Ballistic Silvertip	velocity, fps:	2790	2599	2416	2240	2072
	energy, ft-lb:	2903	2520	2177	1872	1601
	arc, inches:		+1.8	0	−7.8	−22.5
Win. 180 Power-Point	velocity, fps:	2700	2348	2023	1727	1466
	energy, ft-lb:	2913	2203	1635	1192	859
	arc, inches:		+2.7	0	−11.3	−34.4
Win. 180 Power-Point Plus	velocity, fps:	2770	2563	2366	2177	1997
	energy, ft-lb:	3068	2627	2237	1894	1594
	arc, inches:		+1.9	0	−8.1	−23.6
Win. 180 Silvertip	velocity, fps:	2700	2469	2250	2042	1846
	energy, ft-lb:	2913	2436	2023	1666	1362
	arc, inches:		+2.4	0	−9.3	−27.0
Win. 180 Partition Gold	velocity, fps:	2790	2581	2382	2192	2010
	energy, ft-lb:	3112	2664	2269	1920	1615
	arc, inches:		+1.9	0	−8.0	−23.2
Win. 180 Fail Safe	velocity, fps:	2700	2486	2283	2089	1904
	energy, ft-lb:	2914	2472	2083	1744	1450
	arc, inches:		+2.1	0	−8.7	−25.5
.300 H&H Mag.						
Federal 180 Nosler Partition	velocity, fps:	2880	2620	2380	2150	1930
	energy, ft-lb:	3315	2750	2260	1840	1480
	arc, inches:		+1.8	0	−8.0	−23.4
Win. 180 Fail Safe	velocity, fps:	2880	2628	2390	2165	1952
	energy, ft-lb:	3316	2762	2284	1873	1523
	arc, inches:		+1.8	0	−7.9	−23.2
.308 Norma Mag.						
Norma 200 Vulkan	velocity, fps:	2903	2624	2361	2114	
	energy, ft-lb:	3744	3058	2476	1985	
	arc, inches:	0	+1.8	0	−8.0	
.300 Winchester Mag.						
A-Square 180 Dead Tough	velocity, fps:	3120	2756	2420	2108	1820
	energy, ft-lb:	3890	3035	2340	1776	1324
	arc, inches:		+1.6	0	−7.6	−22.9
Federal 150 Sierra Pro Hunt.	velocity, fps:	3280	3030	2800	2570	2360
	energy, ft-lb:	3570	3055	2600	2205	1860
	arc, inches:		+1.1	0	−5.6	−16.4
Federal 150 Trophy Bonded	velocity, fps:	3280	2980	2700	2430	2190
	energy, ft-lb:	3570	2450	2420	1970	1590
	arc, inches:		+1.2	0	−6.0	−17.9
Federal 180 Sierra Pro Hunt.	velocity, fps:	2960	2750	2540	2340	2160
	energy, ft-lb:	3500	3010	2580	2195	1860
	arc, inches:		+1.6	0	−7.0	−20.3
Federal 180 Barnes XLC	velocity, fps:	2960	2780	2600	2430	2260
	energy, ft-lb:	3500	3080	2700	2355	2050
	arc, inches:		+1.5	0	−6.6	−19.2
Federal 180 Trophy Bonded	velocity, fps:	2960	2700	2460	2220	2000
	energy, ft-lb:	3500	2915	2410	1975	1605
	arc, inches:		+1.6	0	−7.4	−21.9

Federal 180 Tr. Bonded HE	velocity, fps:	3100	2830	2580	2340	2110
	energy, ft-lb:	3840	3205	2660	2190	1790
	arc, inches:		+1.4	0	−6.6	−19.7
Federal 180 Nosler Partition	velocity, fps:	2960	2700	2450	2210	1990
	energy, ft-lb:	3500	2905	2395	1955	1585
	arc, inches:		+1.6	0	−7.5	−22.1
Federal 190 Sierra MatchKg. BTHP	velocity, fps:	2900	2730	2560	2400	2240
	energy, ft-lb:	3550	3135	2760	2420	2115
	arc, inches:		+12.9	+22.5	+26.9	+25.1
Federal 200 Sierra GameKing BTSP	velocity, fps:	2830	2680	2530	2380	2240
	energy, ft-lb:	3560	3180	2830	2520	2230
	arc, inches:		+1.7	0	−7.1	−20.4
Federal 200 Nosler Part. HE	velocity, fps:	2930	2740	2550	2370	2200
	energy, ft-lb:	3810	3325	2885	2495	2145
	arc, inches:		+1.6	0	−6.9	−20.1
Federal 200 Trophy Bonded	velocity, fps:	2800	2570	2350	2150	1950
	energy, ft-lb:	3480	2935	2460	2050	1690
	arc, inches:		+1.9	0	−8.2	−23.9
Hornady 150 SP boat-tail	velocity, fps:	3275	2988	2718	2464	2224
	energy, ft-lb:	3573	2974	2461	2023	1648
	arc, inches:		+1.2	0	−6.0	−17.8
Hornady 165 SP boat-tail	velocity, fps:	3100	2877	2665	2462	2269
	energy, ft-lb:	3522	3033	2603	2221	1887
	arc, inches:		+1.3	0	−6.5	−18.5
Hornady 180 SP boat-tail	velocity, fps:	2960	2745	2540	2344	2157
	energy, ft-lb:	3501	3011	2578	2196	1859
	arc, inches:		+1.9	0	−7.3	−20.9
Hornady 180 SPBT HM	velocity, fps:	3100	2879	2668	2467	2275
	energy, ft-lb:	3840	3313	2845	2431	2068
	arc, inches:		+1.4	0	−6.4	−18.7
Hornady 190 SP boat-tail	velocity, fps:	2900	2711	2529	2355	2187
	energy, ft-lb:	3549	3101	2699	2340	2018
	arc, inches:		+1.6	0	−7.1	−20.4
Norma 180 Soft Point	velocity, fps:	3018	2780	2555	2341	
	energy, ft-lb:	3641	3091	2610	2190	
	arc, inches:		+1.5	0	−7.0	
Norma 180 Plastic Point	velocity, fps:	3018	2755	2506	2271	
	energy, ft-lb:	3641	3034	2512	2062	
	arc, inches:		+1.6	0	−7.1	
Norma 180 TXP Swift A-Fr.	velocity, fps:	2920	2688	2467	2256	
	energy, ft-lb:	3409	2888	2432	2035	
	arc, inches:		+1.7	0	−7.4	
Norma 200 Vulkan	velocity, fps:	2887	2609	2347	2100	
	energy, ft-lb:	3702	3023	2447	1960	
	arc, inches:		+1.8	0	−8.2	
Norma 200 Oryx	velocity, fps:	3018	2755	2506	2271	
	energy, ft-lb:	4046	3371	2791	2292	
	arc, inches:		+1.5	0	−7.0	
PMC 150 Barnes X	velocity, fps:	3135	2918	2712	2515	2327
	energy, ft-lb:	3273	2836	2449	2107	1803
	arc, inches:		+1.3	0	−6.1	−17.7
PMC 150 Pointed Soft Point	velocity, fps:	3150	2902	2665	2438	2222
	energy, ft-lb:	3304	2804	2364	1979	1644
	arc, inches:		+1.3	0	−6.2	−18.3
PMC 150 SP boat-tail	velocity, fps:	3250	2987	2739	2504	2281
	energy, ft-lb:	3517	2970	2498	2088	1733
	arc, inches:		+1.2	0	−6.0	−17.4
PMC 180 Barnes X	velocity, fps:	2910	2738	2572	2412	2258
	energy, ft-lb:	3384	2995	2644	2325	2037
	arc, inches:		+1.6	0	−6.9	−19.8
PMC 180 PSP	velocity, fps:	2853	2643	2446	2258	2077
	energy, ft-lb:	3252	2792	2391	2037	1724
	arc, inches:		+1.7	0	−7.5	−21.9
PMC 180 SP boat-tail	velocity, fps:	2900	2714	2536	2365	2200
	energy, ft-lb:	3361	2944	2571	2235	1935
	arc, inches:		+1.6	0	−7.1	−20.3

Rem. 150 PSP Core-Lokt	velocity, fps:	3290	2951	2636	2342	2068
	energy, ft-lb:	3605	2900	2314	1827	1859
	arc, inches:		+1.6	0	−7.0	−20.2
Rem. 180 PSP Core-Lokt	velocity, fps:	2960	2745	2540	2344	2157
	energy, ft-lb:	3501	3011	2578	2196	1424
	arc, inches:		+2.2	+1.9	−3.4	−15.0
Rem. 180 Nosler Partition	velocity, fps:	2960	2725	2503	2291	2089
	energy, ft-lb:	3501	2968	2503	2087	1744
	arc, inches:		+1.6	0	−7.2	−20.9
Rem. 180 Nosler Bal. Tip	velocity, fps:	2960	2774	2595	2424	2259
	energy, ft-lb:	3501	3075	2692	2348	2039
	arc, inches:		+1.5	0	−6.7	−19.3
Rem. 190 PSP boat-tail	velocity, fps:	2885	2691	2506	2327	2156
	energy, ft-lb:	3511	3055	2648	2285	1961
	arc, inches:		+1.6	0	−7.2	−20.8
Rem. 200 Swift A-Frame	velocity, fps:	2825	2595	2376	2167	1970
	energy, ft-lb:	3544	2989	2506	2086	1722
	arc, inches:		+1.8	0	−8.0	−23.5
Speer 180 Grand Slam	velocity, fps:	2950	2735	2530	2334	
	energy, ft-lb:	3478	2989	2558	2176	
	arc, inches:		+1.6	0	−7.0	−20.5
Speer 200 Grand Slam	velocity, fps:	2800	2597	2404	2218	
	energy, ft-lb:	3481	2996	2565	2185	
	arc, inches:		+1.8	0	−7.9	−22.9
Win. 150 Power-Point	velocity, fps:	3290	2951	2636	2342	2068
	energy, ft-lb:	3605	2900	2314	1827	1424
	arc, inches:		+2.6	+2.1	−3.5	−15.4
Win. 150 Fail Safe	velocity, fps:	3260	2943	2647	2370	2110
	energy, ft-lb:	3539	2884	2334	1871	1483
	arc, inches:		+1.3	0	−6.2	−18.7
Win. 165 Fail Safe	velocity, fps:	3120	2807	2515	2242	1985
	energy, ft-lb:	3567	2888	2319	1842	1445
	arc, inches:		+1.5	0	−7.0	−20.0
Win. 180 Power-Point	velocity, fps:	2960	2745	2540	2344	2157
	energy, ft-lb:	3501	3011	2578	2196	1859
	arc, inches:		+1.9	0	−7.3	−20.9
Win. 180 Power-Point Plus	velocity, fps:	3070	2846	2633	2430	2236
	energy, ft-lb:	3768	3239	2772	2361	1999
	arc, inches:		+1.4	0	−6.4	−18.7
Win. 180 Ballistic Silvertip	velocity, fps:	2950	2764	2586	2415	2250
	energy, ft-lb:	3478	3054	2673	2331	2023
	arc, inches:		+1.5	0	−6.7	−19.4
Win. 180 Fail Safe	velocity, fps:	2960	2732	2514	2307	2110
	energy, ft-lb:	3503	2983	2528	2129	1780
	arc, inches:		+1.6	0	−7.1	−20.7
Win. 180 Partition Gold	velocity, fps:	3070	2859	2657	2464	2280
	energy, ft-lb:	3768	3267	2823	2428	2078
	arc, inches:		+1.4	0	−6.3	−18.3
.300 Win. Short Magnum						
Win. 150 Ballistic Silvertip	velocity, fps:	3300	3061	2834	2619	2414
	energy, ft-lb:	3628	3121	2676	2285	1941
	arc, inches:		+1.1	0	−5.4	−15.9
Win. 180 Fail Safe	velocity, fps:	2970	2741	2524	2317	2120
	energy, ft-lb:	3526	3005	2547	2147	1797
	arc, inches:		+1.6	0	−7.0	−20.5
Win. 180 Power-Point	velocity, fps:	2970	2755	2549	2353	2166
	energy, ft-lb:	3526	3034	2598	2214	1875
	arc, inches:		+1.5	0	−6.9	−20.1
.300 Weatherby Mag.						
A-Square 180 Dead Tough	velocity, fps:	3180	2811	2471	2155	1863
	energy, ft-lb:	4041	3158	2440	1856	1387
	arc, inches:		+1.5	0	−7.2	−21.8
A-Square 220 Monolythic Solid	velocity, fps:	2700	2407	2133	1877	1653
	energy, ft-lb:	3561	2830	2223	1721	1334
	arc, inches:		+2.3	0	−9.8	−29.7
Federal 180 Sierra GameKing BTSP	velocity, fps:	3190	3010	2830	2660	2490
	energy, ft-lb:	4065	3610	3195	2820	2480
	arc, inches:		+1.2	0	−5.6	−16.0

Federal 180 Trophy Bonded	velocity, fps:	3190	2950	2720	2500	2290
	energy, ft-lb:	4065	3475	2955	2500	2105
	arc, inches:		+1.3	0	−5.9	−17.5
Federal 180 Tr. Bonded HE	velocity, fps:	3330	3080	2850	2750	2410
	energy, ft-lb:	4430	3795	3235	2750	2320
	arc, inches:		+1.1	0	−5.4	−15.8
Federal 180 Nosler Partition	velocity, fps:	3190	2980	2780	2590	2400
	energy, ft-lb:	4055	3540	3080	2670	2305
	arc, inches:		+1.2	0	−5.7	−16.7
Federal 180 Nosler Part. HE	velocity, fps:	3330	3110	2810	2710	2520
	energy, ft-lb:	4430	3875	3375	2935	2540
	arc, inches:		+1.0	0	−5.2	−15.1
Federal 200 Trophy Bonded	velocity, fps:	2900	2670	2440	2230	2030
	energy, ft-lb:	3735	3150	2645	2200	1820
	arc, inches:		+1.7	0	−7.6	−22.2
Hornady 180 SP	velocity, fps:	3120	2891	2673	2466	2268
	energy, ft-lb:	3890	3340	2856	2430	2055
	arc, inches:		+1.3	0	−6.2	−18.1
Rem. 180 PSP Core-Lokt	velocity, fps:	3120	2866	2627	2400	2184
	energy, ft-lb:	3890	3284	2758	2301	1905
	arc, inches:		+2.4	+2.0	−3.4	−14.9
Rem. 190 PSP boat-tail	velocity, fps:	3030	2830	2638	2455	2279
	energy, ft-lb:	3873	3378	2936	2542	2190
	arc, inches:		+1.4	0	−6.4	−18.6
Rem. 200 Swift A-Frame	velocity, fps:	2925	2690	2467	2254	2052
	energy, ft-lb:	3799	3213	2701	2256	1870
	arc, inches:		+2.8	+2.3	−3.9	−17.0
Speer 180 Grand Slam	velocity, fps:	3185	2948	2722	2508	
	energy, ft-lb:	4054	3472	2962	2514	
	arc, inches:		+1.3	0	−5.9	−17.4
Wby. 150 Pointed Expanding	velocity, fps:	3540	3225	2932	2657	2399
	energy, ft-lb:	4173	3462	2862	2351	1916
	arc, inches:		+2.6	+3.3	0	−8.2
Wby. 150 Nosler Partition	velocity, fps:	3540	3263	3004	2759	2528
	energy, ft-lb:	4173	3547	3005	2536	2128
	arc, inches:		+2.5	+3.2	0	−7.7
Wby. 165 Pointed Expanding	velocity, fps:	3390	3123	2872	2634	2409
	energy, ft-lb:	4210	3573	3021	2542	2126
	arc, inches:		+2.8	+3.5	0	−8.5
Wby. 165 Nosler Bal. Tip	velocity, fps:	3350	3133	2927	2730	2542
	energy, ft-lb:	4111	3596	3138	2730	2367
	arc, inches:		+2.7	+3.4	0	−8.1
Wby. 180 Pointed Expanding	velocity, fps:	3240	3004	2781	2569	2366
	energy, ft-lb:	4195	3607	3091	2637	2237
	arc, inches:		+3.1	+3.8	0	−9.0
Wby. 180 Barnes X	velocity, fps:	3190	2995	2809	2631	2459
	energy, ft-lb:	4067	3586	3154	2766	2417
	arc, inches:		+3.1	+3.8	0	−8.7
Wby. 180 Nosler Partition	velocity, fps:	3240	3028	2826	2634	2449
	energy, ft-lb:	4195	3665	3193	2772	2396
	arc, inches:		+3.0	+3.7	0	−8.6
Wby. 200 Nosler Partition	velocity, fps:	3060	2860	2668	2485	2308
	energy, ft-lb:	4158	3631	3161	2741	2366
	arc, inches:		+3.5	+4.2	0	−9.8
Wby. 220 RN Expanding	velocity, fps:	2845	2543	2260	1996	1751
	energy, ft-lb:	3954	3158	2495	1946	1497
	arc, inches:		+4.9	+5.9	0	−14.6
.300 Dakota						
Dakota 165 Barnes X	velocity, fps:	3200	2979	2769	2569	2377
	energy, ft-lb:	3751	3251	2809	2417	2070
	arc, inches:		+2.1	+1.8	−3.0	−13.2
Dakota 200 Barnes X	velocity, fps:	3000	2824	2656	2493	2336
	energy, ft-lb:	3996	3542	3131	2760	2423
	arc, inches:		+2.2	+1.5	−4.0	−15.2
.300 Pegasus						
A-Square 180 SP boat-tail	velocity, fps:	3500	3319	3145	2978	2817
	energy, ft-lb:	4896	4401	3953	3544	3172
	arc, inches:		+2.3	+2.9	0	−6.8

A-Square 180 Nosler Part.	velocity, fps:	3500	3295	3100	2913	2734
	energy, ft-lb:	4896	4339	3840	3392	2988
	arc, inches:		+2.3	+3.0	0	−7.1
A-Square 180 Dead Tough	velocity, fps:	3500	3103	2740	2405	2095
	energy, ft-lb:	4896	3848	3001	2312	1753
	arc, inches:		+1.1	0	−5.7	−17.5
.300 Remington Ultra Mag						
Federal 180 Trophy Bonded	velocity, fps:	3250	3000	2770	2550	2340
	energy, ft-lb:	4220	3605	3065	2590	2180
	arc, inches:		+1.2	0	−5.7	−16.8
Rem. 150 Swift Scirocco	velocity, fps:	3450	3208	2980	2762	2556
	energy, ft-lb:	3964	3427	2956	2541	2175
	arc, inches:		+1.7	+1.5	−2.6	−11.2
Rem. 180 PSP Core-Lokt	velocity, fps:	3300	3035	2786	2550	2327
	energy, ft-lb:	4352	3682	3102	2599	2163
	arc, inches:		+2.0	+1.8	−3.0	−13.1
Rem. 180 Nosler Partition	velocity, fps:	3250	3037	2834	2640	2454
	energy, ft-lb:	4221	3686	3201	2786	2407
	arc, inches:		+2.4	+1.8	−3.0	−12.7
Rem. 180 Swift Scirocco	velocity, fps:	3250	3048	2856	2672	2495
	energy, ft-lb:	4221	3714	3260	2853	2487
	arc, inches:		+2.0	+1.7	−2.8	−12.3
Rem. 200 Nosler Partition	velocity, fps:	3025	2826	2636	2454	2279
	energy, ft-lb:	4063	3547	3086	2673	2308
	arc, inches:		+2.4	+2.0	−3.4	−14.6
.30–378 Weatherby Mag.						
Wby. 165 Nosler Bal. Tip	velocity, fps:	3500	3275	3062	2859	2665
	energy, ft-lb:	4488	3930	3435	2995	2603
	arc, inches:		+2.4	+3.0	0	−7.4
Wby. 180 Barnes X	velocity, fps:	3450	3243	3046	2858	2678
	energy, ft-lb:	4757	4204	3709	3264	2865
	arc, inches:		+2.4	+3.1	0	−7.4
Wby. 200 Nosler Partition	velocity, fps:	3160	2955	2759	2572	2392
	energy, ft-lb:	4434	3877	3381	2938	2541
	arc, inches:		+3.2	+3.9	0	−9.1
7.82 (.308) Warbird						
Lazzeroni 150 Nosler Part.	velocity, fps:	3680	3432	3197	2975	2764
	energy, ft-lb:	4512	3923	3406	2949	2546
	arc, inches:		+2.1	+2.7	0	−6.6
Lazzeroni 180 Nosler Part.	velocity, fps:	3425	3220	3026	2839	2661
	energy, ft-lb:	4689	4147	3661	3224	2831
	arc, inches:		+2.5	+3.2	0	−7.5
Lazzeroni 200 Swift A-Fr.	velocity, fps:	3290	3105	2928	2758	2594
	energy, ft-lb:	4808	4283	3808	3378	2988
	arc, inches:		+2.7	+3.4	0	−7.9
7.65x53 Argentine						
Norma 180 Soft Point	velocity, fps:	2592	2386	2189	2002	
	energy, ft-lb:	2686	2276	1916	1602	
	arc, inches:		+2.3	0	−9.6	
.303 British						
Federal 150 Hi-Shok	velocity, fps:	2690	2440	2210	1980	1780
	energy, ft-lb:	2400	1980	1620	1310	1055
	arc, inches:		+2.2	0	−9.4	−27.6
Federal 180 Sierra Pro-Hunt.	velocity, fps:	2460	2230	2020	1820	1630
	energy, ft-lb:	2420	1995	1625	1315	1060
	arc, inches:		+2.8	0	−11.3	−33.2
Federal 180 Tr. Bonded HE	velocity, fps:	2590	2350	2120	1900	1700
	energy, ft-lb:	2680	2205	1795	1445	1160
	arc, inches:		+2.4	0	−10.0	−30.0
Hornady 150 Soft Point	velocity, fps:	2685	2441	2210	1992	1787
	energy, ft-lb:	2401	1984	1627	1321	1064
	arc, inches:		+2.2	0	−9.3	−27.4
Hornady 150 SP LM	velocity, fps:	2830	2570	2325	2094	1884
	energy, ft-lb:	2667	2199	1800	1461	1185
	arc, inches:		+2.0	0	−8.4	−24.6
Norma 150 Soft Point	velocity, fps:	2723	2438	2170	1920	
	energy, ft-lb:	2470	1980	1569	1228	
	arc, inches:		+2.2	0	−9.6	

PMC 180 SP boat-tail	velocity, fps:	2450	2276	2110	1951	1799
	energy, ft-lb:	2399	2071	1779	1521	1294
	arc, inches:		+2.6	0	−10.4	−30.1
Rem. 180 SP Core-Lokt	velocity, fps:	2460	2124	1817	1542	1311
	energy, ft-lb:	2418	1803	1319	950	687
	arc, inches, s:		0	−5.8	−23.3	
Win. 180 Power-Point	velocity, fps:	2460	2233	2018	1816	1629
	energy, ft-lb:	2418	1993	1627	1318	1060
	arc, inches, s:		0	−6.1	−20.8	
7.7x58 Japanese Arisaka						
Norma 180 Soft Point	velocity, fps:	2493	2291	2099	1916	
	energy, ft-lb:	2485	2099	1761	1468	
	arc, inches:		+2.6	0	−10.5	
.32–20 Winchester						
Rem. 100 Lead	velocity, fps:	1210	1021	913	834	769
	energy, ft-lb:	325	231	185	154	131
	arc, inches:		0	−31.6	−104.7	
Win. 100 Lead	velocity, fps:	1210	1021	913	834	769
	energy, ft-lb:	325	231	185	154	131
	arc, inches:		0	−32.3	−106.3	
.32 Winchester Special						
Federal 170 Hi-Shok	velocity, fps:	2250	1920	1630	1370	1180
	energy, ft-lb:	1910	1395	1000	710	520
	arc, inches:		0	−8.0	−29.2	
Rem. 170 SP Core-Lokt	velocity, fps:	2250	1921	1626	1372	1175
	energy, ft-lb:	1911	1393	998	710	521
	arc, inches:		0	−8.0	−29.3	
Win. 170 Power-Point	velocity, fps:	2250	1870	1537	1267	1082
	energy, ft-lb:	1911	1320	892	606	442
	arc, inches:		0	−9.2	−33.2	
8mm Mauser (8x57)						
Federal 170 Hi-Shok	velocity, fps:	2360	1970	1620	1330	1120
	energy, ft-lb:	2100	1465	995	670	475
	arc, inches:		0	−7.6	−28.5	
Norma 196 Alaska	velocity, fps:	2395	2112	1850	1611	
	Energy, ft-lb:	2714	2190	1754	1399	
	Arc, inches:		0	−6.3	−22.9	
Norma 196 Soft Point (JS)	velocity, fps:	2526	2244	1981	1737	
	energy, ft-lb:	2778	2192	1708	1314	
	arc, inches:		+2.7	0	−11.6	
Norma 196 Vulkan (JS)	velocity, fps:	2526	2276	2041	1821	
	energy, ft-lb:	2778	2256	1813	1443	
	arc, inches:		+2.6	0	−11.0	
PMC 170 Pointed Soft Point	velocity, fps:	2360	1969	1622	1333	1123
	energy, ft-lb:	2102	1463	993	671	476
	arc, inches:		+1.8	−4.5	−24.3	−63.8
Rem. 170 SP Core-Lokt	velocity, fps:	2360	1969	1622	1333	1123
	energy, ft-lb:	2102	1463	993	671	476
	arc, inches:		0	−7.6	−28.6	
Win. 170 Power-Point	velocity, fps:	2360	1969	1622	1333	1123
	energy, ft-lb:	2102	1463	993	671	476
	arc, inches:		0	−8.2	−29.8	
8mm Remington Mag.						
A-Square 220 Monolythic Solid	velocity, fps:	2800	2501	2221	1959	1718
	energy, ft-lb:	3829	3055	2409	1875	1442
	arc, inches:		+2.1	0	−9.1	−27.6
Rem. 200 Swift A-Frame	velocity, fps:	2900	2623	2361	2115	1885
	energy, ft-lb:	3734	3054	2476	1987	1577
	arc, inches:		+1.8	0	−8.0	−23.9
.338–06						
A-Square 200 Nos. Bal. Tip	velocity, fps:	2750	2553	2364	2184	2011
	energy, ft-lb:	3358	2894	2482	2118	1796
	arc, inches:		+1.9	0	−8.2	−23.6
A-Square 250 SP boat-tail	velocity, fps:	2500	2374	2252	2134	2019
	energy, ft-lb:	3496	3129	2816	2528	2263
	arc, inches:		+2.4	0	−9.3	−26.0
A-Square 250 Dead Tough	velocity, fps:	2500	2222	1963	1724	1507
	energy, ft-lb:	3496	2742	2139	1649	1261
	arc, inches:		+2.8	0	−11.9	−35.5

.338 Winchester Mag.						
A-Square 250 SP boat-tail	velocity, fps:	2700	2568	2439	2314	2193
	energy, ft-lb:	4046	3659	3302	2972	2669
	arc, inches:		+4.4	+5.2	0	−11.7
A-Square 250 Triad	velocity, fps:	2700	2407	2133	1877	1653
	energy, ft-lb:	4046	3216	2526	1956	1516
	arc, inches:		+2.3	0	−9.8	−29.8
Federal 210 Nosler Partition	velocity, fps:	2830	2600	2390	2180	1980
	energy, ft-lb:	3735	3160	2655	2215	1835
	arc, inches:		+1.8	0	−8.0	−23.3
Federal 225 Sierra Pro-Hunt.	velocity, fps:	2780	2570	2360	2170	1980
	energy, ft-lb:	3860	3290	2780	2340	1960
	arc, inches:		+1.9	0	−8.2	−23.7
Federal 225 Trophy Bonded	velocity, fps:	2800	2560	2330	2110	1900
	energy, ft-lb:	3915	3265	2700	2220	1800
	arc, inches:		+1.9	0	−8.4	−24.5
Federal 225 Tr. Bonded HE	velocity, fps:	2940	2690	2450	2230	2010
	energy, ft-lb:	4320	3610	3000	2475	2025
	arc, inches:		+1.7	0	−7.5	−22.0
Federal 250 Nosler Partition	velocity, fps:	2660	2470	2300	2120	1960
	energy, ft-lb:	3925	3395	2925	2505	2130
	arc, inches:		+2.1	0	−8.8	−25.1
Federal 250 Nosler Part HE	velocity, fps:	2800	2610	2420	2250	2080
	energy, ft-lb:	4350	3775	3260	2805	2395
	arc, inches:		+1.8	0	−7.8	−22.5
Hornady 225 Soft Point HM	velocity, fps:	2920	2678	2449	2232	2027
	energy, ft-lb:	4259	3583	2996	2489	2053
	arc, inches:		+1.8	0	−7.6	−22.0
Norma 250 Nosler Partition	velocity, fps:	2657	2470	2290	2118	
	energy, ft-lb:	3920	3387	2912	2490	
	arc, inches:		+2.1	0	−8.7	
PMC 225 Barnes X	velocity, fps:	2780	2619	2464	2313	2168
	energy, ft-lb:	3860	3426	3032	2673	2348
	arc, inches:		+1.8	0	−7.6	−21.6
Rem. 200 Nosler Bal. Tip	velocity, fps:	2950	2724	2509	2303	2108
	energy, ft-lb:	3866	3295	2795	2357	1973
	arc, inches:		+1.6	0	−7.1	−20.8
Rem. 210 Nosler Partition	velocity, fps:	2830	2602	2385	2179	1983
	energy, ft-lb:	3734	3157	2653	2214	1834
	arc, inches:		+1.8	0	−7.9	−23.2
Rem. 225 PSP Core-Lokt	velocity, fps:	2780	2572	2374	2184	2003
	energy, ft-lb:	3860	3305	2815	2383	2004
	arc, inches:		+1.9	0	−8.1	−23.4
Rem. 225 Swift A-Frame	velocity, fps:	2785	2517	2266	2029	1808
	energy, ft-lb:	3871	3165	2565	2057	1633
	arc, inches:		+2.0	0	−8.8	−25.2
Rem. 250 PSP Core-Lokt	velocity, fps:	2660	2456	2261	2075	1898
	energy, ft-lb:	3927	3348	2837	2389	1999
	arc, inches:		+2.1	0	−8.9	−26.0
Speer 250 Grand Slam	velocity, fps:	2645	2442	2247	2062	
	energy, ft-lb:	3883	3309	2803	2360	
	arc, inches:		+2.2	0	−9.1	−26.2
Win. 200 Power-Point	velocity, fps:	2960	2658	2375	2110	1862
	energy, ft-lb:	3890	3137	2505	1977	1539
	arc, inches:		+2.0	0	−8.2	−24.3
Win. 200 Ballistic Silvertip	velocity, fps:	2950	2724	2509	2303	2108
	energy, ft-lb:	3864	3294	2794	2355	1972
	arc, inches:		+1.6	0	−7.1	−20.8
Win. 230 Fail Safe	velocity, fps:	2780	2573	2375	2186	2005
	energy, ft-lb:	3948	3382	2881	2441	2054
	arc, inches:		+1.9	0	−8.1	−23.4
Win. 250 Partition Gold	velocity, fps:	2650	2467	2291	2122	1960
	energy, ft-lb:	3899	3378	2914	2520	2134
	arc, inches:		+2.1	0	−8.7	−25.2
.340 Weatherby Mag.						
A-Square 250 SP boat-tail	velocity, fps:	2820	2684	2552	2424	2299
	energy, ft-lb:	4414	3999	3615	3261	2935
	arc, inches:		+4.0	+4.6	0	−10.6

A-Square 250 Triad	velocity, fps:	2820	2520	2238	1976	1741
	energy, ft-lb:	4414	3524	2781	2166	1683
	arc, inches:		+2.0	0	−9.0	−26.8
Federal 225 Trophy Bonded	velocity, fps:	3100	2840	2600	2370	2150
	energy, ft-lb:	4800	4035	3375	2800	2310
	arc, inches:		+1.4	0	−6.5	−19.4
Wby. 200 Pointed Expanding	velocity, fps:	3221	2946	2688	2444	2213
	energy, ft-lb:	4607	3854	3208	2652	2174
	arc, inches:		+3.3	+4.0	0	−9.9
Wby. 200 Nosler Bal. Tip	velocity, fps:	3221	2980	2753	2536	2329
	energy, ft-lb:	4607	3944	3364	2856	2409
	arc, inches:		+3.1	+3.9	0	−9.2
Wby. 210 Nosler Partition	velocity, fps:	3211	2963	2728	2505	2293
	energy, ft-lb:	4807	4093	3470	2927	2452
	arc, inches:		+3.2	+3.9	0	−9.5
Wby. 225 Pointed Expanding	velocity, fps:	3066	2824	2595	2377	2170
	energy, ft-lb:	4696	3984	3364	2822	2352
	arc, inches:		+3.6	+4.4	0	−10.7
Wby. 225 Barnes X	velocity, fps:	3001	2804	2615	2434	2260
	energy, ft-lb:	4499	3927	3416	2959	2551
	arc, inches:		+3.6	+4.3	0	−10.3
Wby. 250 Pointed Expanding	velocity, fps:	2963	2745	2537	2338	2149
	energy, ft-lb:	4873	4182	3572	3035	2563
	arc, inches:		+3.9	+4.6	0	−11.1
Wby. 250 Nosler Partition	velocity, fps:	2941	2743	2553	2371	2197
	energy, ft-lb:	4801	4176	3618	3120	2678
	arc, inches:		+3.9	+4.6	0	−10.9
.330 Dakota						
Dakota 200 Barnes X	velocity, fps:	3200	2971	2754	2548	2350
	energy, ft-lb:	4547	3920	3369	2882	2452
	arc, inches:		+2.1	+1.8	−3.1	−13.4
Dakota 250 Barnes X	velocity, fps:	2900	2719	2545	2378	2217
	energy, ft-lb:	4668	4103	3595	3138	2727
	arc, inches:		+2.3	+1.3	−5.0	−17.5
.338 Remington Ultra Mag						
Remington 250 Swift A-Fr.	velocity, fps:	2860	2645	2440	2244	2057
	energy, ft-lb:	4540	3882	3303	2794	2347
	arc, inches:		+1.7	0	−7.6	−22.1
.338–378 Weatherby Mag.						
Wby. 200 Nosler Bal. Tip	velocity, fps:	3350	3102	2868	2646	2434
	energy, ft-lb:	4983	4273	3652	3109	2631
	arc, inches:	0	+2.8	+3.5	0	−8.4
Wby. 225 Barnes X	velocity, fps:	3180	2974	2778	2591	2410
	energy, ft-lb:	5052	4420	3856	3353	2902
	arc, inches:	0	+3.1	+3.8	0	−8.9
Wby. 250 Nosler Partition	velocity, fps:	3060	2856	2662	2475	2297
	energy, ft-lb:	5197	4528	3933	3401	2927
	arc, inches:	0	+3.5	+4.2	0	−9.8
8.59 (.338) Titan						
Lazzeroni 200 Nos. Bal. Tip	velocity, fps:	3430	3211	3002	2803	2613
	energy, ft-lb:	5226	4579	4004	3491	3033
	arc, inches:		+2.5	+3.2	0	−7.6
Lazzeroni 225 Nos. Partition	velocity, fps:	3235	3031	2836	2650	2471
	energy, ft-lb:	5229	4591	4021	3510	3052
	arc, inches:		+3.0	+3.6	0	−8.6
Lazzeroni 250 Swift A-Fr.	velocity, fps:	3100	2908	2725	2549	2379
	energy, ft-lb:	5336	4697	4123	3607	3143
	arc, inches:		+3.3	+4.0	0	−9.3
.338 A-Square						
A-Square 200 Nos. Bal. Tip	velocity, fps:	3500	3266	3045	2835	2634
	energy, ft-lb:	5440	4737	4117	3568	3081
	arc, inches:		+2.4	+3.1	0	−7.5
A-Square 250 SP boat-tail	velocity, fps:	3120	2974	2834	2697	2565
	energy, ft-lb:	5403	4911	4457	4038	3652
	arc, inches:		+3.1	+3.7	0	−8.5
A-Square 250 Triad	velocity, fps:	3120	2799	2500	2220	1958
	energy, ft-lb:	5403	4348	3469	2736	2128
	arc, inches:		+1.5	0	−7.1	−20.4

.338 Excaliber						
A-Square 200 Nos. Bal. Tip	velocity, fps:	3600	3361	3134	2920	2715
	energy, ft-lb:	5755	5015	4363	3785	3274
	arc, inches:		+2.2	+2.9	0	−6.7
A-Square 250 SP boat-tail	velocity, fps:	3250	3101	2958	2684	2553
	energy, ft-lb:	5863	5339	4855	4410	3998
	arc, inches:		+2.7	+3.4	0	−7.8
A-Square 250 Triad	velocity, fps:	3250	2922	2618	2333	2066
	energy, ft-lb:	5863	4740	3804	3021	2370
	arc, inches:		+1.3	0	−6.4	−19.2
.348 Winchester						
Win. 200 Silvertip	velocity, fps:	2520	2215	1931	1672	1443
	energy, ft-lb:	2820	2178	1656	1241	925
	arc, inches:		0	−6.2	−21.9	
.357 Magnum						
Federal 180 Hi-Shok HP	velocity, fps:	1550	1160	980	860	770
Hollow Point	energy, ft-lb:	960	535	385	295	235
	arc, inches:		0	−22.8	−77.9	−173.8
Win. 158 Jacketed SP	velocity, fps:	1830	1427	1138	980	883
	energy, ft-lb:	1175	715	454	337	274
	arc, inches:		0	−16.2	−57.0	−128.3
.35 Remington						
Federal 200 Hi-Shok	velocity, fps:	2080	1700	1380	1140	1000
	energy, ft-lb:	1920	1280	840	575	445
	arc, inches:		0	−10.7	−39.3	
Rem. 150 PSP Core-Lokt	velocity, fps:	2300	1874	1506	1218	1039
	energy, ft-lb:	1762	1169	755	494	359
	arc, inches:		0	−8.6	−32.6	
Rem. 200 SP Core-Lokt	velocity, fps:	2080	1698	1376	1140	1001
	energy, ft-lb:	1921	1280	841	577	445
	arc, inches:		0	−10.7	−40.1	
Win. 200 Power-Point	velocity, fps:	2020	1646	1335	1114	985
	energy, ft-lb:	1812	1203	791	551	431
	arc, inches:		0	−12.1	−43.9	
.356 Winchester						
Win. 200 Power-Point	velocity, fps:	2460	2114	1797	1517	1284
	energy, ft-lb:	2688	1985	1434	1022	732
	arc, inches:		+1.6	−3.8	−20.1	−51.2
.358 Winchester						
Win. 200 Silvertip	velocity, fps:	2490	2171	1876	1610	1379
	energy, ft-lb:	2753	2093	1563	1151	844
	arc, inches:		+1.5	−3.6	−18.6	−47.2
.35 Whelen						
Federal 225 Trophy Bonded	velocity, fps:	2600	2400	2200	2020	1840
	energy, ft-lb:	3375	2865	2520	2030	1690
	arc, inches:		+2.3	0	−9.4	−27.3
Rem. 200 Pointed Soft Point	velocity, fps:	2675	2378	2100	1842	1606
	energy, ft-lb:	3177	2510	1958	1506	1145
	arc, inches:		+2.3	0	−10.3	−30.8
Rem. 250 Pointed Soft Point	velocity, fps:	2400	2197	2005	1823	1652
	energy, ft-lb:	3197	2680	2230	1844	1515
	arc, inches:		+1.3	−3.2	−16.6	−40.0
.358 Norma Mag.						
A-Square 275 Triad	velocity, fps:	2700	2394	2108	1842	1653
	energy, ft-lb:	4451	3498	2713	2072	1668
	arc, inches:		+2.3	0	−10.1	−29.8
Norma 250 Woodleigh	velocity, fps:	2799	2442	2112	1810	
	energy, ft-lb:	4350	3312	2478	1819	
	arc, inches:		+2.2	0	−10.0	
.358 STA						
A-Square 275 Triad	velocity, fps:	2850	2562	2292	2039	1764
	energy, ft-lb:	4959	4009	3208	2539	1899
	arc, inches:		+1.9	0	−8.6	−26.1
9.3x57						
Norma 232 Vulkan	velocity, fps:	2329	2031	1757	1512	
	energy, ft-lb:	2795	2126	1591	1178	
	arc, inches:		+3.5	0	−14.9	

Norma 286 Alaska	velocity, fps:	2067	1857	1662	1484	
	energy, ft-lb:	2714	2190	1754	1399	
	arc, inches:		+4.3	0	−17.0	
9.3x62						
A-Square 286 Triad	velocity, fps:	2360	2089	1844	1623	1369
	energy, ft-lb:	3538	2771	2157	1670	1189
	arc, inches:		+3.0	0	−13.1	−42.2
Norma 232 Vulkan	velocity, fps:	2625	2327	2049	1792	
	energy, ft-lb:	3551	2791	2164	1655	
	arc, inches:		+2.5	0	−10.8	
Norma 232 Oryx	velocity, fps:	2625	2294	1988	1708	
	energy, ft-lb:	3535	2700	2028	1497	
	arc, inches:		+2.5	0	−11.4	
Norma 286 Plastic Point	velocity, fps:	2362	2141	1931	1736	
	energy, ft-lb:	3544	2911	2370	1914	
	arc, inches:		+3.1	0	−12.4	
Norma 286 Alaska	velocity, fps:	2362	2135	1920	1720	
	energy, ft-lb:	3544	2894	2342	1879	
	arc, inches:		+3.1	0	−12.5	
9.3x64						
A-Square 286 Triad	velocity, fps:	2700	2391	2103	1835	1602
	energy, ft-lb:	4629	3630	2808	2139	1631
	arc, inches:		+2.3	0	−10.1	−30.8
9.3x74 R						
A-Square 286 Triad	velocity, fps:	2360	2089	1844	1623	
	energy, ft-lb:	3538	2771	2157	1670	
	arc, inches:		+3.6	0	−14.0	
Norma 232 Vulkan	velocity, fps:	2625	2327	2049	1792	
	energy, ft-lb:	3551	2791	2164	1655	
	arc, inches:		+2.5	0	−10.8	
Norma 232 Oryx	velocity, fps:	2526	2191	1883	1605	
	energy, ft-lb:	3274	2463	1819	1322	
	arc, inches:		+2.9	0	−12.8	
Norma 286 Alaska	velocity, fps:	2362	2135	1920	1720	
	energy, ft-lb:	3544	2894	2342	1879	
	arc, inches:		+3.1	0	−12.5	
Norma 286 Plastic Point	velocity, fps:	2362	2135	1920	1720	
	energy, ft-lb:	3544	2894	2342	1879	
	arc, inches:		+3.1	0	−12.5	
.375 Winchester						
Win. 200 Power-Point	velocity, fps:	2200	1841	1526	1268	1089
	energy, ft-lb:	2150	1506	1034	714	
	arc, inches:		0	−9.5	−33.8	
.375 H&H Magnum						
A-Square 300 SP boat-tail	velocity, fps:	2550	2415	2284	2157	2034
	energy, ft-lb:	4331	3884	3474	3098	2755
	arc, inches:		+5.2	+6.0	0	−13.3
A-Square 300 Triad	velocity, fps:	2550	2251	1973	1717	1496
	energy, ft-lb:	4331	3375	2592	1964	1491
	arc, inches:		+2.7	0	−11.7	−35.1
Federal 250 Trophy Bonded	velocity, fps:	2670	2360	2080	1820	1580
	energy, ft-lb:	3955	3100	2400	1830	1380
	arc, inches:		+2.4	0	−10.4	−31.7
Federal 270 Hi-Shok	velocity, fps:	2690	2420	2170	1920	1700
	energy, ft-lb:	4340	3510	2810	2220	1740
	arc, inches:		+2.4	0	−10.9	−33.3
Federal 300 Hi-Shok	velocity, fps:	2530	2270	2020	1790	1580
	energy, ft-lb:	4265	3425	2720	2135	1665
	arc, inches:		+2.6	0	−11.2	−33.3
Federal 300 Nosler Partition	velocity, fps:	2530	2320	2120	1930	1750
	energy, ft-lb:	4265	3585	2995	2475	2040
	arc, inches:		+2.5	0	−10.3	−29.9
Federal 300 Trophy Bonded	velocity, fps:	2530	2280	2040	1810	1610
	energy, ft-lb:	4265	3450	2765	2190	1725
	arc, inches:		+2.6	0	−10.9	−32.8
Federal 300 Tr. Bonded HE	velocity, fps:	2700	2440	2190	1960	1740
	energy, ft-lb:	4855	3960	3195	2550	2020
	arc, inches:		+2.2	0	−9.4	−28.0

Federal 300 Trophy Bonded Sledgehammer Solid	velocity, fps:	2530	2160	1820	1520	1280
	energy, ft-lb:	4265	3105	2210	1550	1090
	arc, inches, s:		0	−6.0	−22.7	−54.6
Hornady 270 SP HM	velocity, fps:	2870	2620	2385	2162	1957
	energy, ft-lb:	4937	4116	3408	2802	2296
	arc, inches:		+2.2	0	−8.4	−23.9
Hornady 300 FMJ RN HM	velocity, fps:	2705	2376	2072	1804	1560
	energy, ft-lb:	4873	3760	2861	2167	1621
	arc, inches:		+2.7	0	−10.8	−32.1
Norma 300 Soft Point	velocity, fps:	2549	2211	1900	1619	
	energy, ft-lb:	4329	3258	2406	1747	
	arc, inches:		+2.8	0	−12.6	
Norma 300 TXP Swift A-Fr.	velocity, fps:	2559	2296	2049	1818	
	energy, ft-lb:	4363	3513	2798	2203	
	arc, inches:		+2.6	0	−10.9	
PMC 270 PSP	velocity, fps:					
	energy, ft-lb:					
	arc, inches:					
PMC 270 Barnes X	velocity, fps:	2690	2528	2372	2221	2076
	energy, ft-lb:	4337	3831	3371	2957	2582
	arc, inches:		+2.0	0	−8.2	−23.4
PMC 300 Barnes X	velocity, fps:	2530	2389	2252	2120	1993
	energy, ft-lb:	4263	3801	3378	2994	2644
	arc, inches:		+2.3	0	−9.2	−26.1
Rem. 270 Soft Point	velocity, fps:	2690	2420	2166	1928	1707
	energy, ft-lb:	4337	3510	2812	2228	1747
	arc, inches:		+2.2	0	−9.7	−28.7
Rem. 300 Swift A-Frame	velocity, fps:	2530	2245	1979	1733	1512
	energy, ft-lb:	4262	3357	2608	2001	1523
	arc, inches:		+2.7	0	−11.7	−35.0
Speer 285 Grand Slam	velocity, fps:	2610	2365	2134	1916	
	energy, ft-lb:	4310	3540	2883	2323	
	arc, inches:		+2.4	0	−9.9	
Speer 300 African GS Tungsten Solid	velocity, fps:	2609	2277	1970	1690	
	energy, ft-lb:	4534	3453	2585	1903	
	arc, inches:		+2.6	0	−11.7	−35.6
Win. 270 Fail Safe	velocity, fps:	2670	2447	2234	2033	1842
	energy, ft-lb:	4275	3590	2994	2478	2035
	arc, inches:		+2.2	0	−9.1	−28.7
Win. 300 Fail Safe	velocity, fps:	2530	2336	2151	1974	1806
	energy, ft-lb:	4265	3636	3082	2596	2173
	arc, inches:		+2.4	0	−10.0	−26.9
.375 Dakota						
Dakota 270 Barnes X	velocity, fps:	2800	2617	2441	2272	2109
	energy, ft-lb:	4699	4104	3571	3093	2666
	arc, inches:		+2.3	+1.0	−6.1	−19.9
Dakota 300 Barnes X	velocity, fps:	2600	2316	2051	1804	1579
	energy, ft-lb:	4502	3573	2800	2167	1661
	arc, inches:		+2.4	−0.1	−11.0	−32.7
.375 Weatherby						
A-Square 300 SP boat-tail	velocity, fps:	2700	2560	2425	2293	2166
	energy, ft-lb:	4856	4366	3916	3503	3125
	arc, inches:		+4.5	+5.2	0	−11.9
A-Square 300 Triad	velocity, fps:	2700	2391	2103	1835	1602
	energy, ft-lb:	4856	3808	2946	2243	1710
	arc, inches:		+2.3	0	−10.1	−30.8
.375 JRS						
A-Square 300 SP boat-tail	velocity, fps:	2700	2560	2425	2293	2166
	energy, ft-lb:	4856	4366	3916	3503	3125
	arc, inches:		+4.5	+5.2	0	−11.9
A-Square 300 Triad	velocity, fps:	2700	2391	2103	1835	1602
	energy, ft-lb:	4856	3808	2946	2243	1710
	arc, inches:		+2.3	0	−10.1	−30.8
.375 Remington Ultra Mag						
Rem. 270 SP	velocity, fps:	2950	2604	2284	1986	1714
	energy, ft-lb:	5216	4066	3126	2365	1761
	arc, inches:		+1.8	0	−8.5	−26

Rem. 300 Swift A-Frame	velocity, fps:	2800	2542	2299	2069	1853
	energy, ft-lb:	5222	4305	3520	2851	2287
	arc, inches:		+1.9	0	−8.6	−25.3
.375 A-Square						
A-Square 300 SP boat-tail	velocity, fps:	2920	2773	2631	2494	2360
	energy, ft-lb:	5679	5123	4611	4142	3710
	arc, inches:		+3.7	+4.4	0	−9.8
A-Square 300 Triad	velocity, fps:	2920	2596	2294	2012	1762
	energy, ft-lb:	5679	4488	3505	2698	2068
	arc, inches:		+1.8	0	−8.5	−25.5
.378 Weatherby						
A-Square 300 SP boat-tail	velocity, fps:	2900	2754	2612	2475	2342
	energy, ft-lb:	5602	5051	4546	4081	3655
	arc, inches:		+3.8	+4.4	0	−10.0
A-Square 300 Triad	velocity, fps:	2900	2577	2276	1997	1747
	energy, ft-lb:	5602	4424	3452	2656	2034
	arc, inches:		+1.9	0	−8.7	−25.9
Wby. 270 Pointed Expanding	velocity, fps:	3180	2921	2677	2445	2225
	energy, ft-lb:	6062	5115	4295	3583	2968
	arc, inches:		+1.3	0	−6.1	−18.1
Wby. 270 Barnes X	velocity, fps:	3150	2954	2767	2587	2415
	energy, ft-lb:	5948	5232	4589	4013	3495
	arc, inches:		+1.2	0	−5.8	−16.7
Wby. 300 RN Expanding	velocity, fps:	2925	2558	2220	1908	1627
	energy, ft-lb:	5699	4360	3283	2424	1764
	arc, inches:		+1.9	0	−9.0	−27.8
Wby. 300 FMJ	velocity, fps:	2925	2591	2280	1991	1725
	energy, ft-lb:	5699	4470	3461	2640	1983
	arc, inches:		+1.8	0	−8.6	−26.1
.38–40 Winchester						
Win. 180 Soft Point	velocity, fps:	1160	999	901	827	
	energy, ft-lb:	538	399	324	273	
	arc, inches:		0	−23.4	−75.2	
.38–55 Winchester						
Win. 255 Soft Point	velocity, fps:	1320	1190	1091	1018	
	energy, ft-lb:	987	802	674	587	
	arc, inches:		0	−33.9	−110.6	
.450/.400 (3")						
A-Square 400 Triad	velocity, fps:	2150	1910	1690	1490	
	energy, ft-lb:	4105	3241	2537	1972	
	arc, inches:		+4.4	0	−16.5	
.450/.400 (3 1/4")						
A-Square 400 Triad	velocity, fps:	2150	1910	1690	1490	
	energy, ft-lb:	4105	3241	2537	1972	
	arc, inches:		+4.4	0	−16.5	
.404 Jeffery						
A-Square 400 Triad	velocity, fps:	2150	1901	1674	1468	1299
	energy, ft-lb:	4105	3211	2489	1915	1499
	arc, inches:		+4.1	0	−16.4	−49.1
.416 Taylor						
A-Square 400 Triad	velocity, fps:	2350	2093	1853	1634	1443
	energy, ft-lb:	4905	3892	3049	2371	1849
	arc, inches:		+3.2	0	−13.6	−39.8
.416 Hoffman						
A-Square 400 Triad	velocity, fps:	2380	2122	1879	1658	1464
	energy, ft-lb:	5031	3998	3136	2440	1903
	arc, inches:		+3.1	0	−13.1	−38.7
.416 Remington Magnum						
A-Square 400 Triad	velocity, fps:	2380	2122	1879	1658	1464
	energy, ft-lb:	5031	3998	3136	2440	1903
	arc, inches:		+3.1	0	−13.2	−38.7
Federal 400 Trophy Bonded Sledgehammer Solid	velocity, fps:	2400	2150	1920	1700	1500
	energy, ft-lb:	5115	4110	3260	2565	2005
	arc, inches:		0	−6.0	−21.6	−49.2
Federal 400 Trophy Bonded	velocity, fps:	2400	2180	1970	1770	1590
	energy, ft-lb:	5115	4215	3440	2785	2245
	arc, inches:		0	−5.8	−20.6	−46.9

Rem. 400 Swift A-Frame	velocity, fps:	2400	2175	1962	1763	1579
	energy, ft-lb:	5115	4201	3419	2760	2214
	arc, inches:		0	−5.9	−20.8	
.416 Rigby						
A-Square 400 Triad	velocity, fps:	2400	2140	1897	1673	1478
	energy, ft-lb:	5115	4069	3194	2487	1940
	arc, inches:		+3.0	0	−12.9	−38.0
Federal 400 Trophy Bonded	velocity, fps:	2370	2150	1940	1750	1570
	energy, ft-lb:	4990	4110	3350	2715	2190
	arc, inches:		0	−6.0	−21.3	−48.1
Federal 400 Trophy Bonded	velocity, fps:	2370	2120	1890	1660	1460
Sledgehammer Solid	energy, ft-lb:	4990	3975	3130	2440	1895
	arc, inches:		0	−6.3	−22.5	−51.5
Federal 410 Woodleigh	velocity, fps:	2370	2110	1870	1640	1440
Weldcore	energy, ft-lb:	5115	4050	3165	2455	1895
	arc, inches:		0	−7.4	−24.8	−55.0
Federal 410 Solid	velocity, fps:	2370	2110	2870	1640	1440
	energy, ft-lb:	5115	4050	3165	2455	1895
	arc, inches:		0	−7.4	−24.8	−55.0
Norma 400 TXP Swift A-Fr.	velocity, fps:	2350	2127	1917	1721	
	energy, ft-lb:	4906	4021	3266	2632	
	arc, inches:		+3.1	0	−12.5	
.416 Rimmed						
A-Square 400 Triad	velocity, fps:	2400	2140	1897	1673	
	energy, ft-lb:	5115	4069	3194	2487	
	arc, inches:		+3.3	0	−13.2	
.416 Dakota						
Dakota 400 Barnes X	velocity, fps:	2450	2294	2143	1998	1859
	energy, ft-lb:	5330	4671	4077	3544	3068
	arc, inches:		+2.5	−0.2	−10.5	−29.4
.416 Weatherby						
A-Square 400 Triad	velocity, fps:	2600	2328	2073	1834	1624
	energy, ft-lb:	6004	4813	3816	2986	2343
	arc, inches:		+2.5	0	−10.5	−31.6
Wby. 350 Barnes X	velocity, fps:	2850	2673	2503	2340	2182
	energy, ft-lb:	6312	5553	4870	4253	3700
	arc, inches:		+1.7	0	−7.2	−20.9
Wby. 400 Swift A-Fr.	velocity, fps:	2650	2426	2213	2011	1820
	energy, ft-lb:	6237	5227	4350	3592	2941
	arc, inches:		+2.2	0	−9.3	−27.1
Wby. 400 RN Expanding	velocity, fps:	2700	2417	2152	1903	1676
	energy, ft-lb:	6474	5189	4113	3216	2493
	arc, inches:		+2.3	0	−9.7	−29.3
Wby. 400 Monolithic Solid	velocity, fps:	2700	2411	2140	1887	1656
	energy, ft-lb:	6474	5162	4068	3161	2435
	arc, inches:		+2.3	0	−9.8	−29.7
10.57 (.416) Meteor						
Lazzeroni 400 Swift A-Fr.	velocity, fps:	2730	2532	2342	2161	1987
	energy, ft-lb:	6621	5695	4874	4147	3508
	arc, inches:		+1.9	0	−8.3	−24.0
.425 Express						
A-Square 400 Triad	velocity, fps:	2400	2136	1888	1662	1465
	energy, ft-lb:	5115	4052	3167	2454	1906
	arc, inches:		+3.0	0	−13.1	−38.3
.44-40 Winchester						
Rem. 200 Soft Point	velocity, fps:	1190	1006	900	822	756
	energy, ft-lb:	629	449	360	300	254
	arc, inches:		0	−33.1	−108.7	−235.2
Win. 200 Soft Point	velocity, fps:	1190	1006	900	822	756
	energy, ft-lb:	629	449	360	300	254
	arc, inches:		0	−33.3	−109.5	−237.4
.44 Remington Magnum						
Federal 240 Hi-Shok HP	velocity, fps:	1760	1380	1090	950	860
	energy, ft-lb:	1650	1015	640	485	395
	arc, inches:		0	−17.4	−60.7	−136.0
Rem. 210 Semi-Jacketed HP	velocity, fps:	1920	1477	1155	982	880
	energy, ft-lb:	1719	1017	622	450	361
	arc, inches:		0	−14.7	−55.5	−131.3

Rem. 240 Soft Point	velocity, fps:	1760	1380	1114	970	878
	energy, ft-lb:	1650	1015	661	501	411
	arc, inches:		0	−17.0	−61.4	−143.0
Rem. 240 Semi-Jacketed Hollow Point	velocity, fps:	1760	1380	1114	970	878
	energy, ft-lb:	1650	1015	661	501	411
	arc, inches:		0	−17.0	−61.4	−143.0
Rem. 275 JHP Core-Lokt	velocity, fps:	1580	1293	1093	976	896
	energy, ft-lb:	1524	1020	730	582	490
	arc, inches:		0	−19.4	−67.5	−210.8
Win. 210 Silvertip HP	velocity, fps:	1580	1198	993	879	795
	energy, ft-lb:	1164	670	460	361	295
	arc, inches:		0	−22.4	−76.1	−168.0
Win. 240 Hollow Soft Point	velocity, fps:	1760	1362	1094	953	861
	energy, ft-lb:	1650	988	638	484	395
	arc, inches:		0	−18.1	−65.1	−150.3
.444 Marlin						
Rem. 240 Soft Point	velocity, fps:	2350	1815	1377	1087	941
	energy, ft-lb:	2942	1755	1010	630	472
	arc, inches:		+2.2	−5.4	−31.4	−86.7
.45–70 Government						
Federal 300 Sierra Pro-Hunt. HP FN	velocity, fps:	1880	1650	1430	1240	1110
	energy, ft-lb:	2355	1815	1355	1015	810
	arc, inches:		0	−11.5	−39.7	−89.1
PMC 350 FNSP	velocity, fps:					
	energy, ft-lb:					
	arc, inches:					
Rem. 300 Jacketed HP	velocity, fps:	1810	1497	1244	1073	969
	energy, ft-lb:	2182	1492	1031	767	625
	arc, inches:		0	−13.8	−50.1	−115.7
Rem. 405 Soft Point	velocity, fps:	1330	1168	1055	977	918
	energy, ft-lb:	1590	1227	1001	858	758
	arc, inches:		0	−24.0	−78.6	−169.4
Win. 300 Jacketed HP	velocity, fps:	1880	1650	1425	1235	1105
	energy, ft-lb:	2355	1815	1355	1015	810
	arc, inches:		0	−12.8	−44.3	−95.5
Win. 300 Partition Gold	velocity, fps:	1880	1558	1292	1103	988
	energy, ft-lb:	2355	1616	1112	811	651
	arc, inches:		0	−12.9	−46.0	−104.9
.450 Nitro Express (3¼")						
A-Square 465 Triad	velocity, fps:	2190	1970	1765	1577	
	energy, ft-lb:	4952	4009	3216	2567	
	arc, inches:		+4.3	0	−15.4	
.450 #2						
A-Square 465 Triad	velocity, fps:	2190	1970	1765	1577	
	energy, ft-lb:	4952	4009	3216	2567	
	arc, inches:		+4.3	0	−15.4	
.458 Winchester Mag.						
A-Square 465 Triad	velocity, fps:	2220	1999	1791	1601	1433
	energy, ft-lb:	5088	4127	3312	2646	2121
	arc, inches:		+3.6	0	−14.7	−42.5
Federal 350 Soft Point	velocity, fps:	2470	1990	1570	1250	1060
	energy, ft-lb:	4740	3065	1915	1205	870
	arc, inches:		0	−7.5	−29.1	−71.1
Federal 400 Trophy Bonded	velocity, fps:	2380	2170	1960	1770	1590
	energy, ft-lb:	5030	4165	3415	2785	2255
	arc, inches:		0	−5.9	−20.9	−47.1
Federal 500 Solid	velocity, fps:	2090	1870	1670	1480	1320
	energy, ft-lb:	4850	3880	3085	2440	1945
	arc, inches:		0	−8.5	−29.5	−66.2
Federal 500 Trophy Bonded	velocity, fps:	2090	1870	1660	1480	1310
	energy, ft-lb:	4850	3870	3065	2420	1915
	arc, inches:		0	−8.5	−29.7	−66.8
Federal 500 Trophy Bonded Sledgehammer Solid	velocity, fps:	2090	1860	1650	1460	1300
	energy, ft-lb:	4850	3845	3025	2365	1865
	arc, inches:		0	−8.6	−30.0	−67.8
Federal 510 Soft Point	velocity, fps:	2090	1820	1570	1360	1190
	energy, ft-lb:	4945	3730	2790	2080	1605
	arc, inches:		0	−9.1	−32.3	−73.9

Hornady 500 FMJ-RN HM	velocity, fps:	2260	1984	1735	1512	
	energy, ft-lb:	5670	4368	3341	2538	
	arc, inches:		0	−7.4	−26.4	
Norma 500 TXP Swift A-Fr.	velocity, fps:	2116	1903	1705	1524	
	energy, ft-lb:	4972	4023	3228	2578	
	arc, inches:		+4.1	0	−16.1	
Rem. 450 Swift A-Frame PSP	velocity, fps:	2150	1901	1671	1465	1289
	energy, ft-lb:	4618	3609	2789	2144	1659
	arc, inches:		0	−8.2	−28.9	
Speer 500 African GS Tungsten Solid	velocity, fps:	2120	1845	1596	1379	
	energy, ft-lb:	4989	3780	2828	2111	
	arc, inches:		0	−8.8	−31.3	
Speer African Grand Slam	velocity, fps:	2120	1853	1609	1396	
	energy, ft-lb:	4989	3810	2875	2163	
	arc, inches:		0	−8.7	−30.8	
Win. 510 Soft Point	velocity, fps:	2040	1770	1527	1319	1157
	energy, ft-lb:	4712	3547	2640	1970	1516
	arc, inches:		0	−10.3	−35.6	
.458 Lott						
A-Square 465 Triad	velocity, fps:	2380	2150	1932	1730	1551
	energy, ft-lb:	5848	4773	3855	3091	2485
	arc, inches:		+3.0	0	−12.5	−36.4
.450 Ackley						
A-Square 465 Triad	velocity, fps:	2400	2169	1950	1747	1567
	energy, ft-lb:	5947	4857	3927	3150	2534
	arc, inches:		+2.9	0	−12.2	−35.8
.460 Short A-Square						
A-Square 500 Triad	velocity, fps:	2420	2198	1987	1789	1613
	energy, ft-lb:	6501	5362	4385	3553	2890
	arc, inches:		+2.9	0	−11.6	−34.2
.450 Dakota						
Dakota 500 Barnes Solid	velocity, fps:	2450	2235	2030	1838	1658
	energy, ft-lb:	6663	5544	4576	3748	3051
	arc, inches:		+2.5	−0.6	−12.0	−33.8
.460 Weatherby Magnum						
A-Square 500 Triad	velocity, fps:	2580	2349	2131	1923	1737
	energy, ft-lb:	7389	6126	5040	4107	3351
	arc, inches:		+2.4	0	−10.0	−29.4
Wby. 450 Barnes X	velocity, fps:	2700	2518	2343	2175	2013
	energy, ft-lb:	7284	6333	5482	4725	4050
	arc, inches:		+2.0	0	−8.4	−24.1
Wby. 500 RN Expanding	velocity, fps:	2600	2301	2022	1764	1533
	energy, ft-lb:	7504	5877	4539	3456	2608
	arc, inches:		+2.6	0	−11.1	−33.5
Wby. 500 FMJ	velocity, fps:	2600	2309	2037	1784	1557
	energy, ft-lb:	7504	5917	4605	3534	2690
	arc, inches:		+2.5	0	−10.9	−33.0
.500/.465						
A-Square 480 Triad	velocity, fps:	2150	1928	1722	1533	
	energy, ft-lb:	4926	3960	3160	2505	
	arc, inches:		+4.3	0	−16.0	
.470 Nitro Express						
A-Square 500 Triad	velocity, fps:	2150	1912	1693	1494	
	energy, ft-lb:	5132	4058	3182	2478	
	arc, inches:		+4.4	0	−16.5	
Federal 500 Woodleigh Weldcore	velocity, fps:	2150	1890	1650	1440	1270
	energy, ft-lb:	5130	3965	3040	2310	1790
	arc, inches:		0	−9.3	−31.3	−69.7
Federal 500 Woodleigh Weldcore Solid	velocity, fps:	2150	1890	1650	1440	1270
	energy, ft-lb:	5130	3965	3040	2310	1790
	arc, inches:		0	−9.3	−31.3	−69.7
Federal 500 Trophy Bonded	velocity, fps:	2150	1940	1740	1560	1400
	energy, ft-lb:	5130	4170	3360	2695	2160
	arc, inches:		0	−7.8	−27.1	−60.8
Federal 500 Trophy Bonded Sledgehammer Solid	velocity, fps:	2150	1940	1740	1560	1400
	ft-lb:	5130	4170	3360	2695	2160
	arc, inches:		0	−7.8	−27.1	−60.8

.470 Capstick						
A-Square 500 Triad	velocity, fps:	2400	2172	1958	1761	1553
	energy, ft-lb:	6394	5236	4255	3445	2678
	arc, inches:		+2.9	0	−11.9	−36.1
.475 #2						
A-Square 480 Triad	velocity, fps:	2200	1964	1744	1544	
	energy, ft-lb:	5158	4109	3240	2539	
	arc, inches:		+4.1	0	−15.6	
.475 #2 Jeffery						
A-Square 500 Triad	velocity, fps:	2200	1966	1748	1550	
	energy, ft-lb:	5373	4291	3392	2666	
	arc, inches:		+4.1	0	−15.6	
.495 A-Square						
A-Square 570 Triad	velocity, fps:	2350	2117	1896	1693	1513
	energy, ft-lb:	6989	5671	4552	3629	2899
	arc, inches:		+3.1	0	−13.0	−37.8
.500 Nitro Express (3″)						
A-Square 570 Triad	velocity, fps:	2150	1928	1722	1533	
	energy, ft-lb:	5850	4703	3752	2975	
	arc, inches:		+4.3	0	−16.1	
.500 A-Square						
A-Square 600 Triad	velocity, fps:	2470	2235	2013	1804	1620
	energy, ft-lb:	8127	6654	5397	4336	3495
	arc, inches:		+2.7	0	−11.3	−33.5
.505 Gibbs						
A-Square 525 Triad	velocity, fps:	2300	2063	1840	1637	
	energy, ft-lb:	6166	4962	3948	3122	
	arc, inches:		+3.6	0	−14.2	
.577 Nitro Express						
A-Square 750 Triad	velocity, fps:	2050	1811	1595	1401	
	energy, ft-lb:	6998	5463	4234	3267	
	arc, inches:		+4.9	0	−18.5	
.577 Tyrannosaur						
A-Square 750 Triad	velocity, fps:	2460	2197	1950	1723	1516
	energy, ft-lb:	10077	8039	6335	4941	3825
	arc, inches:		+2.8	0	−12.1	−36.0
.600 Nitro Express						
A-Square 900 Triad	velocity, fps:	1950	1680	1452	1336	
	energy, ft-lb:	7596	5634	4212	3564	
	arc, inches:		+5.6	0	−20.7	
.700 Nitro Express						
A-Square 1000 Monolithic Solid	velocity, fps:	1900	1669	1461	1288	
	energy, ft-lb:	8015	6188	4740	3685	
	arc, inches:		+5.8	0	−22.2	

Index

Note: Page numbers in italics indicate illustrations

accuracy, 152
acetone, 42
Ackley, P. O., 46, 167
Ackley Improved case, 167, 170
ACTIV, 202
actual range, 131
Adolph, Fred, 24
Agincourt, Battle of, 6
aim point, 112-13, 117
Air-Cushion wad, 201-202
air resistance, 87-90, 97-101, 131
air temperature, 100, 101
Allen, Don, *151*, 190, 192
Alliant, 48, 49
Alphin, Art, 77, 100, 137, 170
altitude, 100-101
aluminum, 35
American Rifleman, The, 26, 52
American Smokeless Powder Company, 40
ammonia dope, 61
ammonium bichromate, 40
ammonium nitrate, 40
ammonium picrate, 40
angles, 130-31
antimony, 63, 65, 202
antimony sulfide, 34, 35
anvils, 33
Any Shot You Want, 77
arquebus, 10
Arrowsmith, George A., 18
Ashley, General W. H., 14
Ashmore locks, 14
A-Square, 175

Bacon, Francis, 6
Bacon, Roger, 9
Baker, Sir Samuel, 59, 207
ballistic charts, 93, 119-21, 150-51, 153-56, 163-65, 175, 181-82, 191-92, 204, 213, 229-71
ballistic coefficient "C," 89-90, 100, 104, 187
  wind and, 117, 120
ballistic pendulum, 87-88
Ballistic Research Institute (BRI), 211
Ballistite, 39, 40
ball powder, 43
barium nitrate, 34, 35
barium oxide, 34
Barnes bullets, 63, 64-65, 164
  Expander MZ, 212
  Original, 70, 147
  X, 70, 73, 85, 162, 175, 181
Barnes shotshells, 212, 213
barrel length, 93-94
  recoil and, 139
barrel temperature, *107*, 109
BB (bulleted breech) Cap, 30
BB guns, 98-99, 113
B. E. Hodgdon, Incorporated, 52
"belted magnum," 31
bench rests, 103-109, 141, 183
Bennett, Thomas, 21-22
Berdan, Hyram, 34
Berdan primer, 33-34
Bernier, Captain, 59, 61
Berthollet, Claude Louis, 11
Beurgless, Edgar, 89
bipods, 114
birdshot, 32
Bitterroot Bullet Company, 64
  Bonded Core, 62
black powder, 9-12, *14*, 34, 37, 39, 41, 48
  drams equivalent in, 32

black powder, *(cont.)*
granule size, 41-42
substitute for, 53-55
Black Powder Express, 31
Blue Mountain Bullets, 73-74
bolt-action rifles, 23-27
bolt face, 81-85
Bonded Core bullets, 62, 64
Boon, Daniel W., 16
Boone, Daniel, 14
bore diameter:
cartridges, 29-31
shotshell, 31-32
boreline, 100, 105-106
bores, 58-62, 93
small, 173-76
boron, 35
BOSS (Ballistic Optimizing Shooting System), 139
Bowman, Les, 27
bows and arrows, 5-8
Boxer, Edward, 33
Boxer-style primers, 33-34
Brady, Mike, 63, 74
braked rifle, 109, 138-39, *180*
breechloading rifles, 17-22, 60
breech pressure, 100
Brennecke, Wilhelm, 208, 209, 212
Bridger, Jim, 16
Brock, Otto, 164
Bronze Point spitzer, 64, 73
Brooks, Randy, 175
Brown, C. Norman, 197
Browning, John Moses, *17*, 21, 22
Browning, Jonathan, 21
Browning, Matt, 22
Browning rifles, 185, 195, 196
Browning shotguns, 209
Brown-Whelen, 197
Brunswick, Colonel, 59
buckshot, 32, 203-206
Buffalo Newton Rifle Corporation, 25
bulk powder, 32, 39
bullet flight, 87-90, 98-101
how not to miss, 111-16
long range shooting and, 145-48
wind and, 117-22
zeroing and, 103-109
bullets, 57-62
bullet-making, 63-68
on leashes, 97-101
for tough game, 69-74
velocity and, 95, 145-48
bullet speed, 91-95, 145-48
Bullseye powder, 41
Bureau of Explosives, 53
burn rate of powder, 42-43, 46
Burrard, Major Gerald, 201
buttpad, 140
buzz-saw effect, 98

"C," 89-90, 100, 104, 117, 120, 187
California Powder Works, 40
cannelures, 63-64
canting, 127-30
cape buffalo hunting, 164
caplock, 15, 16
Carson, Kit, 15, 16
Carter, Jack, 71
cartridges:
.30-06, 161-65
.30 magnums, 177-82
.33s, 187-92
.35s, 193-97
big game, 149-56
bolt action rifles and, 23-27
first, 17-18
headspace and, 81-85
metallic, 18-22, 24
outside the mainstream, 167-72
short and squat, 183-86
small bore, 173-76
case gauges, 82
cases, 81-85
velocity and, 95
CB (conical bullet) Cap, 30
CCI (Cascade Cartridge Industries), 35
cellulose-based powders, 38-42
Center, Warren, 53
chamber, 81-85
charcoal, 38, 41
Chas. Newton Rifle Corporation, 25
"chilled" shot, 202
"Chinese snow," 9
Chisnall, John, 134
chlorate powders, 37-38
chokes, 199-200, 203, 206, 207, 214
chronographs, 88, 91-95
CIL (Canadian Industries, Limited), 48
Clark, Alvin, 57-58
Clark, Homer, 52
collimators, 106
Colt, Samuel, 8
combs, 139, 140
Comfort, Ben, 177
controlled-expansion bullets, 70-72
controlled jerk, 113
copper-jacketed bullets, 102
copper percussion cap, 11
copper units of pressure (CUP), 67, 78
Cordite, 39, 41, 177
Core-Lokt bullets, 64, 73, 175, 176, 194, 195
Coriolis effect, 101
Corzine, Alan, 122
Coxe, Wallace, 89
Crecy, Battle of, 6
crescent buttplate, 140
crimping, 63-64

crusher system, 78
culverins, 10
cupronickel, 61
Curtis, Ted, 52
cut shells, 201, 209

Dakota rifle, *151,* 159, 190, 192
D'Arcy Echols rifle, *157*
datum line, 82
datum point, 82, 84
da Vinci, Leonardo, 6
deceleration, 99, 120, 121
deer hunting, 147, 158-59, 165, 195
  with shotguns, 203-14
deflection, 120-21, 125
Delvigne, Henri Gustave, 58, 59
dense smokeless powders, 32, 39
detonator test, 53
dibutylphthalate, 43
Dittman, Carl, 38
Donaldson Wasp, 168
double-base powders, 47
double rifle, *21,* 22
  recoil and, 138
drag, 87-90, 97-101, 131
drams equivalent, 32, 39
drawn jackets, 63
drift, 117-25
dry-firing, 114
Dualin, 38
duplex loads, 188
Du Pont (E. I. Du Pont de Nemours), 38, 40, 41, 45-49, 89, 208
Du Pont, Eleuthere Irenee, 45
DWM, 64
dynamite, 37

E. C. & Schultz Powder Company, 40
Eddystone Arsenal, 25
Edward II, King, 9
effective range, 131, 146
elephant hunting, 147, 157, 207
Elkhorn Bullets, 73-74
elk hunting, 69-70, 73-74, 112-13, 115-16, 135-36, 147, *152, 159,* 161, 163, 165, 176, 189
Emary, Dave, 65-68, 147
Emrich, Linn, 53, 54
energy measurement, *94*
Enfield, 25, 60, 177
English longbow, 6-7
ether, 42
ethyl acetate, 43
Euler, Leonardo, 6
European cartridge designers, 33-34
EX powder, 46
Express Train rifles, 60
EXPRO, 48, 49
exterior ballistics, 1
extreme spread (ES), 92
extruded propellants, 54

Fabrique Nationale d'Armes de Guerre (FN), 23, 27
false shoulders, 84
FA 70 primers, 162
Federal bullets, 66, 72, 73, 177, 180, 202
  High Energy, 152, 162, 171, 189, 192
  Hi-Shok Softpoint, 72
Federal shotshells, 205, 211
Ferguson, Major Patrick, 17
FH-42 non-mercuric primer, 34
Field & Stream, 175
field gauge, 82
firing range:
  courtesy at, 109
  items needed at, 106
  wind at, 118
first projectiles, 5-8
"Flammable Solid," 52
"flash in the pan," 10
flinching, 107, 137, 138, 140, 152, 161
flintlocks, 10, *11,* 12, *13, 14,* 18
Flobert, 18
flow, 83-84
Forsythe, Alexander John, 11
Foster, Karl, 208, 209
Foster slugs, *207,* 208-14
Fourcroy, Antoine, 11
Frankfort Arsenal, 194
Freudenberg, Rick, 148
full-length sizing, 83
fulminates, 11, 12, 18
Fulton, Bob, 74, 84

gain twist, 60
Galileo, 6, 87, 99
gauge, shotshell, 31-32
Gavre Commission, 89
Gearhardt, Marv, 48
gelatin glue, 34
gelatin powders, 39
Gemmer, J. P., 16
Giant Powder Company, 40
Gibbs, Rocky, 84, 169-70
gilding metal, 162
"go" and "no go" gauges, 82
GOEX, 48
Gold Dust Powder, 40
Goudy, Gary, 193
"Government Pyro," 41
Graham, Richard, 15
"grains," 30
graphite, 42
gravity, 87-90, 97-101, 117, 119, 125, 128, 129-30, 137
Greener, William, 59, 60
Greenhill, Sir Alfred George, 61, 97-98
Griffin & Howe (G&H), 193, 194
groove diameter, 29-30
grooves, barrel, 58-61

group size, 107
guncotton, 37, 39, 41, 42
gunpowder, 37-43, *196*
  for the .30-06, 64
  components of, 9
Gun Servicing, 212
Guthrie, Dr. Samuel, 11
gyroscopic effect, 101

Halidon Hill, Battle of, 6
Hall, Captain John Harris, 17, 18, 19
Hamilton, Alexander, 45
handloading, 215-18
Harpers Ferry muzzleloading rifle, 18
Hastings, Battle of, 6
Hawk bullets, 74, 84
Hawken, Jacob, 14-15
Hawken, Samuel, 14-16
Hawken, William Stewart, 15-16
Hawken rifle, 14-15, 16
Hawker, Colonel, 12
Hawthorne, Lowell, 66
headspace, 81-85, 169
Heckler and Koch, 61
Heise Gauge, 67
Henderson, General, 15
Henry, Benjamin Tyler, 19
Henry, James, 14
Henry rifle, 19
Hercules, 40, 41, *43*
  RL-15 powder, 185
Herter's press, 215, 216
H-48 non-mercuric primer, 34
"high brass," 31
HiVel powder, 41
H-Mantle bullets, 70, 72
Hodgdon, Bruce, 49, 51-55
Hodgdon, J. B., 52, 54
Hodgdon, R. E. (Bob), 52, 53, 54
Hodgdon Powder Company, 49, 51-55
"hold center," 104
Holden, Ron, *188*
holding a rifle still, 114-15
Holland & Holland, 25, 176
  .300, 168, 177-79, 185, 188, 189
hollowpoint bullets, 64
Holt, Wayne, 66, 67
Hopkins, Don, 171, 188
Hornady, Joyce, 65
Hornady, Steve, 73, *112*
Hornady bullets, 64-68
  Interlock, 72-73
  Magnum loads, 152, 162, 171, 180, 189
  Super Shock Tip (SST), 73
Hornady press, 216
Hornady shotshells, 212
Hot-Core bullets, 64
Howard, E. C., 11
"Howard's powder," 11
Howe, James V., 194
Hunt, Walter, 18, 21
hunting bullets, 145-48
  knockdown myth and, 157-59
  military rounds and, 145-46
  shape and weight of, 146-48
hunting rifles, 113
Huntington, Fred, 167

impact extruded jackets, 63
Imperial Chemical Industries Limited, 200, 201
Improved cartridges, 84, 167-72
IMR Powder Company, 48
IMR (Improved Military Rifle) powders, 41, 45-49
Indian bows, 7-8
Infallible powder, 41
Ingalls, Colonel James, 89
Ingalls Tables, 89
interior ballistics, 1
intermittent squeeze, 116
internal combustion, 11
irresponsible hunting, 123, 135
Ithaca shotguns, 209

jacketed bullets, 61-68, 162
Jacobs, General John, 59
jaeger rifle, 12, 13
Jasaitis, Victor, 35
Jeffery, 171, 179, 184, 188, 191, *192*
Jenks, William, 18
Jennie, Fred, 27
Jennings, Lewis, 18
Jensen, Warren, 74
Jensen bullets, 63, 74
JGS Die & Machine, 197
John Hall and Sons, 37
J. Stevens Arms & Tool Company, 114

Keberst rifle, 191
Keith, Elmer, *94*, 171, 188
Kentucky rifle, 12, 14
Kephart, Horace, 15
K-Hornet, 168
Kimble, Fred, 199, 208
Kinchen, Tony, 212
kinetic energy, 137-38, 157-58
King Powder Company, 38
King's Semi-Smokeless Powder, 38
Kleanbore primer, 34, 162
knockdown myth, 157-59
Knox, Neil, 54
Kodiak bullets, 74
Krag-Jorgensen rifle, 162
Krueger, Jerry, *201*
Krupp, 89
Kuisyingen, Battle of, 10
Kynoch, 171

Laflin & Rand, 40
Lancaster, Charles, 207
land diameter, 29-30

L & R Smokeless powder, 40
Lapua cartridges, *188*, 191-92
laser beam sighting, 106
Lazzeroni, John, 159, 181-82, 184-86, 190
Lazzeroni rifles:
  Galaxy, *183*, 185, 190
  Mountain Rifle, 184-85
  Titan, 190, 191
  Warbird, 181-82
leading the target, 135-36
lead pellets, 32
lead styphnate, 34, 35
lead thiocyanate, 34
lead tri-nitro-resorcinate, 34
Lefaucheaux, 207
  needle gun, 11
Lehman, Henry, 14
Leonard Powder Company, 40
Lesmoke, 38
Levenson, Mike, 53
lever-action rifles, 85, 147
Lightfield sabots, 212, 213
Lightning powder, 40
Lindsay, Milton, 38
line of sight, 100, 103-106, 128, 130
lock, 10
Lock a la Miquelet, 10
Lubaloy, 61-62
Ludwig Lowe & Co., 23

M1 bullet, 146, 162
M2 bullet, 146, 162
McCalla, J. M., 19
McMillan actions, 181, 184, 192
magnum, 31
magnum primer, 35, 94-95
magnum shot, 202
Maitland, Stuart, 195
manifest destiny, 13-16
Mannlicker-Schoenauer, *164*, 196
Manton, Joseph, 11
Marlin rifles, 61, 188, 194
Marlin shotguns, 214
Martini-Henry, 60
mass, 137
MAST Technologies, 190
matchlocks, 10, *11*
Mauser, Peter Paul, 23-24, 162
Mauser, Wilhelm, 23
Mauser Bros. and Co., 23-25, 162
Maxim, Hiram, 39
maximum point-blank range, 100, 104, 105
Mayevski, Colonel, 89
Maynard, Edward, 18
Maynard rifle, 193
mean velocity, 92
mercury fulminate, 34
Merrill, J. H., 18
Merrill gun, 11
Metford, William Ellis, 60, 61
Michigan State University, 129
Micro-Groove barrels, 61
military cartridges, 145-46, 161-62
Millcreek Gun Club, 55
millimeter measurements, 30-31
Minie, Captain Claude-Etienne, 59
minute of angle, 108, 164
mirage, 124
missed shots, 111-16
Mobilubricant, 61
"monk's gun," 10
Monterey, battle of, 15
moose hunting, 157, *162*
Morey, Jim, *151*
Moschkau, Johnny, 111-12
Moss, Maxine, 53
Mossberg shotguns, *203*, 214
moving targets, 133-36
MR (military rifle) powder, 45, 47
muzzle break, 109, 138-39, *180*
muzzleloading rifles, 9-16, 18, 63, 65
  twist rates for, 97
mystery shots, 115-16

naming ammunition, 29-32
National Muzzleloading Rifle Association (NMLRA), 53-54
neck sizing, 83-84, 85
needle gun, 18
Nelson, Dick, 124
New Haven Arms Company, 19
New Sporting Powder, 38
Newton, Charles, 24-26, 30, 60-61, 70, 167
Newton, Sir Isaac, 6, 88, 137
Newton Arms Company, 25, 193, 194
Nippes, A. S., 19
nitric acid, 37, 42
nitrocellulose, 38-43, 47
Nitro Express, 31
nitroglycerine, 31, 37-40, 42, 43, 47, 67
Nobel, Alfred, 37, 38, 39, 40, 78
Nobel, Emmanuel, 37
Nock, 11
noise, 138-39, *180*
nomenclature, 29-32
non-corrosive primers, 34-35
non-mercuric primers, 34-35
Norma, 27, 93, 152, 174, 179, 180, 192, 196
Northern Precision, 74
Nosler, John, 70, 147
Nosler bullets, 62, 64, 71, 211, 212
  Ballistic Tip, 73, *158*, 181
  Partition, 63, 70, 72, 73, 85, 95, 147, *159*, 162, 164, *168*, 171, 176, 177, 180, 181, *188*, 189
NRA Foundation, 55
Nyce, Jim, *151*

O'Connor, Jack, 159
Oehler, Dr. Ken, 79, 91
Oehler Ballistic Explorer, 100

offhand shooting, *109,* 140
OKH group, 171, 188
Old Western Scrounger, 72
Olin, 48-49, 66
O'Neil, Charlie, 171, 188
Owens, Harold, 48

Page, Warren, 175
Palma Match bullets, 62, 67
Palmer, Courtland, 18, 19
Palmisano, Dr. Lou, 183, 184, 186
Pape, William, 199
paradox guns, 208
Parkman, Francis, 15
patched ball, 12
Pauly, Johannes, 11
Pawlak, Cathy, 54
Pawlak, Dan, 53, 54
Pearson, Karl, 92
Pentax Lightseeker, *99*
percussion caps, 11-12, 17, 18
Perfect cartridges, 200
petals, 98
Peters, G. M., 38
Peters bullets, 64
PETN, 35
petronel, 10
Peyton Powder, 49
picric acid, 34
piezoelectric gauge, 78-79
Pindell, Ferris, 183
PMC, 212
Pohl, J. J., 38
point-blank range, 100, 104, 105
Poly-Choke, 199, *205*
Polymaq Quik-Shok, *211*
Postman, Jay, 168-69
potassium chlorate, 34, 38
potassium nitrate, 40
Poudre B powder, 38
pounds per square inch (PSI), 78
powder, *see* gunpowder
powdered glass, 34
PPC, 183-84
practice, 114-15
precessional movement, 98
President's Hundred, 66
pressure measurement, 67, 77-79, 94
primers, 33-35, 162
  velocity and, 94-95
Primex, 48-49
progressive powder, 43
pronghorn hunting, 117, *125*
pumpkin balls, 208
punt guns, 31-32
Purdey, James, 60
pyrite, 10
pyrocellulose, 42
Pyrodex, 52-55

Ramshot, 49
ratchet rifling, 61
Rathburg, 34
RCBS, 167, 168, 170, 175, 216
rechambering rifles, 84, 168-69
recoil, 137-41, 150
  light, *152*
  manliness and, 159
  military rounds and, 146, 162
Redding, 170, 216
regressive powder, 43
Reighard, George, 20
Relative Quickness value, 46
Remington, Eliphalet (Lite) II, 13, 18
Remington Arms Co., 25, 27, 47, 49, 170
  bullets, 64, 73, 95, 147, 152, 159, 164
  .30 magnum, 177-81, *191*
  .33 cartridges, 187-88, *191,* 192
  .35 cartridges, 193-96
  7mm magnum, 174-76
Remington shotshells, 205, *208,* 212
Revolutionary War, 9-10, 12
Rheinische-Westphalische Sprengstoff (RWS), 34, 72, 208
rifling, 29-30, 58-62, 97, 164
  formula for rate of twist, 97-98
  wind and, 118
rimless cartridges, 82
Robin Hood Ammunition Company, 40
Robins, Benjamin, 87-88, 91
rocket balls, 18-19, *20*
Rodman, 78
Rogers, Samuel, 40
Rose, J. S., 26
Rostfrei (rust-free) priming, 34
rotational velocity, 98
Rottweil Company, 200
"Ruby N" and "Ruby J" powders, 40
Ruger rifles, 196
Rustless primer, 34

SAAMI (Sporting Arms and Ammunition Manufacturers Institute), 67, 152, 170
sabot slugs, *207,* 208-14
Sako rifles, 181, 184, 192
saltpeter, 11, 38, 41
Saturday Evening Post, 15
Savage rifles, *165, 184,* 196
Savage shotguns, 214
Schultz, 208
Schultz & Larsen, 196
Schwarz, Berthold, 9
scopes, *99,* 113
  adjustments to, *106,* 108
  canting and, 127-30
  zeroing rifles with, 103-109
"secant ogive" nose profile, 68, 122
sectional density, 120, 146-47
semi-rimmed cartridges, 82
semi-smokeless powders, 38

7mm cartridges, *84,* 173-76
Sharpe & Hart, 167-68
Sharps, Christian, 19-20
Sharpshooter powder, 40
Sharps Rifle Manufacturing Company, 19-21
Shaw, Joshua, 11
Shoenbein, Christian, 37
shootability, 152
shooting position, 140
shooting sticks, 114
Shooting Times Alaskan, 196
Shooting Times Westener, 174-76
short magnums, 183-86, 196
shotguns and shotshells, 199-202
  for big game, 207-10
  buckshot, 203-206
  new slugs, 211-14
shotshell nomenclature, 31-32
shotshell primers, 35
Siacci, 89
side-mounted scopes, 130
Sierra Bullets, 63, 64, 67, 69, 89, 90
  ballistics program, 100
  GameKing, 73
  Match King, 192
sightline, 100, 103-106
Simpson, Layne, 174-75
Simpson, Leslie, 193
Sisk, Charlie, *168,* 172
Skipworth, Amy, 51, 52
slings, 114
small bores, 173-76
Smith, Horace, 18-19
Smith, T. D., 100-101
smokeless powders, *14,* 22, 24, 32, 38-43, 52, 147
specific gravity, 61
speed of sound, 88, 90
Speer, Dick, 35
Speer, Vernon, 65
Speer bullets, 64, 79, 164
  Grand Slam, 63, 70-71, 197
  Hot-Cor, 73
Speer primers, 35
Springfield Armory, 24, 161-62
"standard bullet," 88-90
standard deviation, 92-93
stationary game, 135
Staynless primer, 34
steel pellets, 32
Stegall, Keith, 197
stock, 139, 140
Stoeger Arms Co., 23
strain gauge, 79
Subrero, Ascanio, 37
sulfur, 38, 41
sulfuric acid, 34, 37, 42
Super X cartridges, 185-86
Swedish moose target, 133, 134
Swift bullets, 69
  A-Frame, 71, *73,* 95, 162, 164, 175, 192
  Scirocco, 71, 73, *74,* 180

"tangent ogive" nose profile, 68, 122
target rifles, 113
targets for zeroing, 106-107
Taylor, Barry, *151*
Taylor, Stephen, 18
Tennessee rifle, 14
terminal ballistics, 1
tetracene, 35
Theiss breechloader, 17
.30 magnums, 177-82
.30.06, 161-65
.33 cartridges, 193-97
.35 cartridges, 187-92
Thompson-Center, 53, 65
  Contender, *167,* 194
  replicas, *15, 16*
Thouvenin, 59
three-digit bore designations, 29-30
TIG (Torpedo Ideal) bullet, 64, 72
Timmerman, Mike, 65
tin, 41
tin-plated jackets, 61, 162
TNT, 34
tracer bullets, 99
trajectories, *see* bullet flight
trap-door rifle, 18
Trataglia, 87
trigger control, 116
Trophy Bonded bullets, 64, 162, 164, 171, 180, 189
  Bear Claw, 71, 73
trust, 112, 113
Tryon, George, 14
TUG (Torpedo Universal) bullet, 64, 72
.22 rifles, 114-15
twist rates, 59-61, 97, 164
  formula for, 97-98
two-digit bore designations, 30
Tyrolean cheekpieces, 139-40

understudy rifle, 114-15
Union Metallic Cartridge Company, 40
Unique powder, 40
U.S. Army Ballistic Research Laboratories, 89
U.S. Army Ordinances, 146, 161-62
U.S. Department of Transportation, 53
United States Powder Company, 40
U.S. Repeating Arms Company (USRAC), 175, 185
uphill and downhill shooting, 130-31

Valleyfield Chemical Products, 48
Vauquelin, Nicholas Louis, 11
velocity of bullets, 72
  bullet drop and, 90
Vielle, Paul, 38
violence inside the gun, 77-79

Volcanic Repeating Arms Company, 19
Volitional repeater, 18, 21
Volkmann, Frederick, 38-39
von Dreyse, Johann Nikolaus, 18
Von Hersz, 34

Walker, "Bigfoot," 8
Washington, George, 9
Waterbury-Farrell, 65
Weatherby, Roy, 26-27, 29, 35, 93, 152, 169, 174, 178-79
Weatherby rifles, 150-52, 159, 174
  .33 magnum, 189-91
  .300 magnum, 178-81
  wildcat rounds and, 168-71
Weaver Tip-Off scope ring, 129-30
Webernick, Lex, *148*
Welland Chemical, 48
Werner, Larry, 46-49
Wesson, Dan, 18-19
Western Cartridge, 43, 61-62, 162, 177
Western Tool and Copper Works, 64
Westley Richards, 11, 64, 170-71
wheellocks, 10, *11*
Whelen, Colonel Townsend, 30, 194
Whelen cartridges, *193*, 194, 195, 197
Whistler and Aspinwall (WA), 40
White, Stewart Edward, 161
white powder, 38
Whitworth, Joseph, 59-60
Wiggins, O. P., 15
"wildcat" cartridges, 30, 84, 167-72, 174, 175, 179, 189, 190, 195
Wimbledon Match, 177
Winchester Repeating Arms Company, 19, 21, 25, *26*, 49, 64, 147, 159, *169*, 171-72
  .33 cartridges, 187-89
  .35 cartridges, 193-96
  .300 magnum, 177-79
  .300 short magnum, 185-86
  Ballistic Silvertip bullet, 73, 185, 186
  Fail Safe bullet, 71-72, 74, *158*, 162, 164, 177, 185, 186, 192
  long-range big game cartridges, 173-74
  Model 1885, *17*, 21, 22
  Partition Gold bullet, 72, 162
  Power Point bullet, 73
  Silvertip bullet, 73
Winchester shotshells, 205, 209, 211, 212
Winchester-Western, 89
wind, *98*, 99, 100, 117-22
  mirage, luck, and, 123-25
wind flags, 118, 124
windicators, 118, *119*, 124
Wittenmergen, Battle of, 10
Wolfe, Dave, 52, 54
Woodleigh bullets, 72, *181*

yellow prussiate of potash, 38

Zeglin, Fred, *25*, 84
zeroing, 100, 103-109
zero range, 104, 128
zinc, 162, 200